Large Deviations and Adiabatic Transitions for Dynamical Systems and Markov Processes in Fully Coupled Averaging

MEMOIRS
of the
American Mathematical Society

Number 944

Large Deviations and Adiabatic Transitions for Dynamical Systems and Markov Processes in Fully Coupled Averaging

Yuri Kifer

September 2009 • Volume 201 • Number 944 (third of 5 numbers) • ISSN 0065-9266

American Mathematical Society
Providence, Rhode Island

2000 *Mathematics Subject Classification.*
Primary 34C29; Secondary 37D20, 60F10, 60J25.

Library of Congress Cataloging-in-Publication Data

Kifer, Yuri, 1948–
Large deviations and adiabatic transitions for dynamical systems and Markov processes in fully coupled averaging / Yuri Kifer.
p. cm. — (Memoirs of the American Mathematical Society, ISSN 0065-9266 ; no. 944)
"Volume 201, number 944 (third of 5 numbers)."
Includes bibliographical references and index.
ISBN 978-0-8218-4425-0 (alk. paper)
1. Averaging method (Differential equations) 2. Large deviations. 3. Attractors (Mathematics) 4. Differential equations—Qualitative theory. I. Title.

QA372.K536 2009
515′.352—dc22 2009019381

Memoirs of the American Mathematical Society

This journal is devoted entirely to research in pure and applied mathematics.

Subscription information. The 2009 subscription begins with volume 197 and consists of six mailings, each containing one or more numbers. Subscription prices for 2009 are US$709 list, US$567 institutional member. A late charge of 10% of the subscription price will be imposed on orders received from nonmembers after January 1 of the subscription year. Subscribers outside the United States and India must pay a postage surcharge of US$65; subscribers in India must pay a postage surcharge of US$95. Expedited delivery to destinations in North America US$57; elsewhere US$160. Each number may be ordered separately; *please specify number* when ordering an individual number. For prices and titles of recently released numbers, see the New Publications sections of the *Notices of the American Mathematical Society.*

Back number information. For back issues see the *AMS Catalog of Publications.*

Subscriptions and orders should be addressed to the American Mathematical Society, P. O. Box 845904, Boston, MA 02284-5904 USA. *All orders must be accompanied by payment.* Other correspondence should be addressed to 201 Charles Street, Providence, RI 02904-2294 USA.

Memoirs of the American Mathematical Society (ISSN 0065-9266) is published bimonthly (each volume consisting usually of more than one number) by the American Mathematical Society at 201 Charles Street, Providence, RI 02904-2294 USA. Periodicals postage paid at Providence, RI. Postmaster: Send address changes to Memoirs, American Mathematical Society, 201 Charles Street, Providence, RI 02904-2294 USA.

This publication is indexed in *Science Citation Index®*, *SciSearch®*, *Research Alert®*, *CompuMath Citation Index®*, *Current Contents®/Physical, Chemical & Earth Sciences.*
Printed in the United States of America.

♾ The paper used in this book is acid-free and falls within the guidelines established to ensure permanence and durability.
Visit the AMS home page at `http://www.ams.org/`

10 9 8 7 6 5 4 3 2 1 14 13 12 11 10 09

Contents

Abstract

The work treats dynamical systems given by ordinary differential equations in the form $\frac{dX^\varepsilon(t)}{dt} = \varepsilon B(X^\varepsilon(t), Y^\varepsilon(t))$ where fast motions Y^ε depend on the slow motion X^ε (coupled with it) and they are either given by another differential equation $\frac{dY^\varepsilon(t)}{dt} = b(X^\varepsilon(t), Y^\varepsilon(t))$ or perturbations of an appropriate parametric family of Markov processes with freezed slow variables. In the first case we assume that the fast motions are hyperbolic for each freezed slow variable and in the second case we deal with Markov processes such as random evolutions which are combinations of diffusions and continuous time Markov chains. First, we study large deviations of the slow motion X^ε from its averaged (in fast variables Y^ε) approximation $\bar{X}^\varepsilon$. The upper large deviation bound justifies the averaging approximation on the time scale of order $1/\varepsilon$, called the averaging principle, in the sense of convergence in measure (in the first case) or in probability (in the second case) but our real goal is to obtain both the upper and the lower large deviations bounds which together with some Markov property type arguments (in the first case) or with the real Markov property (in the second case) enable us to study (adiabatic) behavior of the slow motion on the much longer exponential in $1/\varepsilon$ time scale, in particular, to describe its fluctuations in a vicinity of an attractor of the averaged motion and its rare (adiabatic) transitions between neighborhoods of such attractors. When the fast motion Y^ε does not depend on the slow one we arrive at a simpler averaging setup studied in numerous papers but the above fully coupled case, which better describes real phenomena, leads to much more complicated problems.

Received by the editor December 4, 2006.

2000 *Mathematics Subject Classification.* Primary: 34C29 Secondary: 37D20, 60F10, 60J25.

Key words and phrases. averaging, hyperbolic attractors, random evolutions,large deviations.

The author was partially supported by US–Israel BSF.

Preface

This work studies the long time behavior of slow motions in two scale fully coupled systems and it consists of two, essentially, independent parts which even have their own introductions. The first part is written having in mind readers with strong backgrounds in smooth dynamical systems and it deals with the case of Axiom A flows as fast motions. The second part is written for probabilists and it studies the case where fast motions are certain Markov processes such as random evolutions and, in particular, diffusions. As we noticed already in [**47**] principal large deviations results for Axiom A systems and Markov processes (satisfying, say, the Doeblin condition) follow from a similar scope of ideas and basic theorems though they rely on quite different machineries and backgrounds. Rate functionals of large deviations turn out to be Legendre transforms of corresponding topological pressures in the dynamical systems case while in the diffusion case they are obtained in the same way from principal eigenvalues of the corresponding infinitesimal generators. This intrinsic connection is further amplified by the fact that in the random diffusion perturbations of dynamical systems setup these principal eigenvalues converge to topological pressures when the perturbation parameter tends to zero (see [**46**]).

Usually, Markov processes are easier to deal with since we can use the Markov property there for free while in the dynamical systems case we have to look for some substitute. We felt that the first part of this work would be quite difficult to follow for most of probabilists in view of its heavy dynamical systems machinery. By this reason the second part is written in the way that it can be read independently of the first one and it relies only on the standard probabilistic background though the strategies of the proof in both parts are similar with the Markov property making arguments easier in the second part which also does not require to deal with geometric pecularities of the hyperbolic deterministic dynamics of the first part. In order to ensure a convenient independent reading of the second part we give full arguments there except for very few references to some general proofs in the first part which do not rely on the specific dynamical systems setup there. Still, the readers having sufficient background both in dynamical systems and Markov processes will certainly benefit from having proofs for both cases in one place and such exposition demonstrates boldly unifying features of these two quite different objects. We observe, that it could be possible to start with some very general (though quite unwieldy) assumptions which would enable us to prove similar results and then verify these assumptions for both cases we are dealing with but we believe that such exposition would make the paper quite difficult to read for both groups of mathematicians this work is addressed to.

Part 1

Hyperbolic Fast Motions

1.1. Introduction

Many real systems can be viewed as a combination of slow and fast motions which leads to complicated double scale equations. Already in the 19th century in applications to celestial mechanics it was well understood (though without rigorous justification) that a good approximation of the slow motion can be obtained by averaging its parameters in fast variables. Later, averaging methods were applied in signal processing and, rather recently, to model climate–weather interactions (see [**36**], [**18**], [**37**] and [**52**]). The classical setup of averaging justified rigorously in [**12**] presumes that the fast motion does not depend on the slow one and most of the work on averaging treats this case only. On the other hand, in real systems both slow and fast motions depend on each other which leads to the more difficult fully coupled case which we study here. This setup emerges, in particular, in perturbations of Hamiltonian systems which leads to fast motions on manifolds of constant energy and slow motions across them.

In this work we consider a system of differential equations for $X^\varepsilon = X^\varepsilon_{x,y}$ and $Y^\varepsilon = Y^\varepsilon_{x,y}$,

$$\text{(1.1.1)} \qquad \frac{dX^\varepsilon(t)}{dt} = \varepsilon B(X^\varepsilon(t), Y^\varepsilon(t)), \quad \frac{dY^\varepsilon(t)}{dt} = b(X^\varepsilon(t), Y^\varepsilon(t))$$

with initial conditions $X^\varepsilon(0) = x$, $Y^\varepsilon(0) = y$ on the product $\mathbb{R}^d \times \mathbf{M}$ where $\mathbf{M}$ is a compact $n_{\mathbf{M}}$-dimensional C^2 Riemannian manifold and $B(x,y)$, $b(x,y)$ are smooth in x, y families of bounded vector fields on $\mathbb{R}^d$ and on $\mathbf{M}$, respectively, so that y serves as a parameter for B and x for b. The solutions of (1.1.1) determine the flow of diffeomorphisms Φ^t_ε on $\mathbb{R}^d \times \mathbf{M}$ acting by $\Phi^t_\varepsilon(x,y) = (X^\varepsilon_{x,y}(t), Y^\varepsilon_{x,y}(t))$. Taking $\varepsilon = 0$ we arrive at the flow $\Phi^t = \Phi^t_0$ acting by $\Phi^t(x,y) = (x, F^t_x y)$ where F^t_x is another family of flows given by $F^t_x y = Y_{x,y}(t)$ with $Y = Y_{x,y} = Y^0_{x,y}$ being the solution of

$$\text{(1.1.2)} \qquad \frac{dY(t)}{dt} = b(x, Y(t)), \quad Y(0) = y.$$

It is natural to view the flow Φ^t as describing an idealised physical system where parameters $x = (x_1, ..., x_d)$ are assumed to be constants (integrals) of motion while the perturbed flow Φ^t_ε is regarded as describing a real system where evolution of these parameters is also taken into consideration. Essentially, the proofs of this paper work also in the slightly more general case when B and b in (1.1.1) together with their derivatives depend Lipschitz continuously on ε (cf. [**55**]) but in order to simplify notations and estimates we do not consider this generalisation here.

Assume that the limit

$$\text{(1.1.3)} \qquad \bar{B}(x) = \bar{B}_y(x) = \lim_{T\to\infty} T^{-1} \int_0^T B(x, F^t_x y) dt$$

exists and it is the same for "many" y's. For instance, suppose that μ_x is an ergodic invariant measure of the flow F^t_x then the limit (1.1.3) exists for μ_x-almost all y and is equal to

$$\bar{B}(x) = \bar{B}_{\mu_x}(x) = \int B(x,y) d\mu_x(y).$$

If $b(x,y)$ does not, in fact, depend on x then $F^t_x = F^t$ and $\mu_x = \mu$ are also independent of x and we arrive at the classical uncoupled setup. Here the Lipschitz continuity of B implies already that $\bar{B}(x)$ is also Lipshitz continuous in x, and so there exists a unique solution $\bar{X} = \bar{X}_x$ of the averaged equation

$$\frac{d\bar{X}^\varepsilon(t)}{dt} = \varepsilon \bar{B}(\bar{X}^\varepsilon(t)), \quad \bar{X}^\varepsilon(0) = x. \tag{1.1.4}$$

In this case the standard averaging principle says (see [**73**]) that for μ-almost all y,

$$\lim_{\varepsilon\to 0} \sup_{0\le t\le T/\varepsilon} |X^\varepsilon_{x,y}(t) - \bar{X}^\varepsilon_x(t)| = 0. \tag{1.1.5}$$

As the main motivation for the study of averaging is the setup of perturbations described above we have to deal in real problems with the fully coupled system (1.1.1) which only in very special situations can be reduced by some change of variables to a much easier uncoupled case where the fast motion does not depend on the slow one. Observe that in the general case (1.1.1) the averaged vector field $\bar{B}(x)$ in (1.1.3) may even not be continuous in x, let alone Lipschitz, and so (1.1.4) may have many solutions or none at all. Moreover, there may exist no natural well dependent on $x \in \mathbb{R}^d$ family of invariant measures μ_x since dynamical systems F^t_x may have rather different properties for different x's. Even when all measures μ_x are the same the averaging principle often does not hold true in the form (1.1.5), for instance, in the presence of resonances (see [**63**] and [**56**]). Thus even basic results on approximation of the slow motion by the averaged one in the fully coupled case cannot be taken for granted and they should be formulated in a different way requiring usually stronger and more specific assumptions.

If convergence in (1.1.3) is uniform in x and y then (see, for instance, [**52**]) any limit point $\bar{Z}(t) = \bar{Z}_x(t)$ as $\varepsilon \to 0$ of $Z^\varepsilon_{x,y}(t) = X^\varepsilon_{x,y}(t/\varepsilon)$ is a solution of the averaged equation

$$\frac{d\bar{Z}(t)}{dt} = \bar{B}(\bar{Z}(t)), \quad \bar{Z}(0) = x. \tag{1.1.6}$$

It is known that the limit in (1.1.3) is uniform in y if and only if the flow F^t_x on $\mathbf{M}$ is uniquely ergodic, i.e. it possesses a unique invariant measure, which occurs rather rarely. Thus, the uniform convergence in (1.1.3) assumption is too restrictive and excludes many interesting cases. Probably, the first relatively general result on fully coupled averaging is due to Anosov [**1**] (see also [**63**] and [**52**]). Relying on the Liouville theorem he showed that if each flow F^t_x preserves a probability measure μ_x on $\mathbf{M}$ having a C^1 dependent on x density with respect to the Riemannian volume m on $\mathbf{M}$ and μ_x is ergodic for Lebesgue almost all x then for any $\delta > 0$,

$$\text{mes}\{(x,y) : \sup_{0\le t\le T/\varepsilon} |X^\varepsilon_{x,y}(t) - \bar{X}^\varepsilon_x(t)| > \delta\} \to 0 \text{ as } \varepsilon \to 0, \tag{1.1.7}$$

where mes is the product of m and the Lebesgue measure in a relatively compact domain $\mathcal{X} \subset \mathbb{R}^d$. An example in Appendix to [**56**] shows that, in general, this convergence in measure cannot be strengthened to the convergence for almost all initial conditions and, moreover, in this example the convergence (1.1.5) does not hold true for any initial condition from a large open domain. Such examples exist due to the presence of resonances , more specifically to the "capture into resonance" phenomenon, which is rather well understood in perturbations of integrable Hamiltonian systems. Resonances lead there to the wealth of ergodic invariant measures and to different time and space averaging. It turns out (see [**11**]) that wealth of ergodic invariant measures with nice properties (such as Gibbs measures) for Axiom A and expanding dynamical systems also yields in the fully coupled averaging setup with the latter fast motions examples of nonconvergence as $\varepsilon \to 0$ for large sets of initial conditions (see Remark 1.2.12).

In Hamiltonian systems, which are a classical object for applications of averaging methods, the whole space is fibered into manifolds of constant energy. For some mechanical systems these manifolds have negative curvature with respect to the natural metric and their motion is described by geodesic flows there. Hyperbolic Hamiltonian systems were discussed, for instance, in [**64**] and a specific example of a particle in a magnetic field leading to such systems was considered recently in [**74**]. Of course, these lead to Hamiltonian systems which are far from integrable. Such situations fall in our framework and they are among main motivations for this work. This suggests to consider the equation (1.1.1) on a (locally trivial) fiber bundle $\mathcal{M} = \{(x, y) : x \in U, y \in M_x\}$ with a base U being an open subset in a Riemannian manifold N and fibers M_x being diffeomorphic compact Riemannian manifolds (see [**75**]). On the other hand, $\mathcal{M}$ has a local product structure and if $\|B\|$ is bounded then the slow motion stays in one chart during time intervals of order Δ/ε with Δ small enough. Hence, studying behavior of solutions of (1.1.1) on each such time interval separately we come back to the product space $\mathbb{R}^d \times \mathbf{M}$ setup and will only have to piece results together to see the picture on a larger time interval of length T/ε.

We assume in the first part of this work that $b(x, y)$ is C^2 in x and y and that for each x in a closure of a relatively compact domain $\mathcal{X}$ the flow F_x^t is Anosov or, more generally, Axiom A in a neighborhood of an attractor Λ_x. Let μ_x^{SRB} be the Sinai-Ruelle-Bowen (SRB) invariant measure of F_x^t on Λ_x and set $\bar{B}(x) = \int B(x, y)d\mu_x^{\mathrm{SRB}}(y)$. It is known (see [**16**]) that the vector field $\bar{B}(x)$ is Lipschitz continuous in x, and so the averaged equations (1.1.4) and (1.1.6) have unique solutions $\bar{X}^\varepsilon(t)$ and $\bar{Z}(t) = \bar{X}^\varepsilon(t/\varepsilon)$. Still, in general, the measures μ_x^{SRB} are singular with respect to the Riemannian volume on $\mathbf{M}$, and so the method of [**1**] cannot be applied here. We proved in [**55**] that, nevertheless, (1.1.7) still holds true in this case, as well, and, moreover, the measure in (1.1.7) can be estimated by $e^{-c/\varepsilon}$ with some $c = c(\delta) > 0$. The convergence (1.1.7) itself without an exponential estimate can be proved by another method (see [**57**]) which can be applied also to some partially hyperbolic fast motions . An extension of the averaging principle in the sense of convergence of Young measures is discussed in Section 1.11.

Once the convergence of $Z_{x,y}^\varepsilon(t) = X_{x,y}^\varepsilon(t/\varepsilon)$ to $\bar{Z}_x(t) = \bar{X}_x^\varepsilon(t/\varepsilon)$ as $\varepsilon \to 0$ is established it is interesting to study the asymptotic behavior of the normalized error

$$(1.1.8) \qquad V_{x,y}^{\varepsilon,\theta}(t) = \varepsilon^{\theta-1}(Z_{x,y}^\varepsilon(t) - \bar{Z}_x(t)), \quad \theta \in [\frac{1}{2}, 1].$$

Namely, in our situation it is natural to study the distributions $m\{y : V_{x,y}^{\varepsilon,\theta}(\cdot) \in A\}$ as $\varepsilon \to 0$ where m is the normalized Riemannian volume on $\mathbf{M}$ and A is a Borel subset in the space C_{0T} of continuous paths $\varphi(t)$, $t \in [0, T]$ on $\mathbb{R}^d$. We will obtain in this work large deviations bounds for $V_{x,y}^\varepsilon = V_{x,y}^{\varepsilon,1}$ which will give, in particular, the result from [**55**] saying that

$$(1.1.9) \qquad m\{y : \|V_{x,y}^\varepsilon\|_{0,T} > \delta\} \to 0 \quad \text{as} \quad \varepsilon \to 0$$

exponentially fast in $1/\varepsilon$ where $\|\cdot\|_{0,T}$ is the uniform norm on C_{0T}. However, the main goal of this work is not to provide another derivation of (1.1.9) but to obtain precise upper and lower large deviations bounds which not only estimate measure of sets of initial conditions for which the slow motion Z^ε exhibits substantially different behavior than the averaged one $\bar{Z}$ but also enable us to go further and to investigate much longer exponential in $1/\varepsilon$ time behavior of Z^ε. Namely, we will be able to study exits of the slow motion from a neighborhood of an attractor of the averaged one and transitions of Z^ε between basins of attractors of $\bar{Z}$. Such evolution, which becomes visible only on much longer than $1/\varepsilon$ time scales, is usually called adiabatic in the framework of averaging. In the simpler case when the fast motion does not depend on the slow one such results were discussed in [**49**]. Still, even in this uncoupled situation descriptions of transitions of the slow motion between attractors of the averaged one were not justified rigorously both in the Markov processes case of [**29**] and in the dynamical systems case of [**49**]. Extending these technique to three scale equations may exhibit stochastic resonance type phenomena producing a nearly periodic motion of the slowest motion which is described in Section 1.10 below. These problems seem to be important in the study of climate–weather interactions and they were discussed in [**18**] and [**37**] in the framework of a model describing transitions between steady climatic states with weather evolving as a fast chaotic system and climate playing the role of the slow motion. Such "very long" time description of the slow motion is usually impossible in the traditional averaging setup which deals with perturbations of integrable Hamiltonian systems. In the fully coupled situation we cannot work just with one hyperbolic flow but have to consider continuously changing fast motions which requires a special technique. In particular, the full flow Φ_ε^t on $\mathbb{R}^d \times \mathbf{M}$ defined above and viewed as a small perturbation of the partially hyperbolic system Φ^t plays an important role in our considerations. It is somewhat surprising that the "very long time" behavior of the slow motion which requires certain "Markov property type" arguments still can be described in the fully coupled setup which involves continuously changing fast hyperbolic motions. It turns out that the perturbed system still possesses semi-invariant expanding cones and foliations and a certain volume lemma type result on expanding leaves plays an important role in our argument for transition from small time were perturbation techniques still works to the long and "very long" time estimates.

It is plausible that moderate deviations type results can be proved for $V_{x,y}^{\varepsilon,\theta}$ when $1/2 < \theta < 1$ and that the distribution of $V_{x,y}^{\varepsilon,1/2}(\cdot)$ in y converges to the distribution of a Gaussian diffusion process in $\mathbb{R}^d$. Still, this requires somewhat different methods and it will not be discussed here. In this regard we can mention limit theorems obtained in [**14**] for a system of two heavy and light particles which leads to an averaging setup for a billiard flow. For the simpler case when b does not depend on x, i.e. when all flows F_x^t are the same, the moderate deviations and Gaussian approximations results were obtained previously in [**50**]. Related results in this uncoupled situation concerning Hasselmann's nonlinear (strong) diffusion approximation of the slow motion X^ε were obtained in [**56**].

We consider also the discrete time case where (1.1.1) is replaced by difference equations for sequences $X^\varepsilon(n) = X_{x,y}^\varepsilon(n)$ and $Y^\varepsilon(n) = Y_{x,y}^\varepsilon(n)$, $n = 0, 1, ...$ so that

$$X^\varepsilon(n+1) - X^\varepsilon(n) = \varepsilon B(X^\varepsilon(n), Y^\varepsilon(n)),$$
$$Y^\varepsilon(n+1) = F_{X^\varepsilon(n)} Y^\varepsilon(n), \ X^\varepsilon(0) = x, Y^\varepsilon(0) = y \tag{1.1.10}$$

where $B : \mathcal{X} \times \mathbf{M} \to \mathbb{R}^d$ is Lipschitz in both variables and the maps $F_x : \mathbf{M} \to \mathbf{M}$ are smooth and depend smoothly on the parameter $x \in \mathbb{R}^d$. Introducing the map

$$\Phi_\varepsilon(x,y) = (X^\varepsilon_{x,y}(1), Y^\varepsilon_{x,y}(1)) = (x + \varepsilon B(x,y), F_x y)$$

we can also view this setup as a perturbation of the map $\Phi(x,y) = (x, F_x y)$ describing an ideal system where parameters $x \in \mathbb{R}^d$ do not change. Assuming that F_x, $x \in \mathbb{R}^d$ are C^2 depending on x families of either C^2 expanding transformations or C^2 Axiom A diffeomorphisms in a neighborhood of an attractor Λ_x we will derive large deviations estimates for the difference $X^\varepsilon_{x,y}(n) - \bar{X}^\varepsilon_x(n)$ where $\bar{X}^\varepsilon = \bar{X}^\varepsilon_x$ solves the equation

$$\frac{d\bar{X}^\varepsilon(t)}{dt} = \varepsilon \bar{B}(\bar{X}^\varepsilon(t)), \;\; \bar{X}^\varepsilon(0) = x \tag{1.1.11}$$

where $\bar{B}(x) = \int B(x,y) d\mu^{\mathrm{SRB}}_x(y)$ and μ^{SRB}_x is the corresponding SRB invariant measure of F_x on Λ_x. The discrete time results are obtained, essentially, by simplifications of the corresponding arguments in the continuous time case which enable us to describe "very long" time behavior of the slow motion also in the discrete time case. Since our methods work not only for fast motions being Axiom A diffeomorphisms but also when they are expanding transformations we can construct simple examples satisfying conditions of our theorems and exhibiting corresponding effects. In particular, we produce in Section 1.9 computational examples which demonstrate transitions of the slow motion between neighborhoods of attractors of the averaged system.

A series of related results for the case when ordinary differential equations in (1.1.1) are replaced by fully coupled stochastic differential equations appeared in [**45**], [**77**]–[**79**], [**66**], and [**5**]. Hasselmann's nonlinear (strong) diffusion approximation of the slow motion in the fully coupled stochastic differential equations setup was justified in [**10**]. When the fast process does not depend on the slow one such results were obtained in [**44**], [**29**], and [**54**]. Especially relevant for our results here is [**78**] and we employ some elements of the probabilistic strategy from this paper. Still, the methods there are quite different from ours and they are based heavily, first, on the Markov property of processes emerging there and, secondly, on uniformity and nondegeneracy of the fast diffusion term assumptions which cannot be satisfied in our circumstances as our deterministic fast motions are very degenerate from this point of view. Note that the proof in [**78**] contains a vicious cycle and substantial gaps which recently were essentially fixed in [**79**]. Some of the dynamical systems technique here resembles [**49**] but the dependence of the fast motion on the slow one complicates the analysis substantially and requires additional machinery. A series of results on Cramer's type asymptotics for fully coupled averaging with Axiom A diffeomorphisms as fast motions appeared recently in [**4**]–[**7**]. Observe that the methods there do not work for continuous time Axiom A dynamical systems considered here, they cannot lead, in principle, to the standard large deviations estimates of our work and they deal with deviations of X^ε from the averaged motion only at the last moment and not of its whole path. Various limit theorems for the difference equations setup (1.1.10) with partially hyperbolic fast motions were obtained recently in [**20**] and [**21**].

The study of deviations from the averaged motion in the fully coupled case seems to be quite important for applications, especially, from phenomenological point of view. In addition to perturbations of Hamiltonian systems mentioned above there are many non Hamiltonian systems which are naturally to consider from the beginning as a combination of fast and slow motions. For instance, Hasselmann [**36**] based his model of weather–climate interaction on the assumption that weather is a fast chaotic motion depending on climate as a slow motion which differs from the corresponding averaged motion mainly by a diffusion term. Though, as shown in [**54**], [**10**] and [**56**], this diffusion error term does not help in the study of large deviations which are responsible for rare transitions of the slow motion between attractors of the averaged one, the latter phenomenon can be described in our framework and it seems to be important in certain models of climate fluctuations (see [**18**] and [**37**]). Very slow nearly periodic motions appearing in the stochastic resonance framework considered in Section 1.10 may also fit into this subject in the discussion on "ice ages". Of course, it is hard to believe that real world chaotic systems can be described precisely by an Anosov or Axiom A flow but one may take comfort in the Chaotic Hypothesis [**32**]: " A chaotic mechanical system can be regarded for practical purposes as a topologically mixing Anosov system".

1.2. Main results

Let F^t be a C^2 flow on a compact Riemannian manifold $\mathbf{M}$ given by a differential equation

$$\frac{dF^ty}{dt}=b(F^ty),\ F^0y=y. \tag{1.2.1}$$

A compact F^t–invariant set $\Lambda\subset\mathbf{M}$ is called hyperbolic if there exists $\kappa>0$ and the splitting $T_\Lambda\mathbf{M}=\Gamma^s\oplus\Gamma^0\oplus\Gamma^u$ into the continuous subbundles $\Gamma^s,\Gamma^0,\Gamma^u$ of the tangent bundle $T\mathbf{M}$ restricted to Λ, the splitting is invariant with respect to the differential DF^t of F^t, Γ^0 is the one dimensional subbundle generated by the vector field b, and there is $t_0>0$ such that for all $\xi\in\Gamma^s$, $\eta\in\Gamma^u$, and $t\geq t_0$,

$$\|DF^t\xi\|\leq e^{-\kappa t}\|\xi\|\quad\text{and}\quad\|DF^{-t}\eta\|\leq e^{-\kappa t}\|\eta\|. \tag{1.2.2}$$

A hyperbolic set Λ is said to be basic hyperbolic if the periodic orbits of $F^t|_\Lambda$ are dense in Λ, $F^t|_\Lambda$ is topologically transitive, and there exists an open set $U\supset\Lambda$ with $\Lambda=\cap_{-\infty<t<\infty}F^tU$. Such a Λ is called a basic hyperbolic attractor if for some open set U and $t_0>0$,

$$F^{t_0}\bar U\subset U\quad\text{and}\quad\cap_{t>0}F^tU=\Lambda$$

where $\bar U$ denotes the closure of U. If $\Lambda=\mathbf{M}$ then F^t is called an Anosov flow.

1.2.1. Assumption. *The family $b(x,\cdot)$ in (1.1.2) consists of C^2 vector fields on a compact $n_{\mathbf{M}}$-dimensional Riemannian manifold $\mathbf{M}$ with uniform C^2 dependence on the parameter x belonging to a neighborhood of the closure $\bar{\mathcal{X}}$ of a relatively compact open connected set $\mathcal{X}\subset\mathbb{R}^d$. Each flow F^t_x, $x\in\bar{\mathcal{X}}$ on $\mathbf{M}$ given by*

$$\frac{dF^t_xy}{dt}=b(x,F^t_xy),\quad F^0_xy=y \tag{1.2.3}$$

possesses a basic hyperbolic attractor Λ_x with a splitting $T_{\Lambda_x}\mathbf{M}=\Gamma^s_x\oplus\Gamma^0_x\oplus\Gamma^u_x$ satisfying (1.2.2) with the same $\kappa>0$ and there exists an open set $\mathcal{W}\subset\mathbf{M}$ and $t_0>0$ such that

$$\Lambda_x\subset\mathcal{W},\ F^t_x\bar{\mathcal{W}}\subset\mathcal{W}\,\forall t\geq t_0,\ \text{and}\ \cap_{t>0}F^t_x\mathcal{W}=\Lambda_x\ \forall x\in\bar{\mathcal{X}}. \tag{1.2.4}$$

Let $J_x^u(t,y)$ be the absolute value of the Jacobian of the linear map $DF_x^t(y): \Gamma_x^u(y) \to \Gamma_x^u(F_x^t y)$ with respect to the Riemannian inner products and set

$$\varphi_x^u(y) = -\frac{dJ_x^u(t,y)}{dt}\Big|_{t=0}. \tag{1.2.5}$$

The function $\varphi_x^u(y)$ is known to be Hölder continuous in y, since the subbundles Γ_x^u are Hölder continuous (see [**13**] and [**60**]), and $\varphi_x^u(y)$ is C^1 in x (see [**16**]).

Let $\mathcal{W}$ satisfy (1.2.4) and set $\mathcal{W}_x^t = \{y \in \mathcal{W} : F_x^s y \in \bar{\mathcal{W}}\ \forall s \in [0,t]\}$. A set $E \subset \mathcal{W}_x^t$ is called $(\delta,t)-$separated for the flow F_x if $y,z \in E$, $y \neq z$ imply $d(F_x^s y, F_x^s z) > \delta$ for some $s \in [0,t]$, where $d(\cdot,\cdot)$ is the distance function on $\mathbf{M}$. For each continuous function ψ on $\mathcal{W}$ set $P_x(\psi,\delta,t) = \sup\{\sum_{y\in E}\exp\int_0^t \psi(F_x^s y)ds : E \subset \mathcal{W}_x^t \text{ is } (\delta,t) - \text{separated for } F_x\}$, $P_x(\psi,\delta,t) = 0$ if $\mathcal{W}_x^t = \emptyset$, and

$$P_x(\psi,\delta) = \limsup_{t\to\infty}\frac{1}{t}\log P_x(\psi,\delta,t).$$

The latter is monotone in δ, and so the limit

$$P_x(\psi) = \lim_{\delta\to 0} P_x(\psi,\delta)$$

exists and it is called the topological pressure of ψ for the flow F_x^t. Let $\mathcal{M}_x$ denotes the space of F_x^t-invariant probability measures on Λ_x then (see, for instance, [**60**]) the following variational principle

$$P_x(\psi) = \sup_{\mu\in\mathcal{M}_x}\Big(\int \psi d\mu + h_\mu(F_x^1)\Big) \tag{1.2.6}$$

holds true where $h_\mu(F_x^1)$ is the Kolmogorov–Sinai entropy of the time-one map F_x^1 with respect to μ. If q is a Hölder continuous function on Λ_x then there exists a unique F_x^t-invariant measure μ_x^q on Λ_x, called the equilibrium state for $\varphi_x^u + q$, such that

$$P_x(\varphi_x^u + q) = \int(\varphi_x^u + q)d\mu_x^q + h_{\mu_x^q}(F_x^1). \tag{1.2.7}$$

We denote μ_x^0 by μ_x^{SRB} since it is usually called the Sinai–Ruelle– Bowen (SRB) measure . Since Λ_x are attractors we have that $P_x(\varphi_x^u) = 0$ (see [**13**]).

For any probability measure ν on $\bar{\mathcal{W}}$ define

$$I_x(\nu) = \begin{cases} -\int \varphi_x^u d\nu - h_\nu(F_x^1) & \text{if } \nu \in \mathcal{M}_x \\ \infty & \text{otherwise.}\end{cases} \tag{1.2.8}$$

Then

$$P_x(\varphi_x^u + q) = \sup_\nu\Big(\int q d\nu - I_x(\nu)\Big).$$

Observe that by the Ruelle inequality (see, for instance, [**60**], Theorem S.2.13), $I_x(\nu) \geq 0$, and so in view of Assumption 1.2.1 for any $\nu \in \mathcal{M}_x$,

$$I_x(\nu) \leq \sup_{y\in\Lambda_x}|\varphi_x^u(y)| \leq \sup_{x\in\mathcal{X},y\in\Lambda_x}|\varphi_x^u(y)| < \infty. \tag{1.2.9}$$

It is known that $h_\nu(F_x^1)$ is upper semicontinuous in ν since hyperbolic flows are entropy expansive (see [**8**]). Thus $I_x(\nu)$ is a lower semicontinuous functional in ν and it is also convex (and affine on $\mathcal{M}_x$) since entropy h_ν is affine in ν (see, for instance, [**80**]). Hence, by the duality theorem (see [**2**], p.201),

$$I_x(\nu) = \sup_{q\in\mathcal{C}(\mathbf{M})}\Big(\int q d\nu - P_x(\varphi_x^u + q)\Big).$$

Observe that this formula can be proved more directly. Namely, if we define $I_x(\nu)$ by it in place of (1.2.8) then (1.2.8) follows for $\nu \in \mathcal{M}_x$ from Theorem 9.12 in [**80**] and it is easy to show directly that $I_x(\nu)$ defined in this way equals ∞ for any finite signed measure ν which is not F_x-invariant.

Since we assume that the vector field B is C^1 in both arguments (here only continuity in y is needed) then for any $x, x' \in \mathcal{X}$ and $\alpha, \beta \in \mathbb{R}^d$ we can define $H(x,x',\beta) = P_x(<\beta, B(x',\cdot)> + \varphi_x^u)$ and $H(x,\beta) = H(x,x,\beta)$. Then

$$(1.2.10) \qquad \begin{aligned} H(x,x',\beta) &= \sup_\nu \big(\int <\beta, B(x',y)> d\nu(y) - I_x(\nu)\big) \\ &= \sup_{\alpha\in\mathbb{R}^d} \big(<\alpha,\beta> - L(x,x',\alpha)\big) \end{aligned}$$

where

$$(1.2.11) \qquad L(x,x',\alpha) = \inf\{I_x(\nu) : \int B(x',y)d\nu(y) = \alpha\}$$

if $\nu \in \mathcal{M}_x$ satisfying the condition in brackets exists and $L(x,x',\alpha) = \infty$, otherwise. Since, $H(x,x',\beta)$ is convex and continuous the duality theorem (see [**2**], p.201) yields that

$$(1.2.12) \qquad L(x,x',\alpha) = \sup_{\beta\in\mathbb{R}^d} \big(<\alpha,\beta> - H(x,x',\beta)\big)$$

provided there exists a probability measure $\nu \in \mathcal{M}_x$ such that $\int B(x',y)d\nu(y) = \alpha$ and $L(x,x',\alpha) = \infty$, otherwise. Clearly, $L(x,x',\alpha)$ is convex and lower semi-continuous in all arguments and, in particular, it is measurable. We set also $L(x,\alpha) = L(x,x,\alpha)$.

Denote by C_{0T} the space of continuous curves $\gamma_t = \gamma(t)$, $t \in [0,T]$ in $\mathcal{X}$ which is the space of continuous maps of $[0,T]$ into $\mathcal{X}$. For each absolutely continuous $\gamma \in C_{0T}$ its velocity $\dot\gamma_t$ can be obtained as the almost everywhere limit of continuous functions $n(\gamma_{t+n^{-1}} - \gamma_t)$ when $n \to \infty$. Hence $\dot\gamma_t$ is measurable in t, and so we can set

$$(1.2.13) \quad S_{0T}(\gamma) = \int_0^T L(\gamma_t,\dot\gamma_t)dt = \int_0^T \inf\{I_{\gamma_t}(\nu) : \dot\gamma_t = \bar B_\nu(\gamma_t),\ \nu \in \mathcal{M}_{\gamma_t}\}dt,$$

where $\bar B_\nu(x) = \int B(x,y)d\nu(y)$, provided for Lebesgue almost all $t \in [0,T]$ there exists $\nu_t \in \mathcal{M}_{\gamma_t}$ for which $\dot\gamma_t = \bar B_{\nu_t}(\gamma_t)$, and $S_{0T}(\gamma) = \infty$ otherwise. It follows from [**13**] and [**16**] that

$$S_{0T}(\gamma) \ge S_{0T}(\gamma^u) = -\int_0^T P_{\gamma_t^u}(\varphi^u_{\gamma_t^u})dt = 0$$

where γ_t^u is the unique solution of the equation

$$(1.2.14) \qquad \dot\gamma_t^u = \bar B(\gamma_t^u), \quad \gamma_0^u = x,$$

where $\bar B(z) = \bar B_{\mu_z^{\mathrm{SRB}}}(z)$, and the equality $S_{0T}(\gamma) = 0$ holds true if and only if $\gamma = \gamma^u$.

Define the uniform metric on C_{0T} by

$$\mathbf{r}_{0T}(\gamma,\eta) = \sup_{0\le t\le T} |\gamma_t - \eta_t|$$

for any $\gamma, \eta \in C_{0T}$. Set

$$\Psi_{0T}^a(x) = \{\gamma \in C_{0T} : \gamma_0 = x,\ S_{0T}(\gamma) \le a\}.$$

Since $L(x,\alpha)$ is lower semicontinuous and convex in α and, in addition, $L(x,\alpha)=\infty$ if $|\alpha|>\sup_{y\in\mathbf{M}}|B(x,y)|$ we conclude that the conditions of Theorem 3 in Ch.9 of [**41**] are satisfied as we can choose a fast growing minorant of $L(x,\alpha)$ required there to be zero in a sufficiently large ball and to be equal, say, $|\alpha|^2$ outside of it. As a result, it follows that S_{0T} is lower semicontinuous functional on C_{0T} with respect to the metric $\mathbf{r}_{0T}$, and so $\Psi^a_{0T}(x)$ is a closed set which plays a crucial role in the large deviations arguments below.

We suppose that the coefficients of (1.1.1) satisfy the following

1.2.2. ASSUMPTION. *There exists $K>0$ such that*

$$\|B(x,y)\|_{C^1(\mathcal{X}\times\mathbf{M})}+\|b(x,y)\|_{C^2(\mathcal{X}\times\mathbf{M})}\le K \tag{1.2.15}$$

where $\|\cdot\|_{C^i(\mathcal{X}\times\mathbf{M})}$ is the C^i norm of the corresponding vector fields on $\mathcal{X}\times\mathbf{M}$.

Set $\mathcal{X}_t=\{x\in\mathcal{X}: X^\varepsilon_{x,y}(s)\in\mathcal{X}$ and $\bar{X}^\varepsilon_x(s)\in\mathcal{X}$ for all $y\in\bar{\mathcal{W}}$, $s\in[0,t/\varepsilon]$, $\varepsilon>0\}$. Clearly, $\mathcal{X}_t\supset\{x\in\mathcal{X}:\inf_{z\in\partial\mathcal{X}}|x-z|\ge 2Kt\}$. The following is one of the main results of this paper.

1.2.3. THEOREM. *Suppose that $x\in\mathcal{X}_T$ and $X^\varepsilon_{x,y}$, $Y^\varepsilon_{x,y}$ are solutions of (1.1.1) with coefficients satisfying Assumptions 1.2.1 and 1.2.2. Set $Z^\varepsilon_{x,y}(t)=X^\varepsilon_{x,y}(t/\varepsilon)$ then for any $a,\delta,\lambda>0$ and every $\gamma\in C_{0T}$, $\gamma_0=x$ there exists $\varepsilon_0=\varepsilon_0(x,\gamma,a,\delta,\lambda)>0$ such that for $\varepsilon<\varepsilon_0$,*

$$m\left\{y\in\mathcal{W}:\ \mathbf{r}_{0T}(Z^\varepsilon_{x,y},\gamma)<\delta\right\}\ge\exp\left\{-\frac{1}{\varepsilon}(S_{0T}(\gamma)+\lambda)\right\} \tag{1.2.16}$$

and

$$m\left\{y\in\mathcal{W}:\ \mathbf{r}_{0T}(Z^\varepsilon_{x,y},\Psi^a_{0T}(x))\ge\delta\right\}\le\exp\left\{-\frac{1}{\varepsilon}(a-\lambda)\right\} \tag{1.2.17}$$

where, recall, m is the normalized Riemannian volume on $\mathbf{M}$. The functional $S_{0T}(\gamma)$ for $\gamma\in C_{0T}$ is finite if and only if $\dot\gamma_t=\bar{B}_{\nu_t}(\gamma_t)$ for $\nu_t\in\mathcal{M}_{\gamma_t}$ and Lebesgue almost all $t\in[0,T]$. Furthermore, $S_{0T}(\gamma)$ achieves its minimum 0 only on γ^u satisfying (1.2.14) for all $t\in[0,T]$. Finally, for any $\delta>0$ there exist $c(\delta)>0$ and $\varepsilon_0>0$ such that for all $\varepsilon<\varepsilon_0$,

$$m\left\{y\in\mathcal{W}:\ \mathbf{r}_{0T}(Z^\varepsilon_{x,y},\bar{Z}_x)\ge\delta\right\}\le\exp\left(-\frac{c(\delta)}{\varepsilon}\right) \tag{1.2.18}$$

where $\bar{Z}_x=\gamma^u$ is the unique solution of (1.2.14).

Observe that (1.2.18) (which was proved already in [**55**] by a less precise large deviations argument) follows from (1.2.17) and the lower semicontinuity of the functional S_{0T} and it says, in particular, that $Z^\varepsilon_{x,\cdot}$ converges to $\bar{Z}_x$ in measure on the space $(\mathcal{W},m)$ with respect to the metric $\mathbf{r}_{0T}$. It is naturally to ask whether we have here also the convergence for m-almost all $y\in\mathcal{W}$. An example due to A.Neishtadt discussed in [**56**] shows that in the classical situation of perturbations of integrable Hamiltonian systems, in general, the averaging principle holds true only in the sense of convergence in measure on the space of intitial conditions but not in the sense of the almost everywhere convergence. This example concerns the simple system $\dot I=\varepsilon(4+8\sin\varphi-I)$, $\dot\varphi=I$ with the one dimensional slow motion I and the fast motion φ evolving on the circle while the corresponding averaged motion $J=\bar I$ satisfies the equation $\dot J=\varepsilon(4-J)$. The resonance occurs here only when $I=0$ but it suffices already to create troubles in the averaging principle.

Namely, it turns out that for any initial condition (I_0, φ_0) with $-2 < I_0 < -1$ there exists a sequence $\varepsilon_n \to 0$ such that $I^{\varepsilon_n}_{I_0,\varphi_0}(1/\varepsilon_n) < J^{\varepsilon_n}_{I_0}(1/\varepsilon_n) - 3/2$ though, of course, convergence in measure holds true here (see [**63**]). Recently (see [**11**] and Remark 1.2.12), such nonconvergence examples were constructed for the difference equations averaging setup (1.1.10) with expanding fast motions and there is no doubt that such examples exist also in the continuous time setup (1.1.1) when fast motions are Axiom A flows as in this paper. Observe also that (1.2.16) and (1.2.17) remain true (with the same proof) if we replace m there by μ^{SRB}_x but as an example in [**11**] shows we cannot, in general, replace m there by an arbitrary Gibbs measure μ_x of F^t_x.

Next, let $V \subset \mathcal{X}$ be a connected open set and put $\tau^\varepsilon_{x,y}(V) = \inf\{t \geq 0 : Z^\varepsilon_{x,y}(t) \notin V\}$ where we take $\tau^\varepsilon_{x,y}(V) = \infty$ if $X^\varepsilon_{x,y}(t) \in V$ for all $t \geq 0$. The following result follows directly from Theorem 1.2.3.

1.2.4. COROLLARY. *Under the conditions of Theorem 1.2.3 for any $T > 0$ and $x \in V$,*

$$\lim_{\varepsilon\to 0} \varepsilon \log m\left\{y \in \mathcal{W} : \tau^\varepsilon_{x,y}(V) < T\right\}$$
$$= -\inf\left\{S_{0t}(\gamma) : \gamma \in C_{0T},\ t \in [0,T],\ \gamma_0 = x,\ \gamma_t \notin V\right\}.$$

Precise large deviations bounds such as (1.2.16) and (1.2.17) of Theorem 1.2.3 (which will be needed uniformly on certain unstable discs) are crucial in our study in Sections 1.7 and 1.8 of the "very long", i.e. exponential in $1/\varepsilon$, time "adiabatic" behaviour of the slow motion which cannot be described usually in the traditional theory of averaging where only perturbations of integrable Hamiltonian systems are considered. Namely, we will describe such long time behavior of Z^ε in terms of the function

$$R(x,z) = \inf_{t\geq 0, \gamma\in C_{0t}} \{S_{0t}(\gamma) : \gamma_0 = x,\ \gamma_t = z\}$$

under various assumptions on the averaged motion $\bar{Z}$. Observe that R satisfies the triangle inequality $R(x_1,x_2) + R(x_2,x_3) \geq R(x_1,x_3)$ for any $x_1,x_2,x_3 \in \mathcal{X}$ and it determines a semi metric on $\mathcal{X}$ which measures "the difficulty'" for the slow motion to move from point to point in terms of the functional S.

Introduce the averaged flow Π^t on $\mathcal{X}_t$ by

$$\frac{d\Pi^t x}{dt} = \bar{B}(\Pi^t x),\ x \in \mathcal{X}_t \tag{1.2.19}$$

where, recall, $\bar{B}(z) = \bar{B}_{\mu^{\mathrm{SRB}}_z}(z)$ and $\bar{B}_\nu(z) = \int B(z,y)d\nu(y)$ for any probability measure ν on $\mathbf{M}$. Call a Π^t-invariant compact set $\mathcal{O} \subset \mathcal{X}$ an S-compact if for any $\eta > 0$ there exist $T_\eta \geq 0$ and an open set $U_\eta \supset \mathcal{O}$ such that whenever $x \in \mathcal{O}$ and $z \in U_\eta$ we can pick up $t \in [0, T_\eta]$ and $\gamma \in C_{0t}$ satisfying

$$\gamma_0 = x,\ \gamma_t = z \text{ and } S_{0t}(\gamma) \leq \eta.$$

It is clear from this definition that $R(x,z) = 0$ for any pair points x, z of an S-compact $\mathcal{O}$ and by the above triangle inequality for R we see that $R(x,z)$ takes on the same value when $z \in \mathcal{X}$ is fixed and x runs over $\mathcal{O}$. We say that the vector field B on $\mathcal{X} \times \mathbf{M}$ is complete at $x \in \mathcal{X}$ if the convex set of vectors $\{\beta\bar{B}_\nu(x) : \beta \in [0,1],\ \nu \in \mathcal{M}_x\}$ contains an open neigborhood of the origin in $\mathbb{R}^d$. In Lemma 1.6.2 we will show that if $\mathcal{O} \subset \mathcal{X}$ is a compact Π^t-invariant set such that B is complete at each $x \in \mathcal{O}$ and either $\mathcal{O}$ contains a dense orbit of the flow Π^t (i.e. Π^t is topologically transitive on $\mathcal{O}$) or $R(x,z) = 0$ for any $x, z \in \mathcal{O}$ then $\mathcal{O}$ is an

S-compact. Moreover, to ensure that $\mathcal{O}$ is an S-compact it suffices to assume that B is complete only at some point of $\mathcal{O}$ and the flow Π^t on $\mathcal{O}$ is minimal , i.e. the Π^t-orbits of all points are dense in $\mathcal{O}$ or, equivalently, for any $\eta > 0$ there exists $T(\eta) > 0$ such that the orbit $\{\Pi^t x,\, t \in [0, T(\eta)]\}$ of length $T(\eta)$ of each point $x \in \mathcal{O}$ forms an η-net in $\mathcal{O}$. The latter condition obviously holds true when $\mathcal{O}$ is a fixed point or a periodic orbit of Π^t but also, more generally, when Π^t on $\mathcal{O}$ is uniquely ergodic (see [**60**], [**65**], [**80**]). Among well known examples of uniquely ergodic flows we can mention irrational translations of tori and horocycle flows on surfaces of negative curvature.

A compact Π^t-invariant set $\mathcal{O} \subset \mathcal{X}$ is called an attractor (for the flow Π^t) if there is an open set $U \supset \mathcal{O}$ and $t_U > 0$ such that

$$\Pi^{t_U}\bar{U} \subset U \text{ and } \lim_{t\to\infty} \text{dist}(\Pi^t z, \mathcal{O}) = 0 \text{ for all } z \in U.$$

For an attractor $\mathcal{O}$ the set $V = \{z \in \mathcal{X} : \lim_{t\to\infty} \text{dist}(\Pi^t z, \mathcal{O}) = 0\}$, which is clearly open, is called the basin (domain of attraction) of $\mathcal{O}$. An attractor which is also an S-compact will be called an S-attractor .

In what follows we will speak about connected open sets V with piecewise smooth boundaries ∂V. The latter can be introduced in various ways but it will be convenient here to adopt the definition from [**17**] saying that ∂V is the closure of a finite union of disjoint, connected, codimension one, extendible C^1 (open or closed) submanifolds of $\mathbb{R}^d$ which are called faces of the boundary. The extendibility condition means that the closure of each face is a part of a larger submanifold of the same dimension which coincides with the face itself if the latter is a compact submanifold. This enables us to extend fields of normal vectors to the boundary of faces and to speak about minimal angles between adjacent faces which we assume to be uniformly bounded away from zero or, in other words, angles between exterior normals to adjacent faces at a point of intersection of their closures are uniformly bounded away from π and $-\pi$. The following result which will be proved in Section 1.7 describes exits of the slow motion from neighborhoods of attractors of the averaged motion.

1.2.5. THEOREM. *Let $\mathcal{O} \subset \mathcal{X}$ be an S-attractor of the flow Π^t whose basin contains the closure $\bar{V}$ of a connected open set V with a piecewise smooth boundary ∂V such that $\bar{V} \subset \mathcal{X}$ and assume that for each $z \in \partial V$ there exists $\varpi = \varpi(z) > 0$ and an F_z^t–invariant probability measure $\nu = \nu_z$ on Λ_z such that*

$$\text{(1.2.20)} \qquad z + s\bar{B}(z) \in V \;\text{ but }\; z + s\bar{B}_\nu(z) \in \mathbb{R}^d \setminus \bar{V} \;\text{ for all } s \in (0, \varpi],$$

i.e. $\bar{B}(z) \neq 0$, $\bar{B}_\nu(z) \neq 0$ and the former vector points out into the interior while the latter into the exterior of V. Set $R_\partial(z) = \inf\{R(z, \tilde{z}) : \tilde{z} \in \partial V\}$ and $\partial_{\min}(z) = \{\tilde{z} \in \partial V : R(z, \tilde{z}) = R_\partial(z)\}$. Then $R_\partial(z)$ takes on the same value R_∂ and $\partial_{\min}(z)$ coincides with the same compact nonempty set $\partial_{\min}$ for all $z \in \mathcal{O}$ while $R_\partial(x) \leq R_\partial$ for all $x \in V$. Furthermore, for any $x \in V$,

$$\text{(1.2.21)} \qquad \lim_{\varepsilon\to 0} \varepsilon \log \int_{\mathcal{W}} \tau^{\varepsilon}_{x,y}(V) dm(y) = R_\partial > 0$$

and for each $\alpha > 0$ there exists $\lambda(\alpha) = \lambda(x, \alpha) > 0$ such that for all small $\varepsilon > 0$,

$$\text{(1.2.22)} \quad m\{y \in \mathcal{W} : e^{(R_\partial - \alpha)/\varepsilon} > \tau^{\varepsilon}_{x,y}(V) \text{ or } \tau^{\varepsilon}_{x,y}(V) > e^{(R_\partial + \alpha)/\varepsilon}\} \leq e^{-\lambda(\alpha)/\varepsilon}.$$

Next, set

$$\Theta_v^\varepsilon(t) = \Theta_v^{\varepsilon,\delta}(t) = \int_0^t \mathbb{I}_{V\setminus U_\delta(\mathcal{O})}(Z_v^\varepsilon(s))ds$$

where $U_\delta(\mathcal{O}) = \{z \in \mathcal{X} : dist(z, \mathcal{O}) < \delta\}$ *and* $\mathbb{I}_\Gamma(z) = 1$ *if* $z \in \Gamma$ *and* $= 0$*, otherwise. Then for any* $x \in V$ *and* $\delta > 0$ *there exists* $\lambda(\delta) = \lambda(x, \delta) > 0$ *such that for all small* $\varepsilon > 0$*,*

$$m\{y \in \mathcal{W} : \Theta_{x,y}^\varepsilon(\tau_{x,y}^\varepsilon(V)) \geq e^{-\lambda(\delta)/\varepsilon}\tau_{x,y}^\varepsilon(V)\} \leq e^{-\lambda(\delta)/\varepsilon}. \tag{1.2.23}$$

Finally, for every $x \in V$ *and* $\delta > 0$*,*

$$\lim_{\varepsilon\to 0} m\{y \in \mathcal{W} : dist(Z_{x,y}^\varepsilon(\tau_{x,y}^\varepsilon(V)), \partial_{\min}) \geq \delta\} = 0 \tag{1.2.24}$$

provided $R_\partial < \infty$ *and the latter holds true if and only if for some* $T > 0$ *there exists* $\gamma \in C_{0T}$*,* $\gamma_0 \in \mathcal{O}$*,* $\gamma_T \in \partial V$ *such that* $\dot{\gamma}_t = \bar{B}_{\nu_t}(\gamma_t)$ *for Lebesgue almost all* $t \in [0, T]$ *with* $\nu_t \in \mathcal{M}_{\gamma_t}$ *then* $R_\partial < \infty$*.*

Theorem 1.2.5 asserts, in particular, that typically the slow motion Z^ε performs rare (adiabatic) fluctuations in the vicinity of an S-attractor $\mathcal{O}$ since it exists from any domain $U \supset \mathcal{O}$ with $\bar{U} \subset V$ for the time much smaller than $\tau^\varepsilon(V)$ (as the corresponding number $R_\partial = R_{\partial U}$ will be smaller) and by (1.2.23) it can spend in $V \setminus U_\delta(\mathcal{O})$ only small proportion of time which implies that Z^ε exits from U and returns to $U_\delta(\mathcal{O})$ (exponentially in $1/\varepsilon$) many times before it finally exits V. We observe that in the much simpler uncoupled setup corresponding results in the case of $\mathcal{O}$ being an attracting point were obtained for a continuous time Markov chain and an Axiom A flow as fast motions in [**29**] and [**49**], respectively, but the proofs there rely on the lower semicontinuity of the function R which does not hold true in general, and so extra conditions like S-compactness of $\mathcal{O}$ or, more specifically, the completness of B at $\mathcal{O}$ should be assumed there, as well. It is important to observe that the intuition based on diffusion type small random perturbations of dynamical systems should be applied with caution to problems of large deviations in averaging since the S-functional of Theorem 1.2.3 describing them is more complex and have rather different properties than the corresponding functional emerging in diffusion type random perturbations of dynamical systems (see [**31**]). The reason for this is the deterministic nature of the slow motion Z^ε which unlike a diffusion can move only with a bounded speed, and moreover, even in order to ensure its "diffusive like" local behaviour (i.e. to let it go in many directions) some extra nondegeneracy type conditions on the vector field B are required.

Our next result describes rare (adiabatic) transitions of the slow motion Z^ε between basins of attractors of the averaged flow Π^t which we consider now in the whole $\mathbb{R}^d$ and impose certain conditions on the structure of its ω-limit set .

1.2.6. Assumption. *Assumptions 1.2.1 and 1.2.2 hold true for* $\mathcal{X} = \mathbb{R}^d$*, the family* $\{F_x^t,\ t \leq 1,\ x \in \mathbb{R}^d\}$ *is a compact set of diffeomorphisms in the* C^2 *topology,*

$$\|B(x, y)\|_{C^2(\mathbb{R}^d\times\mathbf{M})} \leq K \tag{1.2.25}$$

for some $K > 0$ *independent of* x, y *and there exists* $r_0 > 0$ *such that*

$$(x, B(x, y)) \leq -K^{-1} \text{ for any } y \in \mathcal{W} \text{ and } |x| \geq r_0. \tag{1.2.26}$$

The condition (1.2.26) means that outside of some ball all vectors $B(x, y)$ have a bounded away from zero projection on the radial direction which points out to the origin. This condition can be weakened, for instance, it suffices that

$$\lim_{d\to\infty} \inf\{R(x,z) : \operatorname{dist}(x,z) \geq d\} = \infty$$

but, anyway, we have to make some assumption which ensure that the slow motion stays (at least, for "most" initial points $y \in \mathcal{W}$) in a compact region where really interesting dynamics takes place.

Next, suppose that the ω-limit set of the averaged flow Π^t is compact and it consists of two parts, so that the first part is a finite number of S-attractors $\mathcal{O}_1, ..., \mathcal{O}_\ell$ whose basins $V_1, ..., V_\ell$ have piecewise smooth boundaries $\partial V_1, ..., \partial V_\ell$ and the remaining part of the ω-limit set is contained in $\cup_{1\leq j\leq \ell}\partial V_j$. We assume also that for any $z \in \cap_{1\leq i\leq k}\partial V_{j_i}$, $k \leq \ell$ there exist $\varpi = \varpi(z) > 0$ and an F_z^t-invariant measures $\nu_1, ..., \nu_k$ such that

$$z + s\bar{B}_{\nu_i}(z) \in V_{j_i} \text{ for all } s \in (0, \varpi] \text{ and } i = 1, ..., k, \tag{1.2.27}$$

i.e. $\bar{B}_{\nu_i}(z) \neq 0$ and it points out into the interior of V_{j_i} which means that from any boundary point it is possible to go to any adjacent basin along a curve with an arbitrarily small S-functional. Let $\delta > 0$ be so small that the δ-neighborhood $U_\delta(\mathcal{O}_i) = \{z \in \mathcal{X} : \operatorname{dist}(z, \mathcal{O}_i) < \delta\}$ of each $\mathcal{O}_i$ is contained with its closure in the corresponding basin V_i. For any $x \in V_i$ set

$$\tau^\varepsilon_{x,y}(i) = \inf\big\{t \geq 0 : Z^\varepsilon_{x,y}(t) \in \cup_{j\neq i} U_\delta(\mathcal{O}_j)\big\}.$$

In Section 1.8 we will derive the following result.

1.2.7. THEOREM. *The function $R_{ij}(x) = \inf_{z\in V_j} R(x,z)$ takes on the same value R_{ij} for all $x \in \mathcal{O}_i$, $i \neq j$. Let $R_i = \min_{j\neq i, j\leq \ell} R_{ij}$. Then for any $x \in V_i$,*

$$\lim_{\varepsilon\to 0} \varepsilon \log \int_{\mathcal{W}} \tau^\varepsilon_{x,y}(i) dm(y) = R_i > 0 \tag{1.2.28}$$

and for any $\alpha > 0$ there exists $\lambda(\alpha) = \lambda(x, \alpha) > 0$ such that for all small $\varepsilon > 0$,

$$m\big\{y \in \mathcal{W} : e^{(R_i-\alpha)/\varepsilon} > \tau^\varepsilon_{x,y}(i) \text{ or } \tau^\varepsilon_{x,y}(i) > e^{(R_i+\alpha)/\varepsilon}\big\} \leq e^{-\lambda(\alpha)/\varepsilon}. \tag{1.2.29}$$

Next, set

$$\Theta^{\varepsilon,i}_v(t) = \Theta^{\varepsilon,i,\delta}_v(t) = \int_0^t \mathbb{I}_{V_i\setminus U_\delta(\mathcal{O}_i)}(Z^\varepsilon_v(s))ds.$$

Then for any $x \in V_i$ and $\delta > 0$ there exists $\lambda(\delta) = \lambda(x, \delta) > 0$ such that for all small $\varepsilon > 0$,

$$m\{y \in \mathcal{W} : \Theta^{\varepsilon,i}_{x,y}(\tau^\varepsilon_{x,y}(i)) \geq e^{-\lambda(\delta)/\varepsilon}\tau^\varepsilon_{x,y}(i)\} \leq e^{-\lambda(\delta)/\varepsilon}. \tag{1.2.30}$$

Now, suppose that the vector field B is complete on ∂V_i for some $i \leq \ell$ (which strengthens (1.2.27) there) and the restriction of the ω-limit set of Π^t to ∂V_i consists of a finite number of S-compacts. Assume also that there is a unique index $\iota(i) \leq \ell$, $\iota(i) \neq i$ such that $R_i = R_{i\iota(i)}$. Then for some $\lambda = \lambda(x) > 0$ and all small $\varepsilon > 0$,

$$m\big\{y \in \mathcal{W} : Z^\varepsilon_{x,y}(\tau^\varepsilon_{x,y}(i)) \notin V_{\iota(i)}\big\} \leq e^{-\lambda/\varepsilon}. \tag{1.2.31}$$

Finally, suppose that the above conditions hold true for all $i = 1, ..., \ell$. Define $\iota_0(i) = i$, $\tau^\varepsilon_v(i,1) = \tau^\varepsilon_v(i)$ and recursively,

$$\iota_k(i) = \iota(\iota_{k-1}(i)) \text{ and } \tau^\varepsilon_v(i,k) = \tau^\varepsilon_v(i,k-1) + \tau^\varepsilon_{v_\varepsilon(k-1)}\big(j(v_\varepsilon(k-1))\big),$$

where $v_\varepsilon(k) = \Phi_\varepsilon^{\varepsilon^{-1}\tau_v^\varepsilon(i,k)}v$, $j((x,y)) = j$ *if* $x \in V_j$, *and set* $\Sigma_i^\varepsilon(k,a) = \sum_{l=1}^k \exp\big((R_{\iota_{l-1}(i),\iota_l(i)} + a)/\varepsilon\big)$. *Then for any* $x \in V_i$ *and* $\alpha > 0$ *there exists* $\lambda(\alpha) = \lambda(x,\alpha) > 0$ *such that for all* $n \in \mathbb{N}$ *and sufficiently small* $\varepsilon > 0$,

$$(1.2.32) \qquad m\{y \in \mathcal{W} : \Sigma_i^\varepsilon(k,-\alpha) > \tau_{x,y}^\varepsilon(i,k) \text{ or } \tau_{x,y}^\varepsilon(i,k) > \Sigma_i^\varepsilon(k,\alpha) \text{ for some } k \le n\} \le ne^{-\lambda(\alpha)/\varepsilon}$$

and for some $\lambda = \lambda(x) > 0$,

$$(1.2.33) \qquad m\{y \in \mathcal{W} : Z_{x,y}^\varepsilon(\tau_{x,y}^\varepsilon(i,k)) \notin V_{\iota_k(i)} \text{ for some } k \le n\} \le ne^{-\lambda/\varepsilon}.$$

Generically there exists only one index $\iota(i)$ such that $R_i = R_{i\iota(i)}$ and in this case Theorem 1.2.7 asserts that $Z_{x,y}^\varepsilon$, $x \in V_i$ arrives (for "most" $y \in \mathcal{W}$) at $V_{\iota(i)}$ after it leaves V_i. If $\mathcal{I}(i) = \{j : R_i = R_{ij}\}$ contains more than one index then the method of the proof of Theorem 1.2.7 enables us to conclude that in this case $Z_{x,y}^\varepsilon$, $x \in V_i$ arrives (for "most" $y \in \mathcal{W}$) at $\cup_{j\in\mathcal{I}(i)}V_j$ after leaving V_i but now we cannot specify the unique basin of attraction of one of $\mathcal{O}_j$'s where $Z_{x,y}^\varepsilon$ exits from V_i. If the succession function ι is uniquely defined then it determines an order of transitions of the slow motion Z^ε between basins of attractors of $\bar{Z}$ and because of their finite number Z^ε passes them in certain cyclic order going around such cycle exponentially many in $1/\varepsilon$ times while spending the total time in a basin V_i which is approximately proportional to $e^{R_i/\varepsilon}$. If there exist several cycles of indices $i_0, i_1, ..., i_{k-1}, i_k = i_0$ where $i_j \le \ell$ and $i_{j+1} = \iota(i_j)$ then transitions between different cycles may also be possible. In the uncoupled case with fast motions being continuous time Markov chains a description of such transitions via certain hierarchy of cycles appeared in [**29**] and [**31**] without detailed proofs but relying on some heuristic arguments. In our fully coupled deterministic setup a rigorous justification of the corresponding description seems to be difficult in a more or less general situation though for some specific simple examples (as, for instance, those which are considered in Section 1.9) this looks feasible while it is not clear whether it is possible to describe in our situation a limiting as $t \to \infty$ behaviour of the slow motion $Z^\varepsilon(t)$ when ε is small but fixed.

The proof of Theorems 1.2.5, 1.2.7 and to certain extent also of Theorem 1.2.3 rely, in particular, on certain "Markov property type" arguments which enable us to extend estimates on relatively short time intervals to very long time intervals by, essentially, iterating them where the crucial role is played by a volume lemma type result of Section 1.3 together with the technique of (t,δ)-separated sets and Bowen's (t,δ)-balls on unstable leaves of the perturbed flow Φ_ε^t. Moreover, the proof of (1.2.32) and (1.2.33) require certain rough strong Markov property type arguments which enable us to study the slow motion at subsequent hitting times $\tau_{x,y}^\varepsilon(i,n)$ of small neighborhoods of attractors of the averaged motion.

In order to produce a wide class of systems satisfying the conditions of Theorem 1.2.7 we can choose, for instance, a vector field $\tilde{B}(x)$ on $\mathbb{R}^d$ whose ω-limit set satisfies the conditions stated above for the averaged system together with a family of vector fields $\hat{B}(x,y)$ on $\mathbb{R}^d$ (parametrized by $y \in \mathbf{M}$) such that $\int_{\mathbf{M}} \hat{B}(x,y)d\mu_x^{\mathrm{SRB}}(y) \equiv 0$ and then set $B(x,y) = \tilde{B}(x) + \hat{B}(x,y)$. As a specific example we can take the flows F_x^t, $x \in \mathbb{R}^1_- = (-\infty, 0)$ to be geodesic flows on the manifold $\mathbf{M}$ with (changing) constant negative curvature x, $\tilde{B}$ to be a one dimensional vector field on $\mathbb{R}^1$ and $\hat{B}(x,y)$ can be just a function $\hat{B}(y)$ on $\mathbf{M}$ with zero integral with respect to the Lebesgue measure there.

In Section 1.9 we will derive similar results for the discrete time case where differential equations (1.1.1) are replaced by difference equations (1.1.10). Namely, recall that a compact subset Λ of a compact Riemannian manifold $\mathbf{M}$ is called hyperbolic if it is F-invariant and there exists $\kappa > 0$ and the splitting $T_\Lambda \mathbf{M} = \Gamma^s \bigoplus \Gamma^u$ into the continuous subbundles Γ^s, Γ^u of the tangent bundle $T\mathbf{M}$ restricted to Λ, the splitting is invariant with respect to the differential DF of F, and there is $n_0 > 0$ such that for all $\xi \in \Gamma^s$, $\eta \in \Gamma^u$, and $n \geq n_0$ the inequalities (1.2.2) with t replaced by n hold true. A hyperbolic set Λ is said to be basic hyperbolic if the periodic orbits of $F|_\Lambda$ are dense in Λ, $F|_\Lambda$ is topologically transitive, and there exists an open set $U \supset \Lambda$ with $\Lambda = \bigcap_{-\infty<n<\infty} F^n U$. Such a Λ is called a basic hyperbolic attractor if for some open set U and $n_0 > 0$,

$$F^{n_0}\bar{U} \subset U \quad \text{and} \quad \bigcap_{n>0} F^t U = \Lambda$$

where $\bar{U}$ denotes the closure of U. If $\Lambda = \mathbf{M}$ then F^t is called an Anosov flow. If F is a C^2 endomorphism of $\mathbf{M}$ and there exists $\kappa > 0$ such that $\|DF\xi\| \geq e^\kappa \|\xi\|$ for all $\xi \in T\mathbf{M}$ then F is called an expanding map (or transformation) of $\mathbf{M}$. It will be convenient for our exposition to use the notation of the expanding subbundle Γ^u also in the case of expanding maps where, of course, $\Gamma^u = T\mathbf{M}$. We replace now Assumption 1.2.1 by the following one.

1.2.8. ASSUMPTION. *The family $F_x = \Phi(x, \cdot)$ in (1.1.10) consists of C^2-diffeomorphisms or endomorphisms of a compact $n_{\mathbf{M}}$-dimensional Riemannian manifold $\mathbf{M}$ with uniform C^2 dependence on the parameter x belonging to a neighborhood of the closure $\bar{\mathcal{X}}$ of a relatively compact open connected set $\mathcal{X} \subset \mathbb{R}^d$. All F_x, $x \in \bar{\mathcal{X}}$ are either expanding maps of $\mathbf{M}$ or diffeomorphisms possessesing basic hyperbolic attractors Λ_x with hyperbolic splittings satisfying (1.2.2) with the same $\kappa > 0$ and there exists an open set $\mathcal{W} \subset \mathbf{M}$ and $n_0 > 0$ satisfying (1.2.4) with n in place of t.*

Let $J^u_x(y)$ be the absolute value of the Jacobian of the linear map $DF_x(y) : \Gamma^u_x(y) \to \Gamma^u_x(F_x y)$ with respect to the Riemannian inner products and set

$$\varphi^u_x(y) = -\log J^u_x(y). \tag{1.2.34}$$

The function $\varphi^u_x(y)$ is known to be Hölder continuous in y, since the subbundles Γ^u_x are Hölder continuous (see [**60**]), and $\varphi^u_x(y)$ is C^1 in x (see [**16**]). The topological pressure $P_x(\psi)$ of a function ψ for F is defined similarly to the continuous time (flow) case above but now time should run only over integers and the integral $\int_0^t \psi(F^s_x y)ds$ should be replaced by the sum $\sum_{k=0}^{n-1} \psi(F^k y)$ (see [**60**]). Again the variational principle (1.2.6) holds true and if q is a Hölder continuous function on Λ_x there exists a unique F_x–invariant measure μ^q_x on Λ_x, called the equilibrium state for $\varphi^u_x + q$ which satisfies (1.2.7). In particular, $\mu^0_x = \mu^{\mathrm{SRB}}_x$ is usually called the Sinai–Ruelle–Bowen (SRB) measure. Since Λ_x are attractors we have that $P_x(\varphi^u_x) = 0$ (see [**13**]) and the same holds true in the expanding case, as well. Next, we define $I_x(\nu)$, $H(x, x', \beta)$, $H(x, \beta)$, $L(x, x', \alpha)$, $L(x, \alpha)$, $S_{0T}(\gamma)$, and γ^u as in (1.2.8) and (1.2.10)–(1.2.14). In place of Assumption 1.2.2 we will rely now on the similar one concerning the equation (1.1.10).

1.2.9. ASSUMPTION. *There exists $K > 0$ such that*

$$\|B(x,y)\|_{C^1(\mathcal{X}\times\mathbf{M})} + \|\Phi(x,y)\|_{C^2(\mathcal{X}\times\mathbf{M})} \leq K \tag{1.2.35}$$

where the first $\|\cdot\|_{C^1(\mathcal{X}\times\mathbf{M})}$ *is the* C^1 *norm of the corresponding vector fields on* $\mathcal{X}\times\mathbf{M}$ *and the second expression is the* C^2 *norm (with respect to the corresponding Riemannian metrics) of the map* $\Phi:\mathcal{X}\times\mathbf{M}\to\mathcal{X}\times\mathbf{M}$ *acting by* $\Phi(x,y)=(x,F_xy)$.

1.2.10. THEOREM. *Assume that Assumptions 1.2.8 and 1.2.9 are satisfied and that* $X^\varepsilon(n)=X^\varepsilon_{x,y}(n)$, $n=0,1,2,...$ *is obtained by (1.1.10). For* $t\in[n,n+1]$ *define* $X^\varepsilon(t)=(t-n)X^\varepsilon(n+1)+(n+1-t)X^\varepsilon(n)$ *and set* $Z^\varepsilon_{x,y}(t)=X^\varepsilon_{x,y}(t/\varepsilon)$. *Then Theorem 1.2.3 and Corollary 1.2.4 hold true with the corresponding functionals* S_{0t}. *Theorems 1.2.5 and 1.2.7 hold true, as well, under the corresponding assumptions about the family* $\{F_x,\,x\in\mathcal{X}\}$ *(with* $\mathcal{X}=\mathbb{R}^d$ *in the case of Theorem 1.2.7) and about the averaged system (1.1.11) (in particular, about its attractors) in place of the system (1.1.6).*

In Section 1.9 we exhibit computations which demonstrate the phenomenon of Theorem 1.2.7 in the discrete time case for two simple examples where F_xy are one dimensional maps $y\to 3y+x$ (mod 1) and the averaged equation has three attracting fixed points.

In the last Section 1.10 we discuss a stochastic resonance type phenomenon which can be exhibited in three scale systems where fast motions are hyperbolic flows (hyperbolic diffeomorphisms, expanding transformations) as above depending on the intermediate and slow motions while the intermediate motion performs rare transitions between attracting fixed points of corresponding averaged systems which under certain conditions creates a nearly periodic motion of the slow one dimensional motion.

1.2.11. REMARK. *Computation or even estimates of functionals* $S_{0T}(\gamma)$ *seem to be quite difficult already for simple discrete (and, of course, more for continuous) time examples since this leads to complicated nonclassical variational problems. This is crucial in order to estimate numbers* R_{ij} *which according to Theorem 1.2.7 are responsible for transitions of the slow motion between basins of attractors of the averaged system.*

1.2.12. REMARK. *The estimate (1.2.18) shows that* $Z^\varepsilon_{x,y}$ *tends as* $\varepsilon\to 0$ *to* $\bar Z_x$ *uniformly on* $[0,T]$ *in the sense of convergence in measure* m *considered on the space of initial conditions* $y\in\mathbf{M}$. *A natural question to ask is whether the convergence for almost all (fixed) initial conditions also takes place in our circumstances. In* [**11**] *we give a negative answer to this question, in paricular, for the following simple discrete time example*

$$\begin{aligned}&\big(X^\varepsilon_{x,y}(n+1),Y^\varepsilon_{x,y}(n+1)\big)\\&=\big(X^\varepsilon_{x,y}(n)+\varepsilon\sin\big(2\pi Y^\varepsilon_{x,y}(n)\big),\,2Y^\varepsilon_{x,y}(n)+X^\varepsilon_{x,y}(n)\ (mod\ 1)\big).\end{aligned}$$

Identifying 0 and 1 we view y variable as belonging to the circle in order to fit into our setup where the fast motion runs on a compact manifold. The averaged equation (1.1.11) has here zero in the right hand side so the averaged motion stays forever at the initial point. The discrete time version of (1.2.18) asserted by Theorem 1.2.10 implies that

$$\max_{0\le n\le 1/\varepsilon}|X^\varepsilon_{x,y}(n)-x|\to 0 \text{ as } \varepsilon\to 0 \tag{1.2.36}$$

in the sense of convergence in (the Lebesgue) measure on the circle but we show in [**11**] *that for each* x *there is a set* Γ_x *of full Lebesgue measure on the circle such* *that if* $y \in \Gamma_x$ *then* $\limsup$ *as* $\varepsilon \to 0$ *of the left hand side in (1.2.36) is positive, i.e. there is no convergence for Lebesgue almost all* y *there. Namely, it turns out that for almost all initial conditions there exists a sequence* $\varepsilon_i \to 0$ *such that the fast motion* $Y_{x,y}^{\varepsilon_i}(n)$ *stays for a time of order* $1/\varepsilon_i$ *close to an orbit* $\{2^n v \text{ (mod 1)}, n \geq 0\}$ *of the doubling map with* v *being a generic point with respect to a Gibbs invariant measure* μ *of this map satisfying* $\int_0^1 \sin(2\pi v)d\mu(v) \neq 0$ *which prevents (1.2.36).*

1.3. Dynamics of Φ_ε^t

For readers convenience we exhibit, first, in this section the setup and necessary technical results from [**55**] and though their proofs can can be found in [**55**] we provide for completness and readers' convenience their slightly modified and corrected version also here.

Any vector $\xi \in T(\mathbb{R}^d \times \mathbf{M}) = \mathbb{R}^d \oplus T\mathbf{M}$ can be uniquely written as $\xi = \xi^{\mathcal{X}} + \xi^{\mathcal{W}}$ where $\xi^{\mathcal{X}} \in T\mathbb{R}^d$ and $\xi^{\mathcal{W}} \in T\mathbf{M}$ and it has the Riemannian norm $|||\xi||| = |\xi^{\mathcal{X}}| + \|\xi^{\mathcal{W}}\|$ where $|\cdot|$ is the usual Euclidean norm on $\mathbb{R}^d$ and $\|\cdot\|$ is the Riemannian norm on $T\mathbf{M}$. The corresponding metrics on $\mathbf{M}$ and on $\mathbb{R}^d \times \mathbf{M}$ will be denoted by $d_{\mathbf{M}}$ and *dist*, respectively, so that if $z_1 = (x_1, w_1)$, $z_2 = (x_2, w_2) \in \mathbb{R}^d \times \mathbf{M}$ then $dist(z_1, z_2) = |x_1 - x_2| + d_{\mathbf{M}}(w_1, w_2)$. It is known (see [**69**]) that the hyperbolic splitting $T_{\Lambda_x}\mathbf{M} = \Gamma_x^s \oplus \Gamma_x^0 \oplus \Gamma_x^u$ over Λ_x can be continuously extended to the splitting $T_{\mathcal{W}}\mathbf{M} = \Gamma_x^s \oplus \Gamma_x^0 \oplus \Gamma_x^u$ over $\mathcal{W}$ which is forward invariant with respect to DF_x^s and satisfies exponential estimates with a uniform in $x \in \mathcal{X}$ positive exponent which we denote again by $\kappa > 0$, i.e. we assume now that

$$\|DF_x^t \xi\| \leq e^{-\kappa t}\|\xi\| \quad \text{and} \quad \|DF_x^{-t}\eta\| \leq e^{-\kappa t}\|\eta\| \tag{1.3.1}$$

provided $\xi \in \Gamma_x^s(w)$, $\eta \in \Gamma_x^u(F_x^t w)$, $t \geq t_0$, and $w \in \mathcal{W}$. Moreover, by [**16**] (see also [**71**]) we can choose these extensions so that $\Gamma_x^s(w)$ and $\Gamma_x^u(w)$ will be Hölder continuous in w and C^1 in x in the corresponding Grassmann bundle. Actually, since $\mathcal{W}$ is contained in the basin of each attractor Λ_x, any point $w \in \mathcal{W}$ belongs to the stable manifold $W_x^s(v)$ of some point $v \in \Lambda_x$ (see [**13**]), and so we choose naturally $\Gamma_x^s(w)$ to be the tangent space to $W_x^s(v)$ at w. Now each vector $\xi \in T_{x,w}(\mathcal{X} \times \mathcal{W}) = T_x\mathcal{X} \oplus T_w\mathcal{W}$ can be represented uniquely in the form $\xi = \xi^{\mathcal{X}} + \xi^s + \xi^0 + \xi^u$ with $\xi^{\mathcal{X}} \in T_x\mathcal{X}$, $\xi^s \in \Gamma_x^s(w)$, $\xi^0 \in \Gamma_x^0(w)$ and $\xi^u \in \Gamma_x^u(w)$. We denote also $\xi^{0s} = \xi^s + \xi^0$ and $\xi^{0u} = \xi^u + \xi^0$. For each small $\varepsilon, \alpha > 0$ set $\mathcal{C}^u(\varepsilon, \alpha) = \{\xi \in T(\mathcal{X} \times \mathcal{W}) : \|\xi^{0s}\| \leq \varepsilon\alpha^{-2}\|\xi^u\| \text{ and } \|\xi^{\mathcal{X}}\| \leq \varepsilon\alpha^{-1}\|\xi^u\|\}$ and $\mathcal{C}_{x,w}^u(\varepsilon, \alpha) = \mathcal{C}^u(\varepsilon, \alpha) \cap T_{x,w}(\mathcal{X} \times \mathcal{W})$ which are unstable cones around Γ^u and $\Gamma_x^u(w)$, respectively. Similarly, we define $\mathcal{C}^s(\varepsilon, \alpha) = \{\xi \in T(\mathcal{X} \times \mathcal{W}) : \|\xi^{0u}\| \leq \varepsilon\alpha^{-2}\|\xi^s\| \text{ and } \|\xi^{\mathcal{X}}\| \leq \varepsilon\alpha^{-1}\|\xi^s\|\}$ and $\mathcal{C}_{x,w}^s(\varepsilon, \alpha) = \mathcal{C}^s(\varepsilon, \alpha) \cap T_{x,w}(\mathcal{X} \times \mathcal{W})$ which are stable cones around Γ^s and $\Gamma_x^s(w)$, respectively. Put $(\mathcal{X} \times \mathcal{W})_t = \{(x, w) : \Phi_\varepsilon^u(x, w) \in (\mathcal{X} \times \mathcal{W})\ \forall u \in [0, t]\}$, where, recall, Φ_ε^t is the flow determined by the equations (1.1.1).

1.3.1. LEMMA. . *There exist* α_0, $\varepsilon(\alpha)$, $t_1 > 0$ *such that if* $z = (x, y) \in (\mathcal{X} \times \mathcal{W})_t$ *and* $t \geq t_1$, $\alpha \leq \alpha_0$, $\varepsilon \leq \varepsilon(\alpha)$ *then*

$$D_z\Phi_\varepsilon^t \mathcal{C}_z^u(\varepsilon, \alpha) \subset \mathcal{C}_{\Phi_\varepsilon^t z}^u(\varepsilon, \alpha), \quad \mathcal{C}_z^s(\varepsilon, \alpha) \supset D_z\Phi_\varepsilon^{-t}\mathcal{C}_{\Phi_\varepsilon^t z}^s(\varepsilon, \alpha), \tag{1.3.2}$$

and for any $\xi \in \mathcal{C}_z^u(\varepsilon, \alpha)$, $\eta \in \mathcal{C}_{\Phi_\varepsilon^t z}^s(\varepsilon, \alpha)$,

$$|||D_z\Phi_\varepsilon^t \xi||| \geq e^{\frac{1}{2}\kappa t}|||\xi|||, \quad |||D_z\Phi_\varepsilon^{-t}\eta||| \geq e^{\frac{1}{2}\kappa t}|||\eta|||. \tag{1.3.3}$$

PROOF. Let $\xi = \xi^{\mathcal{X}} + \xi^u + \xi^{0s} \in T_z(\mathcal{X} \times \mathbf{M})$, $D_z\Phi^t\xi^{\mathcal{X}} = \zeta = \zeta^{\mathcal{X}} + \zeta^u + \zeta^{0s} \in T_{\Phi^t z}(\mathcal{X} \times \mathbf{M})$, $z = (x, y)$, $D_y F_x^t \xi^u = \eta^u$, and $D_y F_x^t \xi^{0s} = \eta^{0s}$. Then $D_z\Phi^t\xi =$

$\zeta^{\mathcal{X}}+(\zeta^u+\eta^u)+(\zeta^{0s}+\eta^{0s})$ and $\|\xi^{\mathcal{X}}\|=\|\zeta^{\mathcal{X}}\|$, $\|\zeta^u\|\le Ce^{Ct}\|\xi^{\mathcal{X}}\|$, $\|\zeta^{0s}\|\le Ce^{Ct}\|\xi^{\mathcal{X}}\|$, $\|\eta^{0s}\|\le C\|\xi^{0s}\|$ for some $C\ge 1$ independent of ξ and $\|\eta^u\|\ge e^{\kappa t}\|\xi^u\|$ if $t\ge t_0$. Hence, for $t\ge t_0$,

$$\|\zeta^u+\eta^u\|\ge\|\eta^u\|-\|\zeta^u\|\ge e^{\kappa t}\|\xi^u\|-Ce^{Ct}\|\xi^{\mathcal{X}}\|$$

and

$$\|\zeta^{0s}+\eta^{0s}\|\le\|\zeta^{0s}\|+\|\eta^{0s}\|\le Ce^{Ct}\|\xi^{\mathcal{X}}\|+C\|\xi^{0s}\|.$$

If $\xi\in\mathcal{C}^u_z(\varepsilon,\alpha)$ then $\|\xi^u\|\ge\alpha\varepsilon^{-1}\|\xi^{\mathcal{X}}\|$ and $\|\xi^u\|\ge\alpha^2\varepsilon^{-1}\|\xi^{0s}\|$. Hence, by the above,

$$\|\zeta^u+\eta^u\|\ge\alpha\varepsilon^{-1}(\frac{1}{2}e^{\kappa t}-\varepsilon\alpha^{-1}Ce^{Ct})\|\xi^{\mathcal{X}}\|+\frac{1}{2}e^{\kappa t}\alpha^2\varepsilon^{-1}\|\xi^{0s}\|.$$

Set $t_1=\kappa^{-1}\ln 6$, choose $\alpha\le 6^{-2C/\kappa}$ and $\varepsilon=\varepsilon(\alpha)\le\alpha^2/4C$. Then we obtain that $D_z\Phi^t\xi\in\mathcal{C}^u_{\Phi^t z}(\varepsilon,2\alpha)$ for all $t\in[t_1,t_2]$, and so by continuity of the splitting $\Gamma^s_x\oplus\Gamma^0_x\oplus\Gamma^u_x$ and by perturbation arguments it follows that $D_z\Phi^t_\varepsilon\xi\in\mathcal{C}^u_{\Phi^t_\varepsilon z}(\varepsilon,\alpha)$ for all $t\in[t_1,2t_1]$ provided ε is small enough. Repeating this argument for $t\in[it_1,(i+1)t_1]$, $i=2,3,..$ we conclude the proof of the first part of (1.3.2) and its second part follows in the same way.

Next, for $\xi\in\mathcal{C}^u_z(\varepsilon,\alpha)$ and $t\ge t_0$,

$$|||D_z\Phi^t\xi|||\ge\|\eta^u\|-\|\zeta^{\mathcal{X}}\|-\|\zeta^u\|-\|\zeta^{0s}\|-\|\eta^{0s}\|\ge e^{\kappa t}\|\xi^u\|$$
$$-(1+2Ce^{Ct})\|\xi^{\mathcal{X}}\|-C\|\xi^{0s}\|\ge(e^{\kappa t}-\alpha^{-1}\varepsilon(1+2Ce^{Ct})-\alpha^{-2}\varepsilon C)\|\xi^u\|$$
$$\ge(e^{\kappa t}-\alpha^2(1+2Ce^{Ct})-\alpha^{-2}\varepsilon C)(1+\varepsilon\alpha^{-1}+\varepsilon\alpha^{-2})^{-1}|||\xi|||.$$

Choose α_0, $\varepsilon(\alpha)$ so small (for instance, $\varepsilon(\alpha)=\alpha^3$) that for all $\alpha\le\alpha_0$ and $\varepsilon\le\varepsilon(\alpha)$,

$$e^{\kappa t}-\varepsilon\alpha^{-1}(1+2Ce^{Ct})-\varepsilon\alpha^{-2}C\ge(1+\varepsilon\alpha^{-1}+\varepsilon\alpha^{-2})e^{\frac{2}{3}\kappa t}$$

for all $t\in[t_1,2t_1]$. Then, $|||D_z\Phi^t\xi|||\ge e^{\frac{2}{3}\kappa t}|||\xi|||$ for all such t, and so if ε small enough we have also $\|D_z\Phi^t_\varepsilon\xi\|\ge e^{\frac{1}{2}\kappa t}\|\xi\|$. Using (1.3.2) and repeating this argument for $D_z\Phi^{it_1}_\varepsilon\xi$, $i=1,2,...$ in place of ξ we derive (ii) for all $t\ge t_1$. The proof for stable cones $\mathcal{C}^s_\varepsilon(\varepsilon,\alpha)$ is similar. □

For any linear subspace Ξ of $T_z(\mathbb{R}^d\times\mathbf{M})$ denote by $J^\Xi_\varepsilon(t,z)$ absolute value of the Jacobian of the linear map $D_z\Phi^t_\varepsilon:\Xi\to D_z\Phi^t_\varepsilon\Xi$ with respect to inner products induced by the Riemannian metric. For each $z=(x,y)\in\mathbb{R}^d\times\mathbf{M}$ set also

$$J^u_\varepsilon(t,z)=\exp\big(-\int_0^t\varphi^u_{X^\varepsilon_{x,y}(s)}(Y^\varepsilon_{x,y}(s))ds\big).\tag{1.3.4}$$

Let n_u be the dimension of $\Gamma^u_x(w)$ which does not depend on x and w by continuity considerations. If Ξ is an n^u-dimensional subspace of $T_z(\mathcal{X}\times\mathcal{W})$, $z=(x,y)$, and $\Xi\subset\mathcal{C}^u_{x,y}(\varepsilon,\alpha)$ then it follows easily from Assumption 1.2.2 and Lemma 1.3.1 that there exists a constant $C_1>0$ independent of $z\in\mathcal{X}\times\mathcal{W}$ and of a small ε such that for any $t\ge 0$,

$$(1-C_1\varepsilon)^t\le J^\Xi_\varepsilon(t,z)(J^u_\varepsilon(t,z))^{-1}\le(1+C_1\varepsilon)^t.\tag{1.3.5}$$

Recall, that an embedded C^k, $k=1,2$ l-dimensional disc D in $\mathbb{R}^d\times\mathbf{M}$, $l\le d+n_{\mathbf{M}}$ is the image of an l-dimensional disc (ball) K in $\mathbb{R}^{d+n_{\mathbf{M}}}$ centered at 0 under a C^k diffeomorphism of a neighborhood of 0 in $\mathbb{R}^{d+n_{\mathbf{M}}}$ into $\mathbb{R}^d\times\mathbf{M}$ and we define the boundary ∂D of D as the image of the boundary ∂K of K considered in the corresponding l-dimensional Euclidean subspace of $\mathbb{R}^{d+n_{\mathbf{M}}}$. Denote by $U(z,\rho)$ the ball in $\mathbb{R}^d\times\mathbf{M}$ centered at z and let $\mathcal{D}^u_\varepsilon(z,\alpha,\rho,C)$, $C\ge 1$ be the set of all C^1

embedded n_u–dimensional closed discs $D \subset \mathcal{X}\times\mathcal{W}$ such that $z \in D$, $TD \subset \mathcal{C}^u(\varepsilon,\alpha)$ and if $v \in \partial D$ then $C\rho \le d_D(v,z) \le C^2\rho$ where TD is the tangent bundle over D and d_D is the interior metric on D. Each disc $D \in \mathcal{D}^u_\varepsilon(z,\alpha,\rho,C)$ will be called unstable or expanding and, clearly, $D \subset U(z, C^2\rho)$ and if ε/α^2 and ρ are small enough and $C > 1$ then $\text{dist}(v,z) \ge \rho$ for any $v \in \partial D$. Let $D \in \mathcal{D}^u_\varepsilon((x,y),\alpha,\rho,C)$ and $z \in D \subset \mathcal{X}\times\mathcal{W}$. Set $U^\varepsilon_D(t,z,L) = \{\tilde z \in D : d_{\Phi^s_\varepsilon D}(\Phi^s_\varepsilon z, \Phi^s_\varepsilon \tilde z) \le L \quad \forall s \in [0,t]\}$ and let $\pi_1 : \mathcal{X}\times\mathcal{W} \to \mathcal{X}$ and $\pi_2 : \mathcal{X}\times\mathcal{W} \to \mathcal{W}$ be natural projections on the first and second factors, respectively.

1.3.2. LEMMA. *Let ε, α, t, (x,y) be as in Lemma 1.3.1 and $T > 0$. There exist $\rho_0, c, c_{\rho,T}, C \ge 1$ such that if $\rho \le \rho_0$, $D \in \mathcal{D}^u_\varepsilon((x,y),\alpha,\rho,C)$, $z \in D$, $V_{s,t}(z) = \Phi^s_\varepsilon U^\varepsilon_D(t,z,C\rho)$, $V_t(z) = V_{t,t}(z)$, $V = V_{0,t}(z) \subset D$ and $t \ge 0$ then*

(i) $d_{V_{s,t}(z)}(\Phi^s_\varepsilon v, \Phi^s_\varepsilon z) \le c^{-1}e^{-\frac12\kappa(t-s)}d_{V_t(z)}(\Phi^t_\varepsilon v,\Phi^t_\varepsilon z) \le c^{-1}C\rho e^{-\frac12\kappa(t-s)}$ for any $v \in V$ and $s \in [0,t]$, where d_U is the interior distance on U;

(ii) $TV_{s,t}(z) \subset \mathcal{C}^u(\varepsilon,\alpha)$ and $V_t(z) \in \mathcal{D}^u_\varepsilon(\Phi^t_\varepsilon z,\alpha,\rho,\sqrt{C})$ provided $\partial D \cap U^\varepsilon_D(t,z,C\rho) = \emptyset$;

(iii) For all $v \in V$ and $0 \le s \le t$,

$$|\pi_1\Phi^s_\varepsilon v - \pi_1\Phi^s_\varepsilon z| \le Cc^{-1}\rho\varepsilon\alpha^{-1}(1-\varepsilon\alpha^{-1}-\varepsilon\alpha^{-2})^{-1}e^{-\frac12\kappa(t-s)}.$$

(iv) $c_{\rho,T} \le m_D(V)J^u_\varepsilon(t,z) \le c^{-1}_{\rho,T}$ provided $t \le T/\varepsilon$, where m_D is the induced (not normalized) Riemannian volume on D.

PROOF. (i) Let γ be a smooth curve on $V_t(z)$ connecting $a = \Phi^t_\varepsilon z$ and $b = \Phi^t_\varepsilon v$. then $\tilde\gamma = \Phi^{s-t}_\varepsilon\gamma$ is a smooth curve on $V_{s,t}(z)$ connecting $\Phi^s_\varepsilon z$ and $\Phi^s_\varepsilon v$. Since $T\tilde\gamma \subset TV_s(z) \subset \mathcal{C}^u(\varepsilon,\alpha)$ then by (1.3.3), $\text{length}(\gamma) \ge e^{\kappa(t-s)/2}\text{length}(\tilde\gamma)$ if $t-s \ge t_1$. Then for such t and s,

$$d_{V_s}(\Phi^s_\varepsilon v,\Phi^s_\varepsilon z) \le \text{length}(\tilde\gamma) \le e^{-\kappa(t-s)/2}\text{length}(\gamma).$$

Observe that (i) is nontrivial only for large $t-s$, so minimizing in γ in the above inequality we derive the assertion (i).

Next, we derive (ii). Its first part follows from (1.3.2). By the definition of V, $d_{V_t(z)}(w,z) \le C\rho$ for any $w \in \partial V_t(z)$. It remains to show that $d_{V_t(z)}(w,z) \ge \sqrt{C}\rho$ for any $w \in \partial V_t(z)$. Indeed, suppose $d_{V_t(z)}(w,z) < \sqrt{C}\rho$. Set

$$d_1 = \sup_{-t_1\le s\le 0, v\in\mathcal{X}\times\mathcal{W}} \|D_v\Phi^s_\varepsilon\|.$$

Next, we conclude via perturbation arguments that $d_1 > 0$ provided ε is small enough. Let $w = \Phi^t_\varepsilon v$. It follows from Lemma 1.3.1 that $d_{V_s(z)}(\Phi^s_\varepsilon v,\Phi^s_\varepsilon z) < d_1\sqrt{C}\rho$ for all $s \in [0,t]$. Hence, if $\sqrt{C} \ge d_1$ then $v \notin \partial V$, and so $w \notin \partial V_t(z)$.

In order to derive (iii) take an arbitrary smooth curve γ on $V_{s,t}(z)$ connecting $\Phi^s_\varepsilon v$ and $\Phi^s_\varepsilon z$. Then $\frac{d\gamma(s)}{ds} = \dot\gamma(s) \in \mathcal{C}^u_{\gamma(s)}(\varepsilon,\alpha)$. It follows that if $\gamma(s) = (\gamma^{\mathcal{X}}(s),\gamma^{\mathbf{M}}(s))$ with $\gamma^{\mathcal{X}}(s) \in \mathcal{X}$ and $\gamma^{\mathbf{M}}(s) \in \mathbf{M}$ then $\dot\gamma^{\mathbf{M}}(s) = \dot\gamma^{0s}(s) + \dot\gamma^u(s)$, $\|\dot\gamma^{\mathcal{X}}(s)\| \le \varepsilon\alpha^{-1}\|\dot\gamma^u(s)\|$, $\|\dot\gamma^{0s}(s)\| \le \varepsilon\alpha^{-2}\|\dot\gamma^u(s)\|$, and so

$$|||\dot\gamma(s)||| \ge \|\dot\gamma^u(s)\| - \|\dot\gamma^{0s}(s)\| - \|\dot\gamma^{\mathcal{X}}(s)\| \ge \|\dot\gamma^u(s)\|(1-\varepsilon\alpha^{-1}-\varepsilon\alpha^{-2}).$$

Hence

$$\|\dot\gamma^{\mathcal{X}}(s)\| \le \varepsilon\alpha^{-1}(1-\varepsilon\alpha^{-1}-\varepsilon\alpha^{-2})^{-1}|||\dot\gamma(s)|||,$$

and so

$$|\pi_1\Phi^s_\varepsilon v - \pi_1\Phi^s_\varepsilon z| \le \varepsilon\alpha^{-1}(1-\varepsilon\alpha^{-1}-\varepsilon\alpha^{-2})^{-1}\text{length}(\gamma).$$

Minimizing the right hand side here over such γ we obtain (iii) using (i). Finally, (iv) follows from (1.3.5), (i), (ii), and the Hölder continuity of φ^u (as a function on $\mathcal{X} \times W$). □

For each $y \in \Lambda_x$ and $\varrho > 0$ small enough set $W_x^s(y,\varrho) = \{\tilde{y} \in \mathcal{W} : d_{\mathbf{M}}(F_x^t y, F_x^t \tilde{y}) \le \varrho\ \forall t \ge 0\}$ and $W_x^u(y,\varrho) = \{\tilde{y} \in \mathcal{W} : d_{\mathbf{M}}(F_x^t y, F_x^t \tilde{y}) \le \varrho\ \forall t \le 0\}$ which are local stable and unstable manifolds for F_x at y. According to [**40**] and [**69**] these families can be included into continuous families of n^s and n^u–dimensional stable and unstable C^1 discs $W_x^s(y,\varrho)$ and $W_x^u(y,\varrho)$, respectively, defined for all $y \in \mathcal{W}$ and such that $W_x^s(y,\varrho)$ is tangent to Γ_x^s, $W_x^u(y,\varrho)$ is tangent to Γ_x^u, $F_x^t W_x^s(y,\varrho) \subset W_x^s(F_x^t y,\varrho)$, and $W_x^u(y,\varrho) \supset F_x^{-t} W_x^u(F_x^t y,\varrho)$. Actually, as we noted it above if $y \in \mathcal{W}$ then y belongs to a stable manifold $W_x^s(\tilde{y})$ of some $\tilde{y} \in \Lambda_x$ and we choose $W_x^s(y,\varrho)$ to be the subset of $W_x^s(\tilde{y})$.

1.3.3. LEMMA. *For any $0 < \rho_1 < \rho_0$ small enough and a continuous function g on $\mathcal{X} \times \mathcal{W}$ uniformly in $D \in \mathcal{D}_\varepsilon^u(z,\alpha,\rho,C)$, $x' \in \mathcal{X}$, $z \in \mathcal{X} \times \mathcal{W}$ and $\rho \in [\rho_1,\rho_0]$,*

$$\text{(1.3.6)}\quad \lim_{t\to\infty} \frac{1}{t} \log \int_D \exp\Big(\int_0^t g(x', F^r_{\pi_1 z}\pi_2 v)dr\Big)\, dm_D(v) = P_{\pi_1 z}(g(x',\cdot) + \varphi^u_{\pi_1 z})$$

where m_D is the induced Riemannian volume on D.

PROOF. Set $\tilde{W} = \pi_2 D$, $x = \pi_1 z$ and $y \in \pi_2 z$. By standard transversality considerations we can define a one-to-one map $\tilde{\pi} : \tilde{W} \to \tilde{\pi}\tilde{W} \subset W_x^u(y,\tilde{C}\rho)$ by $\tilde{\pi}(\tilde{w}) = w \in F_x^\tau W_x^s(\tilde{w},\tilde{C}\rho)$ provided $\alpha, \rho, |\tau| > 0$ are small and $\tilde{C} > 0$ is sufficiently large. By the absolute continuity of the stable foliation arguments (see, for instance, [**65**], Section 3.3) we conclude that $\tilde{\pi}$ and its inverse have bounded Jacobians. It follows that it suffices to establish (1.3.6) for $D = \{x\} \times W$ where $W = W_x^u(y,\gamma)$ uniformly in $\gamma \in [\gamma_0, \gamma_0^{-1}]$, $\gamma_0 > 0$.

Set $W_\tau = F_x^\tau W$, $W_{r,q} = \cup_{r \le \tau \le q} F_x^\tau W$ and

$$I_{x,x'}^V(t) = \int_V \exp\big(\int_0^t g(x', F_x^\tau v)d\tau\big)dm_V(v).$$

Then

$$I_{x,x'}^{W_\tau}(t) = \textstyle\int_{W_\tau} \exp\big(\int_0^t g(x', F_x^\theta v)d\theta\big)dm_{W_\tau}(v)$$
$$= \textstyle\int_W \exp\big(\int_\tau^{t+\tau} g(x', F_x^\theta v)d\theta\big)J_x^u(\tau,v)dm_W(v),$$

and so

$$re^{-2(\tau+r)\|g\|} I_{x,x'}^W(t) \le I_{x,x'}^{W_{\tau,\tau+r}}(t) \le re^{(\tau+r)(2\|g\|+\|\varphi^u\|)} I_{x,x'}^W(t)$$

where $\|\cdot\|$ is the supremum norm on $\mathcal{X} \times \mathcal{W}$. Since $\mathcal{W} \subset \cup_{v\in\Lambda_x} W_x^s(v)$ by [**13**] then given $\eta > 0$ there exist $\gamma(\eta), \tau(\eta) > 0$, $v \in \Lambda_x$, and $U \subset \cup_{-r\le\theta\le r} F_x^\theta W_x^u(v)$ such that for any $\gamma \le \gamma(\eta)$ we can define a one-to-one map $\phi : U \to W_{\tau(\eta),\tau(\eta)+r}$ by $\phi(w) \in W_x^s(w,\eta)$. By standard absolute continuity of the stable foliation considerations (see [**65**], Section 3.3) it follows that ϕ and its inverse have bounded Jacobians which together with the above arguments yield that it suffices to prove Lemma 1.3.3 only when $y \in \Lambda_x$, and so (see [**13**]), $W = W_x^u(y,\gamma) \subset \Lambda_x$. We observe also that without loss of generality we can assume γ to be sufficiently small since we can always cover $W_x^u(y,\gamma)$ by $W_x^u(y_i,\tilde{\gamma})$, $i = 1,...,k$ with $y_1 = y$, $k = k(\gamma,\tilde{\gamma})$ and small $\tilde{\gamma} \le \gamma$, so proving Lemma 1.3.3 for all such $W_x^u(y_i,\tilde{\gamma})$ will imply it for $W_x^u(y,\gamma)$ itself.

So now assume that $W = W_x^u(y,\gamma) \subset \Lambda_x$ and we claim that for any $\eta > 0$ there exists $\tau(\eta,\gamma) > 0$ such that $W_{0,\tau}$ forms an η–net in Λ_x for any $\tau \ge \tau(\eta,\gamma)$. Indeed, by topological transitivity there exists $v \in \Lambda_x$ whose orbit is dense in Λ_x, and so by standard ergodicity considerations with respect to any ergodic invariant measure with full support on Λ_x (take, for instance, the SRB measure) we conclude that for any τ already $\{F_x^r v,\ r \geq \tau\}$ is dense in Λ_x. Then by transversality of W_x^s and $\cup_\theta F_x^\theta W_x^u$ there exists $r > 0$ such that $F_x^r v \in W_x^s(w, \gamma)$ for some $w \in W$, and so the forward orbit of w is dense in Λ_x, whence our claim holds true. By compactness and structural stability considerations it follows that we can choose the same $\tau(\eta, \gamma)$ for all $y \in \mathcal{W}$ and $x \in \mathcal{X}$.

For any set $V \subset \Lambda_x$ put $U_V(t, y, \zeta) = \{v \in V : d(F_x^r y, F_x^r v) \leq \zeta\ \ \forall r \in [0, t]\}$. Recall, that a finite set $E \subset \Lambda_x$ is called $(\zeta, t)-$separated for the flow F_x^t if $y, \tilde{y} \in E$, $y \neq \tilde{y}$ implies $\tilde{y} \notin U_{\Lambda_x}(t, y, \zeta)$. A set $E \subset \Lambda_x$ is called $(\zeta, t)-$spanning if for any $y \in \Lambda_x$ there is $\tilde{y} \in E$ such that $y \in U_{\Lambda_x}(t, \tilde{y}, \zeta)$. Let $W_{0,\tau}$ be an $\eta-$net in Λ_x and E be a maximal $(\zeta, t)-$separated subset of $W_{0,\tau}$. Then $U_{W_{0,\tau}}(t, y, \zeta/2)$, $y \in E$ are disjoint sets. By transversality of $W_{0,\tau}$ and W_x^s there exists $C_1 > 0$ such that for any $y \in \Lambda_x$ we can find $v(y) \in W_{0,\tau}$ such that $y \in W_x^s(v(y), C_1\eta)$, and so for some $w(y) \in E$, $y \in U_{\Lambda_x}(t, w(y), C_2(\zeta + \eta))$ with some $C_2 > 0$ large enough but independent of x, y, ζ, η. Hence, E is $(C_2(\zeta + \eta), t)-$spanning, and so $W_{0,\tau} \subset \cup_{y \in E} U_{W_{0,\tau}}(t, y, C_2(\zeta + \eta))$. Assume, first, that $g = g(x', v)$ in (1.3.6) is Hölder continuous in v. Then by standard volume lemma arguments (see [**13**]) we obtain for $V = U_{W_{0,\tau}}(t, y, \gamma)$, $\gamma > 0$ and $y \in W_{0,\tau}$ that

$$\begin{aligned}&\big|\log \textstyle\int_V \exp\big(\int_0^t g(x', F_x^r v)dr\big)dm_V(v)\\ &\quad - \textstyle\int_0^t \big(g(x', F_x^r y) + \varphi_x^u(F_x^r y)\big)dr\big| \leq C(\gamma)\end{aligned}$$

where $C(\gamma) > 0$ does not depend on x, x', y, t. Now (1.3.6) follows from the above integral estimates and the uniform in $(\gamma, t)-$separated and $(\gamma, t)-$spanning sets approximation of the topological pressure (see, for instance, [**8**] and [**28**]). For a general continuous g approximate it uniformly by Hölder continuous functions and (1.3.6) will follow in this case again. The limit (1.3.6) is uniform in x' and in z since $P_{F_{\pi_1 z}}(g(x', \cdot) + \varphi^u_{\pi_1 z})$ uniformly continuous in x' (easy to see) and it is uniformly continuous in z (see [**16**]) and, furthermore, it follows from Lemma 5.1 from [**16**] that the family $\frac{1}{t} \log \int_D \exp\left(\int_0^t g(x', F^r_{\pi_1 z}\pi_2 v)dr\right) dm_D(v)$, $t \geq 1$ is equicontinuous in z. □

1.3.4. PROPOSITION. *For any $\rho, C, b > 0$ with C large and $C\rho$ small enough there exists a positive function $\zeta_{b,\rho,T}(\Delta, s, \varepsilon)$ such that*

$$\limsup_{\Delta \to 0} \limsup_{\varepsilon \to 0} \limsup_{s \to \infty} \zeta_{b,\rho,T}(\Delta, s, \varepsilon) = 0 \tag{1.3.7}$$

and for any $x, x' \in \mathcal{X}$, $y \in \mathcal{W}$, $t \geq t_1$, $\tau \leq \frac{T}{\varepsilon} - t$, $\beta \in \mathbb{R}^d$, $|\beta| \leq b$, $D \in \mathcal{D}^u_\varepsilon((x, y), \alpha, \rho, C)$, $z \in D$ and $V = U^\varepsilon_D(t, z, C\rho)$ satisfying $V \cap \partial D = \emptyset$ we have

$$\begin{aligned}&\Big|\tfrac{1}{\tau} \log \textstyle\int_V \exp\langle\beta, \int_t^{t+\tau} B(x', Y_v^\varepsilon(s))ds\rangle dm_D(v) + \tfrac{1}{\tau}\log J^u_\varepsilon(t, z)\\ &\quad - P_{\pi_1 z_t}(\langle\beta, B(x', \cdot)\rangle + \varphi^u_{\pi_1 z_t})\Big| \leq \zeta_{b,\rho,T}(\varepsilon\tau, \min(\tau, (\log \tfrac{1}{\varepsilon})^\lambda), \varepsilon)\end{aligned} \tag{1.3.8}$$

where $\langle\cdot,\cdot\rangle$ is the inner product, $z_t = \Phi^t_\varepsilon z$, $\lambda \in (0, 1)$, and m_D is the induced Riemannian volume on D.

PROOF. By (1.1.1) and (1.2.15) for any $w, \tilde{w} \in \mathbb{R}^d \times \mathcal{W}$,

$$d(\Phi_\varepsilon^s w, \Phi^s \tilde{w}) \le d(w, \tilde{w}) + \int_0^s \|b(\Phi_\varepsilon^u w) - b(\Phi^u \tilde{w})\| du$$
$$\le d(w, \tilde{w}) + K \int_0^s d(\Phi_\varepsilon^u w, \Phi^u \tilde{w}) du,$$

where, $d = dist$ and, recall, $\Phi = \Phi_0$. Then, by Gronwall's inequality

$$d(\Phi_\varepsilon^s w, \Phi^s \tilde{w}) \le e^{Ks} d(w, \tilde{w}).$$

Hence,

$$d_{\mathbf{M}}(\pi_2 \Phi_\varepsilon^s w, F_{\pi_1 z_r}^s \pi_2 w) \le d(\Phi_\varepsilon^s w, \Phi^s(\pi_1 z_r, \pi_2 w)) \le e^{Ks} |\pi_1 w - \pi_1 z_r|.$$

Recall, that by Lemma 1.3.2(iii) for any $w \in V_r(z)$, $r \le t$,

$$|\pi_1 z_r - \pi_1 w| \le C c^{-1} \rho \varepsilon \alpha^{-1} (1 - \varepsilon \alpha^{-1} - \varepsilon \alpha^{-2})^{-1}.$$

Set

$$I_{x'}^\varepsilon(v, r, q) = \exp\langle \beta, \int_r^q B(x', \pi_2(\Phi_\varepsilon^s v)) ds \rangle.$$

Then

$$\int_{V_{0,r}(z)} I_{x'}^\varepsilon(v, r, r+s) dm_D(v) = \int_{V_r(z)} I_{x'}^\varepsilon(w, 0, s) J_\varepsilon^{\Xi_w}(-r, w) dm_{V_r(z)}(w) \tag{1.3.9}$$

where $V_{0,r}(z) = U_D^\varepsilon(r, z, C\rho)$, Ξ_w is the tangent space to $V_r(z)$ at w and $m_{V_r(z)}$ is the induced Riemannian volume on $V_r(z)$. By (1.3.4), (1.3.5), Lemma 1.3.2(i), and the Hölder continuity of the function φ^u,

$$C_2^{-1}(1 + C_1\varepsilon)^{-r} \le J_\varepsilon^{\Xi_w}(-r, w) J_\varepsilon^u(r, z) \le C_2(1 - C_1\varepsilon)^{-r} \tag{1.3.10}$$

for some $C_2 > 0$ independent of ε, r, $z \in D$ and $w \in V_r(z)$. Since $V_r(z) \in \mathcal{D}_\varepsilon^u(z_r, \alpha, \rho, \sqrt{C})$ by Lemma 1.3.2(ii), it follows from (1.2.15) and the above estimates that

$$(\nu_b(\varepsilon, s))^{-1} \le \int_{V_r(z)} I_{x'}^\varepsilon(w, 0, s) dm_{V_r(z)}(w) \tag{1.3.11}$$
$$\times \left(\int_{V_r(z)} \exp\langle \beta, \int_0^s B(x', F_{\pi_1 z_r}^\sigma \pi_2 w) d\sigma \rangle dm_{V_r(z)}(w) \right)^{-1} \le \nu_b(\varepsilon, s)$$

where

$$\nu_b(\varepsilon, s) = 2 + C_3 \exp(C_3 b e^{Ks} \varepsilon),$$

$C_3 > 0$ is a constant independent of $\varepsilon, \rho, b, r, z, x'$ and $\beta \in \mathbb{R}^d$ with $|\beta| \le b$.

Next, choose $\lambda \in (0,1)$ and $\theta(\varepsilon) \in [(\log \frac{1}{\varepsilon})^\lambda, 2(\log \frac{1}{\varepsilon})^\lambda]$ so that $n = \tau/\theta(\varepsilon)$ is an integer. If $n \le 1$ then (1.3.8) follows from (1.3.9)–(1.3.11) and Lemma 1.3.3. Now, let $n > 1$, $k < n$ and $v \in V_{0,t}(z)$. Then by (1.2.15) and Lemma 1.3.2(i) for any $w \in V_{0,t+k\theta(\varepsilon)}(v)$,

$$C_4^{-1} \le I_{x'}^\varepsilon(w, t, t + (k+1)\theta(\varepsilon))$$
$$\times \left(I_{x'}^\varepsilon(v, t, t + k\theta(\varepsilon)) I_{x'}^\varepsilon(w, t + k\theta(\varepsilon), t + (k+1)\theta(\varepsilon)) \right)^{-1} \le C_4$$

and

$$C_4^{-1} \le I_{x'}^\varepsilon(v, t, t + k\theta(\varepsilon)) \left(I_{x'}^\varepsilon(w, t, t + k\theta(\varepsilon)) \right)^{-1} \le C_4 \tag{1.3.12}$$

where $C_4 = C_4(b) = e^{4bKC\rho c^{-1}\kappa^{-1}}$. Integrating the inequalities above we obtain

$$(1.3.13)\, C_4^{-1} \le \int_{V_{0,t+k\theta(\varepsilon)}(v)} I^\varepsilon_{x'}(w,t,t+(k+1)\theta(\varepsilon))dm_D(w)\Big(I^\varepsilon_{x'}(v,t,t+k\theta(\varepsilon)) \times \int_{V_{0,t+k\theta(\varepsilon)}(v)} I^\varepsilon_{x'}(w,t+k\theta(\varepsilon),t+(k+1)\theta(\varepsilon))dm_D(w)\Big)^{-1} \le C_4.$$

From the estimates (1.3.9)–(1.3.11) together with Lemma 1.3.3 we conclude that for some $C_5 > 0$ independent of $t, k, \varepsilon, \rho, v$, and x',

$$(1.3.14)\quad C_5^{-1}e^{-\theta(\varepsilon)\eta_{b,\rho}(\varepsilon)}\big(\nu_b(\varepsilon,\theta(\varepsilon))\big)^{-1} \le \int_{V_{0,t+k\theta(\varepsilon)}(v)} I^\varepsilon_{x'}(w,t+k\theta(\varepsilon), t+(k+1)\theta(\varepsilon))dm_D(w)J^u_\varepsilon(t+k\theta(\varepsilon),v)\exp\big(-\theta(\varepsilon)P_{F_{\pi_1 v_{t+k\theta(\varepsilon)}}}(\langle\beta,B(x',\cdot)\rangle + \varphi^u_{\pi_1 v_{t+k\theta(\varepsilon)}})\big) \le e^{\theta(\varepsilon)\eta_{b,\rho}(\varepsilon)}\nu_b(\varepsilon,\theta(\varepsilon))$$

where $v_s = \Phi^s_\varepsilon v$ and $\eta_{b,\rho}(\varepsilon) > 0$, $\eta_{b,\rho}(\varepsilon) \to 0$ as $\varepsilon \to 0$. Observe that by (1.1.1), (1.2.15) and Lemma 1.3.2(iii),

$$|\pi_1 v_{t+k\theta(\varepsilon)} - \pi_1 z_t| \le K\varepsilon\tau + Cc^{-1}\rho\varepsilon\alpha^{-1}(1-\varepsilon\alpha^{-1}-\varepsilon\alpha^{-2})^{-1},$$

and so setting $P = P_{F_{\pi_1 z_t}}(\langle\beta,B(x',\cdot)\rangle + \varphi^u_{\pi_1 z_t})$ we obtain by (1.2.15) and [**16**] (see also [**61**] and [**71**]) that

$$(1.3.15)\qquad \big|P - P_{F_{\pi_1 v_{t+k\theta(\varepsilon)}}}(\langle\beta,B(x',\cdot)\rangle + \varphi^u_{\pi_1 v_{t+k\theta(\varepsilon)}})\big| \le C_6\varepsilon\tau$$

for some $C_6 = C_6(b) > 0$ independent of $v, k \le n, \varepsilon, t, z, x'$, and $\beta \in \mathbb{R}^d$ with $|\beta| \le b$ provided, say, $\tau \ge 1$ which we can assume without loss of generality.

A finite set $E \subset D$ will be called (s,γ,ε,D)–separated if $v_i, v_j \in E$, $v_i \ne v_j$ implies that $v_i \notin U^\varepsilon_D(s,v_j,\gamma)$. Let E_k be a maximal $(t+k\theta(\varepsilon), C\rho,\varepsilon,D)$–separated set in D and define

$$E^U_k = \{v \in E_k : U^\varepsilon_D(t+k\theta(\varepsilon),v,C\rho)\cap U \ne \emptyset\}.$$

Then for $k \ge 1$,

$$(1.3.16)\quad U^\varepsilon_D(t,z,\gamma+a_kC\rho) \supset \cup_{v\in E_k^{U^\varepsilon_D(t,z,\gamma)}} U^\varepsilon_D(t+k\theta(\varepsilon),v,C\rho) \supset U^\varepsilon_D(t,z,\gamma)$$

where $a_k = c^{-1}e^{-\frac12\kappa k\theta(\varepsilon)}$ and the left hand side of (1.3.16) follows from Lemma 1.3.2(i) assuming that ε is small enough. Observe also that $U^\varepsilon_D(t+k\theta(\varepsilon),v,C\rho/2)$ are disjoint for different $v \in E_k$. For $k = 1,2,...,n-1$ set $V(k,\rho) = U^\varepsilon_D(t,z,C\rho(1+\sum_{j=k}^{n-1}a_j))$ and $V(-k,\rho) = U^\varepsilon_D(t,z,C\rho(1-\sum_{j=k}^{n-1}a_j))$ with $V(n,\rho) = V = V_{0,t}(z)$. Then by (1.3.12)–(1.3.16) and Lemma 1.3.2(iii),

$$\begin{aligned}
&\int_{V(k+1,\rho)} I^\varepsilon_{x'}(v,t,t+(k+1)\theta(\varepsilon))dm_D(w)\\
&\le \sum_{v\in E_k^{V(k+1,\rho)}}\int_{V_{0,t+k\theta(\varepsilon)}(v)} I^\varepsilon_{x'}(w,t,t+(k+1)\theta(\varepsilon))dm_D(w)\\
&\le C_4\sum_{v\in E_k^{V(k+1,\rho)}} I^\varepsilon_{x'}(v,t,t+k\theta(\varepsilon))\int_{V_{0,t+k\theta(\varepsilon)}(v)} I^\varepsilon_{x'}(w,t+k\theta(\varepsilon),t\\
&+(k+1)\theta(\varepsilon))dm_D(w) \le C_5C_4e^{\theta(\varepsilon)(\eta_{b,\rho}(\varepsilon)+C_6\varepsilon\tau+P)}\nu_b(\varepsilon,\theta(\varepsilon))\sum_{v\in E_k^{V(k+1,\rho)}}\\
&\big(J^u_\varepsilon(t+k\theta(\varepsilon),v)\big)^{-1}I^\varepsilon_{x'}(v,t,t+k\theta(\varepsilon))\\
&\le C_5C_4^2c^{-1}_{\rho/2,T}e^{\theta(\varepsilon)(\eta_{b,\rho}(\varepsilon)+C_6\varepsilon\tau+P)}\nu_b(\varepsilon,\theta(\varepsilon))\\
&\times\sum_{v\in E_k^{V(k+1,\rho)}}\int_{U^\varepsilon_D(t+k\theta(\varepsilon),v,C\rho/2)} I^\varepsilon_{x'}(w,t,t+k\theta(\varepsilon))dm_D(w)\\
&\le C_5C_4^2c^{-1}_{\rho/2,T}e^{\theta(\varepsilon)(\eta_{b,\rho}(\varepsilon)+C_6\varepsilon\tau+P)}\nu_b(\varepsilon,\theta(\varepsilon))\int_{V(k,\rho)} I^\varepsilon_{x'}(w,t,t+k\theta(\varepsilon))dm_D(w).
\end{aligned}$$

Similarly, we obtain

$$\int_{V(-(k+1),\rho)} I_{x'}^\varepsilon(w,t,t+(k+1)\theta(\varepsilon))dm_D(w)$$
$$\geq C_5^{-1}C_4^{-2}c_{\rho,T}e^{-\theta(\varepsilon)(\eta_{b,\rho/2}(\varepsilon)+C_6\varepsilon\tau+P)}(\nu_b(\varepsilon,\theta(\varepsilon))^{-1}$$
$$\times \int_{V(-k,\rho)} I_{x'}^\varepsilon(w,t,t+k\theta(\varepsilon))dm_D(w).$$

Emloying these estimates recursively for $k=n-1,n-2,...,1$ and estimating $\int_{V(\pm 1,\rho)} I_{x'}^\varepsilon(w,t,t+\theta(\varepsilon))dm_D(w)$ by (1.3.14) with $k=0$ and with $V(\pm 1,\rho)$ in place of $V_{0,t}(v)$ we derive (1.3.8) with $\zeta_{b,\rho,T}(\Delta,s,\varepsilon)=s^{-1}\log\big(C_5C_4^2(c_{\rho,T}^{-1}+c_{\rho/2,T}^{-1})\big)+\eta_{b,\rho}(\varepsilon)+\eta_{b,\rho/2}(\varepsilon)+2(\theta(\varepsilon))^{-1}\log\nu_b(\varepsilon,\theta(\varepsilon))+C_6\Delta$ provided $n\geq 1$. □

Next, under Assumption 1.2.6 we derive a volume lemma type assertion (see [**13**]) which will be needed in Sections 1.7 and 1.8 and which will hold true on any time intervals and not just on time intervals of order $1/\varepsilon$ as in Lemma 1.3.2(iv). In order to do so we will consider a subset of embedded C^2 discs from $\mathcal{D}_\varepsilon^u(z,\alpha,\rho,C)$ taking special care of their C^2 bounds.

Namely, let $\mathrm{Exp}_y : T_y\mathbf{M}\to\mathbf{M}$ be the exponential map which is a diffeomorphism of $V_\delta^y=\{\xi\in T_y\mathbf{M} : \|\xi\|<\delta\}$ onto the open δ−neighborhood $U_\delta(y)$ of y provided $\delta>0$ is small enough. Given $x,\tilde{x}\in\mathbb{R}^d$, $y\in\mathbf{M}$ and $\xi\in T_y\mathbf{M}$ set

$$\chi_{x,y}(\tilde{x},\xi)=(x+\tilde{x},\ \mathrm{Exp}_y\xi)$$

which is a diffeomorphism of $\mathbb{R}^d\times V_\delta^y$ onto $\mathbb{R}^d\times U_\delta(y)$. Let $D\in\mathcal{D}_\varepsilon^u(z,\alpha,\rho,C)$, $z=(x,y)$, $y\in\mathcal{W}$ be an embedded C^2 disc. Assuming that $C^2\rho<\delta$ we can define $\hat{D}=\chi_{x,y}^{-1}(D)$ which is a C^2 hypersurface in $\{\tilde{x} : |\tilde{x}\|<\delta\}\times V_\delta^y$. If δ is sufficiently small then the tangent subbubndle $T\hat{D}$ over $\hat{D}$ still stays close to $\Gamma_x^u(y)$, and so we can represent $\hat{D}$ as a parametric set $(\eta,\varphi(\eta),x(\eta))$ where $\eta\in\Gamma_x^u(y)$, $\varphi(\eta)\in\Gamma_x^{0s}(y)$ and $x(\eta)\in\mathbb{R}^d$. We will write that $D\in\hat{\mathcal{D}}_\varepsilon^u(z,\alpha,\rho,C,L)$ if the parametric representation of the corresponding $\hat{D}$ as above satisfies

$$\max_{i,j,k,l}\max\big(\big|\frac{\partial^2\varphi_i(\eta)}{\partial\eta_k\partial\eta_l}\big|,\ \frac{\partial^2 x_j(\eta)}{\partial\eta_k\partial\eta_l}\big|\big)\leq L.$$

1.3.5. Lemma. *There exists $t_1\geq t_0$ such that for any $t\geq t_1$ we can choose $\delta>0$ small enough and $L>0$ large enough so that if $D\in\hat{\mathcal{D}}_\varepsilon^u(z,\alpha,\rho,C,L)$ and $C^2\rho<\delta$ then*

$$\{v\in\Phi_\varepsilon^tD : d_{\Phi_\varepsilon^tD}(\Phi_\varepsilon^tz,v)\leq C^2\rho\}\in\hat{\mathcal{D}}_\varepsilon^u(\Phi_\varepsilon^tz,\alpha,\rho,C,L).$$

Proof. Since the differential $D_0\mathrm{Exp}_y$ of the exponential map at zero is the identity map it follows from the definition of $\mathcal{D}_\varepsilon^u(z,\alpha,\rho,C)$ that

$$\max_{i,j,k}\max\big(\big|\frac{\partial\varphi_i(\eta)}{\partial\eta_k}\big|,\ \frac{\partial x_j(\eta)}{\partial\eta_k}\big|\big)\leq c(\varepsilon,\delta)$$

where $c(\varepsilon,\delta)\to 0$ (uniformly in all D as above) as $\varepsilon,\delta\to 0$. Let $z=(x,y)$ and set

$$f_{x,y,\varepsilon}^t=\chi_{\Phi_\varepsilon^t(x,y)}^{-1}\circ\Phi_\varepsilon^t\circ\chi_{x,y}$$

which for each fixed $t>0$ and a sufficiently small $\delta>0$ (depending on t) defines a diffeomorphism of $\mathbb{R}^d\times V_\delta^y$ onto its image. By (1.3.2) the tangent subbundle over Φ_ε^tD is contained in $\mathcal{C}^u(\varepsilon,\alpha)$, and so for small $\delta>0$ the tangent subbundle $T(f_{x,y,\varepsilon}^t\hat{D})$ over $f_{x,y,\varepsilon}^t\hat{D}$ stays close to $\Gamma_{\tilde{x}}^u(\tilde{y})$ where $\tilde{x}=\pi_1\Phi_\varepsilon^t(z)$ and $\tilde{y}=\pi_2\Phi_\varepsilon^t(z)$. Hence, we can represent $f_{x,y,\varepsilon}^t(\hat{D})$ in a parametric form $\big(\tilde{\eta},\tilde{\varphi}(\tilde{\eta}),\tilde{x}(\tilde{\eta})\big)$ where $\tilde{\eta}\in$

$\Gamma^u_{\tilde x}(\tilde y)$, $\tilde\varphi(\tilde\eta)\in\Gamma^{0s}_{\tilde x}(\tilde y)$ and $\tilde x(\tilde\eta)\in\mathbb{R}^d$. Fix some $t>t_0$ so that (1.3.3) holds true. Write $f^t_{x,y,\varepsilon}(\eta,\varphi,x)=(\tilde\eta,\tilde\varphi,\tilde x)$, so that, in particular,

$$f^t_{x,y,\varepsilon}(\eta,\varphi(\eta),x(\eta))=\big(\tilde\eta,\tilde\varphi(\tilde\eta),\tilde x(\tilde\eta)\big).$$

Then

$$\tilde\eta=A\eta+a_{\varepsilon,\delta}(\eta,\varphi,x),\ \tilde\varphi=B\varphi+b_{\varepsilon,\delta}(\eta,\varphi,x),\ \tilde x=x+c_{\varepsilon,\delta}(\eta,\varphi,x) \tag{1.3.17}$$

where $\eta\in\Gamma^u_x(y)$, $\varphi\in\Gamma^{0s}_x(y)$, $x\in\mathbb{R}^d$, A is an $n^u\times n^u$–matrix, B is an $n^{0s}\times n^{0s}$–matrix with $n^{0s}=n_{\mathbf{M}}-n^u$ and

$$\|a_{\varepsilon,\delta}(\eta,\varphi,x)\|_{C^1}+\|b_{\varepsilon,\delta}(\eta,\varphi,x)\|_{C^1}+\|c_{\varepsilon,\delta}(\eta,\varphi,x)\|_{C^1}\le c(\varepsilon,\delta) \tag{1.3.18}$$

for all $(\eta,\varphi)\in V^y_\delta$ and $|x|<\delta$ where $c(\varepsilon,\delta)\to 0$ as $\varepsilon,\delta\to 0$. By (1.2.15), (1.3.1) and Assumption 1.2.6 it follows that there exists a constant $R>0$ such that for any $\xi\in\Gamma^{0s}_x(y)$, $x\in\mathbb{R}^d$, $y\in\mathbf{M}$,

$$\|DF^t_x\xi\|\le R\|\xi\|. \tag{1.3.19}$$

By (1.3.1) we can choose $t>0$ large enough and then ε and δ small enough so that for all $\eta\in\Gamma^u_x(y)$,

$$\|A\eta\|\ge(1+R)\|\eta\|. \tag{1.3.20}$$

Now t is fixed and we can choose ε and δ so small that (1.3.19) implies that,

$$\|B\|<1+R. \tag{1.3.21}$$

In order to shorten notations for every vector function $f(\zeta)=(f_1(\zeta),...,f_l(\zeta))$, $\zeta=(\zeta_1,...,\zeta_k)$ we denote by $\frac{\partial f}{\partial\zeta}$ the Jacobi matrix $(\partial f_i(\zeta)/\partial\zeta_j)$ and by $\frac{\partial^2 f}{\partial\zeta^2}$ we denote the collection $(\partial^2 f_i(\zeta)/\partial\zeta_j\partial\zeta_k)$. We set also

$$\|\frac{\partial f}{\partial\zeta}\|=\max_{i,j}|\frac{\partial f_i(\zeta)}{\partial\zeta_j}|\ \text{and}\ \|\frac{\partial^2 f}{\partial\zeta^2}\|=\max_{i,j,k}|\frac{\partial^2 f_i}{\partial\zeta_j\partial\zeta_k}|.$$

Observe that by Assumptions 1.2.1, 1.2.2 and 1.2.6 for any $t>0$ there exists $\hat R=\hat R_t>0$ such that

$$\sup_{\varepsilon\le 1}\sup_{|u|\le t}\|\Phi^u_\varepsilon\|_{C^2}\le\hat R \tag{1.3.22}$$

and

$$\max\big(\|a_{\varepsilon,\delta}(\eta,\varphi,x)\|_{C^2},\|b_{\varepsilon,\delta}(\eta,\varphi,x)\|_{C^2},\|c_{\varepsilon,\delta}(\eta,\varphi,x)\|_{C^2}\big)\le 2\hat R+1. \tag{1.3.23}$$

It follows by (1.3.17)–(1.3.23) (with natural product notations) that

$$\|\frac{\partial^2\tilde\varphi}{\partial\tilde\eta^2}\|=\|\frac{\partial^2\tilde\varphi}{\partial\eta^2}(\frac{\partial\eta}{\partial\tilde\eta})^2+\frac{\partial\tilde\varphi}{\partial\eta}\frac{\partial^2\eta}{\partial\tilde\eta^2}\|\le\|\frac{\partial^2\tilde\varphi}{\partial\eta^2}\|(1+R)^{-2}+c(\varepsilon,\delta)\hat R$$

and

$$\|\tfrac{\partial^2\tilde\varphi}{\partial\eta^2}\|=\|B\tfrac{\partial^2\varphi}{\partial\eta^2}+\tfrac{\partial^2 b_{\varepsilon,\delta}}{\partial\eta^2}+2\tfrac{\partial^2 b_{\varepsilon,\delta}}{\partial\eta\partial\varphi}\tfrac{\partial\varphi}{\partial\eta}+2\tfrac{\partial^2 b_{\varepsilon,\delta}}{\partial\eta\partial x}\tfrac{\partial x}{\partial\eta}\tfrac{\partial^2 b_{\varepsilon,\delta}}{\partial\varphi^2}\big(\tfrac{\partial\varphi}{\partial\eta}\big)^2$$
$$+2\tfrac{\partial^2 b_{\varepsilon,\delta}}{\partial\varphi\partial x}\tfrac{\partial\varphi}{\partial\eta}\tfrac{\partial x}{\partial\eta}+\tfrac{\partial^2 b_{\varepsilon,\delta}}{\partial x^2}\big(\tfrac{\partial x}{\partial\eta}\big)^2\|\le(1+R)L+(2\hat R+1)(1+c(\varepsilon,\delta))^2.$$

Similarly,

$$\|\frac{\partial^2\tilde x}{\partial\eta^2}\|\le(1+R)L+(2\hat R+1)(1+R)^{-2}(1+c(\varepsilon,\delta))^2+c(\varepsilon,\delta)\hat R.$$

Choosing $L \geq R^{-1}(2\hat{R} + R + 2)$ we obtain that if

$$\max\left\{\left\|\frac{\partial^2 \varphi}{\partial \eta^2}\right\|, \left\|\frac{\partial^2 x}{\partial \eta^2}\right\|\right\} \leq L$$

then

$$\max\left\{\left\|\frac{\partial^2 \tilde{\varphi}}{\partial \tilde{\eta}^2}\right\|, \left\|\frac{\partial^2 \tilde{x}}{\partial \tilde{\eta}^2}\right\|\right\} \leq L$$

and the assertion of Lemma 1.3.5 follows. □

The main purpose of the previous result is to derive the following volume lemma type assertion which plays an essential role in Section 1.6.

1.3.6. Lemma. *For any $\beta \in (0, C^2\rho)$ there exists $c_\beta > 0$ such that if $D \in \hat{D}^u_\varepsilon(z, \alpha, \rho, C, L)$ and L is large enough then for any $t > 0$ and $v, w \in D$ satisfying $w \in U^\varepsilon_D(t, v, \beta) \subset D$,*

$$c_\beta \leq m_D\big(U^\varepsilon_D(t, v, \beta)\big) J^{T_w D}_\varepsilon(t, w) \leq c_\beta^{-1}. \tag{1.3.24}$$

Proof. Set $V_{s,t} = \Phi^s_\varepsilon U^\varepsilon_D(t, v, \beta)$ and $V_t = V_{t,t}$. Similarly to Lemma 1.3.2(ii), $V_t \in \mathcal{D}^u_\varepsilon(\Phi^t_\varepsilon v, \alpha, \beta C^{-1}, \sqrt{C})$, and so by uniformity considerations there exists $\tilde{c}_\beta > 0$ independent of v, t and D as above such that

$$\tilde{c}_\beta \leq m_{V_t}(V_t) = \int_{U^\varepsilon_D(t,v,\beta)} J^{T_w D}_\varepsilon(t, w) dm_D(w) \leq \tilde{c}_\beta^{-1}. \tag{1.3.25}$$

Choose $l \in \mathbb{N}$ so that $t_2 = t/l \in [t_1, 2t_1)$ and set $w_k = \Phi^{kt_2}_\varepsilon w$, $v_k = \Phi^{kt_2}_\varepsilon v$. Then for any $w \in U^\varepsilon_D(t, v, \beta)$,

$$J^{T_w D}_\varepsilon(t, w) = \prod_{k=0}^{l-1} J^{T_{w_k} V_{kt_2,t}}_\varepsilon(t_2, w_k) \tag{1.3.26}$$

and by Lemma 1.3.2(i),

$$d_{V_{kt_2,t}}(w_k, v_k) \leq c^{-1}\beta e^{-\frac{1}{2}\kappa(l-k)t_2}. \tag{1.3.27}$$

By (1.3.5), (1.3.22), (1.3.26), and (1.3.27) together with Lemma 1.3.5 we conclude that there exists a constant $\tilde{C} > 0$ such that

$$\big|\ln J^{T_{w_k} V_{kt_2,t}}_\varepsilon(t_2, w_k) - \ln J^{T_{v_k} V_{kt_2,t}}_\varepsilon(t_2, v_k)\big| \leq \tilde{C} e^{-\frac{1}{2}\kappa(l-k)t_2}. \tag{1.3.28}$$

Now (1.3.24) follows from (1.3.25), (1.3.26), and (1.3.28) with

$$c_\beta = \tilde{c}_\beta \exp\big(-2\tilde{C}(1 - e^{-\frac{1}{2}\kappa t_2})^{-1}\big).$$

□

1.4. Large deviations: preliminaries

We will need the following version of general large deviations bounds when usual assumptions hold true with errors. An upper bound similar to (1.4.3) below appeared previously in [**55**]. For simplicity we will formulate the result for $\mathbb{R}^d$–valued random vectors though the same arguments work for random variables with values in a Banach space. The proof is a strightforward modification of the standard one (cf. [**47**]) but still we exhibit it here for readers' convenience.

1.4.1. LEMMA. *Let $H = H(\beta)$, $\eta = \eta(\beta)$ be uniformly bounded on compact sets functions on $\mathbb{R}^d$ and $\{\Xi_\tau, \tau \geq 1\}$ be a family of $\mathbb{R}^d$−valued random vectors on a probability space $(\Omega, \mathcal{F}, P)$ such that $|\Xi_\tau| \leq C < \infty$ with probability one for some constant C and all $\tau \geq 1$. For any $a > 0$ and $\alpha, \beta_0 \in \mathbb{R}^d$ set*

$$(1.4.1)\quad L_a^{\beta_0}(\alpha) = \sup_{\beta \in \mathbb{R}^d, |\beta+\beta_0| \leq a} (\langle \beta, \alpha \rangle - H(\beta)),\ L_a(\alpha) = L_a^0(\alpha),\ L(\alpha) = L_\infty(\alpha).$$

(i) For any $\lambda, a > 0$ there exists $\tau_0 = \tau(\lambda, a, C)$ such that whenever for some $\tau \geq \tau_0$, $\beta_0 \in \mathbb{R}^d$ and each $\beta \in \mathbb{R}^d$ with $|\beta + \beta_0| \leq a$,

$$(1.4.2)\qquad H_\tau(\beta) = \tau^{-1} \log E e^{\tau \langle \beta, \Xi_\tau \rangle} \leq H(\beta) + \eta(\beta)$$

then for any compact set $\mathcal{K} \subset \mathbb{R}^d$,

$$(1.4.3)\qquad P\{\Xi_\tau \in \mathcal{K}\} \leq \exp\left(-\tau(L_a^{\beta_0}(\mathcal{K}) - \eta_a^{\beta_0} - \lambda|\beta_0| - \lambda)\right)$$

where

$$(1.4.4)\qquad \eta_a^{\beta_0} = \sup\{\eta(\beta) : |\beta + \beta_0| \leq a\} \text{ and } L_a^{\beta_0}(\mathcal{K}) = \inf_{\alpha \in \mathcal{K}} L_a^{\beta_0}(\alpha).$$

(ii) Suppose that $\alpha_0 \in \mathbb{R}^d$, $0 < a \leq \infty$ and there exists $\beta_0 \in \mathbb{R}^d$ such that $|\beta_0| \leq a$ and

$$(1.4.5)\qquad H(\beta_0) = \langle \beta_0, \alpha_0 \rangle - L_a(\alpha_0).$$

If (1.4.2) holds true then for any $\delta > 0$,

$$(1.4.6)\qquad P\{|\Xi_\tau - \alpha_0| \leq \delta\} \leq \exp\left(-\tau(L_a(\alpha_0) - \eta(\beta_0) - \delta|\beta_0|)\right).$$

(iii) Assume that $\alpha_0, \beta_0 \in \mathbb{R}^d$ satisfy (1.4.5). For any $\lambda, a > 0$ there exists $\tau_0 = \tau(\lambda, a, C)$ such that whenever for some $\tau \geq \tau_0$ and each $\beta \in \mathbb{R}^d$ with $|\beta| \leq a$ the inequality (1.4.2) holds true together with

$$(1.4.7)\qquad \tau^{-1} \log E e^{\tau \langle \beta, \Xi_\tau \rangle} \geq H(\beta) - \eta(\beta)$$

then for any $\gamma, \delta > 0$, $\gamma \leq \delta$,

$$(1.4.8)\qquad \begin{aligned} P\{|\Xi_\tau - \alpha_0| < \delta\} &\geq \exp\left(-\tau(L(\alpha_0) + \eta(\beta_0) + \gamma|\beta_0|)\right) \\ &\times \Big(1 - \exp\big(-\tau(\tilde{L}_a^{\beta_0}(\mathcal{K}_{\gamma,C}(\alpha_0)) - \eta_a - \eta(\beta_0) - \lambda|\beta_0| - \lambda)\big)\Big) \end{aligned}$$

where

$$\tilde{L}_a^{\beta_0}(\alpha) = L_a(\alpha) - \langle \beta_0, \alpha \rangle + H(\beta_0),$$

$\tilde{L}_a^{\beta_0}(\mathcal{K}) = \inf_{\alpha \in \mathcal{K}} \tilde{L}_a^{\beta_0}(\alpha)$, $\eta_a = \eta_a^0$, $\mathcal{K}_{\gamma,C}(\alpha_0) = \overline{U_C(0)} \setminus U_\gamma(\alpha_0)$, $U_\gamma(\alpha) = \{\tilde{\alpha} : |\tilde{\alpha} - \alpha| < \gamma\}$ and $\bar{U}$ denotes the closure of U.

PROOF. (i) By (1.4.1) for any $\alpha \in \mathcal{K}_C = \mathcal{K} \cap \overline{U_C(0)}$ and $\lambda > 0$ there exists $\beta_\lambda(\alpha) \in \mathbb{R}^d$ such that

$$(1.4.9)\qquad |\beta_\lambda(\alpha) + \beta_0| \leq a \text{ and } \langle \beta_\lambda(\alpha), \alpha \rangle - H(\beta_\lambda(\alpha)) > L_a^{\beta_0}(\alpha) - \lambda/2.$$

Set $\gamma_{a,\lambda}(\alpha) = \frac{\lambda}{2}\min(1, a^{-1})$ and cover the compact set $\mathcal{K}_C$ by open balls $U_{\gamma_{a,\lambda}}(\alpha)$, $\alpha \in \mathcal{K}_C$. Let $U_{\gamma_{a,\lambda}(\alpha_1)}, ..., U_{\gamma_{a,\lambda}(\alpha_n)}$ be a finite subcover with a minimal number n of elements. Observe that n does not exceed the maximal number of points in $\overline{U_C(0)}$ with pairwise distances at least $\frac{1}{2}\gamma_{a,\lambda}$ and the latter number depends only on C, a and λ. By (1.4.2) and (1.4.9) for each $i = 1, ..., n$,

$$e^{\tau\eta_a^{\beta_0}} \geq E\mathbb{I}_{\Xi_\tau \in U_{\gamma_{a,\lambda}(\alpha_i)}(\alpha_i)} e^{\tau(\langle\beta_\lambda(\alpha_i),\Xi_\tau\rangle - H(\beta_\lambda(\alpha_i)))}$$

$$\geq e^{-\tau\lambda/2} E\mathbb{I}_{\Xi_\tau \in U_{\gamma_{a,\lambda}(\alpha_i)}(\alpha_i)} e^{\tau\left(\langle\beta_\lambda(\alpha_i),\alpha_i\rangle - |\beta_0|\lambda/2 - H(\beta_\lambda(\alpha_i))\right)}$$

$$e^{\tau\left(L_a^{\beta_0}(\alpha_i) - |\beta_0|\lambda/2 - \lambda/2\right)} P\{\Xi_\tau \in U_{\gamma_{a,\lambda}(\alpha_i)}(\alpha_i)\}.$$

Since $L_a^{\beta_0}(\alpha_i) \geq L_a^{\beta_0}(\mathcal{K})$ and $|\Xi_\tau| \leq C$ then summing these inequalities in $i = 1, ..., n$ we obtain

$$(1.4.10) \qquad P\{\Xi_\tau \in \mathcal{K}\} = P\{\Xi_\tau \in \mathcal{K}_C\} \leq e^{-\tau\left(L_a^{\beta_0}(\mathcal{K}) - \eta_a^{\beta_0} - |\beta_0|\lambda/2 - \tau^{-1}\log n - \lambda/2\right)}.$$

Since n is bounded by a number depending only on λ, a, and C we can choose $\tau_0 = \tau_0(\lambda, a, C)$ so that $\tau_0^{-1}\log n \leq \lambda/2$ which together with (1.4.10) yield (1.4.3).

(ii) By (1.4.2) and (1.4.5),

$$e^{\tau\eta(\beta_0)} \geq E\mathbb{I}_{\Xi_\tau \in U_\delta(\alpha_0)} e^{\tau\left(\langle\beta_0,\Xi_\tau\rangle - H(\beta_0)\right)} \geq e^{\tau\left(L_a(\alpha_0) - \delta|\beta_0|\right)} P\{\Xi_\tau \in U_\delta(\alpha_0)\}$$

and (1.4.6) follows.

(iii) By (1.4.5) and (1.4.7) for any $\gamma \leq \delta$,

$$(1.4.11) \qquad P\{|\Xi_\tau - \alpha_0| < \delta\} \geq P\{|\Xi_\tau - \alpha_0| < \gamma\}$$

$$= E_\tau^{\beta_0}\mathbb{I}_{|\Xi_\tau - \alpha_0| < \gamma} e^{-\tau\left(\langle\beta_0,\Xi_\tau\rangle - H_\tau(\beta_0)\right)}$$

$$\geq e^{-\tau\left(L(\alpha_0) + |\beta_0|\gamma + \eta(\beta_0)\right)} P_\tau^{\beta_0}\{|\Xi_\tau - \alpha_0| < \gamma\}$$

where $E_\tau^{\beta_0}$ is the expectation with respect to the probability measure P^{β_0} on $(\Omega, \mathcal{F})$ such that

$$\frac{dP_\tau^{\beta_0}}{dP} = e^{\tau(\langle\beta_0,\Xi_\tau\rangle - H_\tau(\beta_0))}.$$

Now by (1.4.2) and (1.4.5) for any $\beta \in \mathbb{R}^d$ with $|\beta + \beta_0| \leq a$ we obtain that

$$(1.4.12) \qquad \tau^{-1}\log E_\tau^{\beta_0} e^{\tau\langle\beta,\Xi_\tau\rangle} = H_\tau(\beta + \beta_0) - H_\tau(\beta_0)) \leq \tilde{H}^{\beta_0}(\beta) + \tilde{\eta}^{\beta_0}(\beta)$$

where $\tilde{H}^{\beta_0}(\beta) = H(\beta + \beta_0) - H(\beta_0)$ and $\tilde{\eta}^{\beta_0}(\beta) = \eta(\beta + \beta_0) + \eta(\beta_0)$. Observe that

$$(1.4.13) \qquad \sup_{\beta\in\mathbb{R}^d, |\beta+\beta_0|\leq a} \left(\langle\beta, \alpha\rangle - \tilde{H}^{\beta_0}(\beta)\right) = L_a(\alpha) - \langle\beta_0, \alpha\rangle + H(\beta_0) = \tilde{L}_a^{\beta_0}(\alpha).$$

Thus, applying (i) on the probability space $(\Omega, \mathcal{F}, P_\tau^{\beta_0})$ we derive that

$$(1.4.14) \quad P_\tau^{\beta_0}\{|\Xi_\tau - \alpha_0| \geq \gamma\} \leq \exp\left(-\tau\left(\tilde{L}_a^{\beta_0}(\mathcal{K}_{\gamma,C}(\alpha_0)) - \eta_a - \eta(\beta_0) - \lambda|\beta_0| - \lambda\right)\right)$$

provided $\tau \geq \tau_0$ for a sufficiently large $\tau_0 = \tau_0(\lambda, a, C)$. This together with (1.4.11) yield (1.4.8). □

1.4.2. LEMMA. *Let S_n, $n = 1, 2, ...$ be a nondecreasing sequence of lower semicontinuous functions on a metric space M and let $S = \lim_{n\to\infty} S_n$. Assume that S is also lower semicontinuous and for any compact set $\mathcal{K} \subset M$ denote*

$$S_n(\mathcal{K}) = \inf_{\gamma\in\mathcal{K}} S_n(\gamma) \text{ and } S(\mathcal{K}) = \inf_{\gamma\in\mathcal{K}} S(\gamma).$$

Then

$$(1.4.15) \qquad \lim_{n\to\infty} S_n(\mathcal{K}) = S(\mathcal{K}).$$

PROOF. By the lower semicontinuity of S_n and S and by compactness of $\mathcal{K}$ it follows that there exist $\hat{\gamma}_n, \hat{\gamma} \in \mathcal{K}$ such that $S_n(\hat{\gamma}_n) = S_n(\mathcal{K})$ and $S(\hat{\gamma}) = S(\mathcal{K})$. Passing if needed to a subsequence assume that $\hat{\gamma}_n \to \tilde{\gamma} \in \mathcal{K}$ as $n \to \infty$. Since

$$S_n(\mathcal{K}) = S_n(\hat{\gamma}_n) \le S_n(\hat{\gamma}) \tag{1.4.16}$$

then

$$\limsup_{n\to\infty} S_n(\mathcal{K}) \le S(\hat{\gamma}) = S(\mathcal{K}). \tag{1.4.17}$$

Assume now that $S(\mathcal{K}) < \infty$. Since

$$S(\mathcal{K}) = S(\hat{\gamma}) \le S(\tilde{\gamma})$$

then for any $\varepsilon > 0$ there exists $n(\varepsilon)$ such that

$$S(\tilde{\gamma}) \le S_{n(\varepsilon)}(\tilde{\gamma}) + \varepsilon. \tag{1.4.18}$$

By the lower semicontinuity of $S_{n(\varepsilon)}(\gamma)$ it follows that for $m \ge n(\varepsilon)$ large enough

$$S(\tilde{\gamma}) \le S_{n(\varepsilon)}(\hat{\gamma}_m) + 2\varepsilon \le S_m(\hat{\gamma}_m) + 2\varepsilon \tag{1.4.19}$$

where we use also that $S_m, m = 1, 2, ...$ is a nondecreasing sequence. Since (1.4.19) holds true for any $m \ge n(\varepsilon)$ large enough and for each $\varepsilon > 0$ we can pass there to the limit so that, first, $m \to \infty$ and then $\varepsilon \to 0$ yielding that

$$S(\mathcal{K}) \le \liminf_{m\to\infty} S_n(\mathcal{K})$$

which together with (1.4.17) give (1.4.15) under the condition $S(\mathcal{K}) < \infty$. If $S(\mathcal{K}) = \infty$ then $S(\tilde{\gamma}) = \infty$ and for any $A > 0$ there exists $n(A)$ such that $S_n(\tilde{\gamma}) > A$ for any $n \ge n(A)$. By the lower semicontinuity of S_n we conclude that $S_n(\hat{\gamma}_m) > A$ for $m \ge n$ large enough which implies that $S_m(\hat{\gamma}_m) > A$ for all sufficiently large m. Hence

$$\liminf_{m\to\infty} S_m(\mathcal{K}) = \liminf_{m\to\infty} S_m(\hat{\gamma}_m) > A \tag{1.4.20}$$

and since A is arbitrary the left hand side of (1.4.20) equals infinity, i.e. again (1.4.15) holds trues with both parts of it being equal ∞. □

In the next section we will employ the following general result which will enable us to subdivide time into small intervals freezing the slow variable on each of them so that the estimate (1.3.8) of Proposition 1.3.4 becomes sufficiently precise and, on the other hand, we will not change much the corresponding functionals S_{0T} appearing in required large deviations estimates. This result is certainly not new, it is cited in [**79**] as a folklore fact and a version of it can be found in [**59**], p.67 but for readers convenience we give its proof here.

1.4.3. LEMMA. *Let $f = f(t)$ be a measurable function on $\mathbb{R}^1$ equal zero outside of $[0, T]$ and such that $\int_0^T |f(t)|dt < \infty$. For each positive integer m and $c \in [0, T]$ define $f_m(t, c) = f([(t + c)\Delta^{-1}]\Delta - c)$ where $\Delta = T/m$ and $[\cdot]$ denotes the integral part. Then there exists a sequence $m_i \to \infty$ such that for Lebesgue almost all $c \in [0, T]$,*

$$\lim_{i\to\infty} \int_0^T |f(t) - f_{m_i}(t, c)|dt = 0. \tag{1.4.21}$$

PROOF. For each $\delta > 0$ there exists a C^1 function g on $\mathbb{R}^1$ equal zero outside of $[0, T]$ and such that

$$\int_0^T |g(t) - f(t)|dt \le \delta/T. \tag{1.4.22}$$

Define $g_m(t, c)$ as above with g in place of f. Then

$$\begin{aligned}(1.4.23)\ &\textstyle\int_0^T dc \int_0^T |g_m(t,c) - f_m(t,c)|dt \le \int_0^T dc \sum_{i=0}^{\infty} |g(i\Delta - c) - f(i\Delta - c)|\Delta \\ &= \textstyle\sum_{i=1}^m \Delta \int_0^{i\Delta} |g(u) - f(u)|du = \sum_{i=1}^m \Delta \sum_{k=0}^{i-1} \int_{k\Delta}^{(k+1)\Delta} |g(u) - f(u)|du \\ &= \textstyle\Delta \sum_{k=0}^{m-1} (m-k) \int_{k\Delta}^{(k+1)\Delta} |g(u) - f(u)|du \le T \int_0^T |g(u) - f(u)|du \le \delta.\end{aligned}$$

We have also

$$\int_0^T \int_0^T |g_m(t,c) - g(t)|dtdc \le \Delta \sup_{0\le t\le T} g'(t). \tag{1.4.24}$$

Since

$$|f(t) - f_m(t,c)| \le |f(t) - g(t)| + |g(t) - g_m(t,c)| + |g_m(t,c) - f_m(t,c)|$$

it follows from (1.4.22)–(1.4.24) that

$$\lim_{m\to\infty} \int_0^T \int_0^T |f(t) - f_m(t,c)|dtdc = 0.$$

This together with the Chebyshev inequality and the Borel–Cantelli lemma yield (1.4.21) for some sequence $m_i \to \infty$ and Lebesgue almost all $c \in [0, T]$. □

1.5. Large deviations: Proof of Theorem 1.2.3

1.5.1. LEMMA. *Let $x_i, \tilde{x}_i \in \mathcal{X}$, $i = 0, 1, ..., N$, $0 = t_0 < t_1 < ... < t_{N-1} < t_N = T$, $\Delta = \max_{0\le i\le N-1}(t_{i+1} - t_i)$, $\xi_i = (x_i - x_{i-1})(t_i - t_{i-1})^{-1}$, $n(t) = \max\{j \ge 0 : t \ge t_j\}$, $\psi(t) = \tilde{x}_{n(t)}$, $v \in \mathcal{X} \times \mathbf{M}$,*

$$\Xi_j^\varepsilon(v, x) = (t_j - t_{j-1})^{-1} \int_{t_{j-1}}^{t_j} B(x, Y_v^\varepsilon(s/\varepsilon))ds,$$

and for $t \in [0, T]$,

$$Z_{v,x}^{\varepsilon,\psi}(t) = x + \int_0^t B(\psi(s), Y_v^\varepsilon(s/\varepsilon))ds. \tag{1.5.1}$$

Then

$$\begin{aligned}(1.5.2)\qquad &\left|\Xi_j^\varepsilon(v, x_{j-1}) - (t_j - t_{j-1})^{-1}(Z_v^\varepsilon(t_j) - Z_v^\varepsilon(t_{j-1}))\right| \\ &\le K\left|Z_v^\varepsilon(t_{j-1}) - x_{j-1}\right| + \tfrac{1}{2}K^2(t_j - t_{j-1}),\end{aligned}$$

$$\begin{aligned}(1.5.3)\qquad &\textstyle\sup_{0\le s\le t}\left|Z_{v,x}^{\varepsilon,\psi}(s) - \psi(s)\right| \le |x - x_0| + \max_{0\le j\le n(t)} |x_j - \tilde{x}_j| \\ &\textstyle+K\Delta + n(t)\Delta \max_{1\le j\le n(t)} \left|\Xi_j^\varepsilon(v, \tilde{x}_{j-1}) - \xi_j\right|\end{aligned}$$

and

$$\sup_{0\le s\le t} \left|Z_v^\varepsilon(s) - Z_{v,x}^{\varepsilon,\psi}(s)\right| \le e^{Kt}\left(|\pi_1 v - x| + Kt \sup_{0\le s\le t} \left|Z_{v,x}^{\varepsilon,\psi}(s) - \psi(s)\right|\right) \tag{1.5.4}$$

where, recall, $Z_v^\varepsilon(s) = X_v^\varepsilon(s/\varepsilon)$ and $\pi_1 v = z \in \mathcal{X}$ if $v = (z, y) \in \mathcal{X} \times \mathbf{M}$.

PROOF. By (1.2.12),

$$\big|(t_j-t_{j-1})\Xi_j^\varepsilon(v,x_{j-1})-(Z_v^\varepsilon(t_j)-Z_v^\varepsilon(t_{j-1}))\big|\le\int_{t_{j-1}}^{t_j}\big|B(x_{j-1},Y_v^\varepsilon(\tfrac{s}{\varepsilon}))$$
$$-B(Z_v^\varepsilon(s),Y_v^\varepsilon(\tfrac{s}{\varepsilon}))\big|ds\le\int_{t_{j-1}}^{t_j}\big(\big|B(x_{j-1},Y_v^\varepsilon(\tfrac{s}{\varepsilon}))-B(Z_v^\varepsilon(t_{j-1}),Y_v^\varepsilon(\tfrac{s}{\varepsilon}))\big|$$
$$+\big|B(Z_v^\varepsilon(t_{j-1}),Y_v^\varepsilon(\tfrac{s}{\varepsilon}))-B(Z_v^\varepsilon(s),Y_v^\varepsilon(\tfrac{s}{\varepsilon}))\big|\big)ds\le K(t_j-t_{j-1})|Z_v^\varepsilon(t_{j-1})$$
$$-x_{j-1}|+K\int_{t_{j-1}}^{t_j}|Z_v^\varepsilon(s)-Z_v^\varepsilon(t_{j-1})|ds\le K(t_j-t_{j-1})|Z_v^\varepsilon(t_{j-1})-x_{j-1}|$$
$$+K^2\int_{t_{j-1}}^{t_j}(s-t_{j-1})ds\le K(t_j-t_{j-1})|Z_v^\varepsilon(t_{j-1})-x_{j-1}|+\tfrac{1}{2}K^2(t_j-t_{j-1})^2$$

and (1.5.2) follows.

Observe, that

$$Z_{v,x}^{\varepsilon,\psi}(s)-\psi(s)=x-x_0+(x_{n(s)}-\tilde{x}_{n(s)})$$
$$+\sum_{j=1}^{n(s)}(t_j-t_{j-1})\big(\Xi_j^\varepsilon(v,\tilde{x}_{j-1})-\xi_j\big)+\int_{t_{n(s)}}^{s}B\big(\tilde{x}_{n(s)},Y_v^\varepsilon(\tfrac{u}{\varepsilon})\big)du$$

and (1.5.3) follows in view of (1.2.15). Next, by (1.2.15),

$$\big|Z_v^\varepsilon(s)-Z_{v,x}^{\varepsilon,\psi}(s)\big|\le|\pi_1v-x|+\int_0^s\big|B(Z_v^\varepsilon(u),Y_v^\varepsilon(\tfrac{u}{\varepsilon}))$$
$$-B(Z_{v,x}^{\varepsilon,\psi}(u),Y_v^\varepsilon(\tfrac{u}{\varepsilon}))\big|du+\int_0^s\big|B(Z_{v,x}^{\varepsilon,\psi}(u),Y_v^\varepsilon(\tfrac{u}{\varepsilon}))-B(\psi(u),Y_v^\varepsilon(\tfrac{u}{\varepsilon}))\big|du$$
$$\le|\pi_1v-x|+K\int_0^s\big|Z_{v,x}^{\varepsilon,\psi}(u)-\psi(u)\big|du+K\int_0^s\big|Z_v^\varepsilon(u)-Z_{v,x}^{\varepsilon,\psi}(u)\big|du$$

and (1.5.4) follows from the Gronwall inequality . □

For any $x',x''\in\mathcal{X}$ and $\beta,\xi\in\mathbb{R}^d$ set

$$L_b(x',x'',\xi)=\sup_{\beta\in\mathbb{R}^d,|\beta|\le b}\big(\langle\beta,\xi\rangle-H(x',x'',\beta)\big),$$

and $L_b(x,\xi)=L_b(x,x,\xi)$ with $H(x',x'',\beta)$ given by (1.2.10).

1.5.2. PROPOSITION. *Let x_j,t_j,ξ_j,N,Δ,T and Ξ_j^ε be the same as in Lemma 1.5.1 and assume that*

$$\hat{\Delta}=\min_{0\le i\le N-1}(t_{i+1}-t_i)\ge\Delta/3. \tag{1.5.5}$$

Fix also $\rho>0$ so that Proposition 1.3.4 holds true.

(i) There exist $\delta_0>0,\varepsilon_0(\Delta)>0$ and $C_T(b)>0$ independent of $x,x_j,\tilde{x}_j,\xi_j$ such that if $\delta\le\delta_0$ and $\varepsilon\le\varepsilon_0(\Delta)$ then for any $b>0$,

$$m\{y\in\mathcal{W}:\max\nolimits_{1\le j\le N}\big|\Xi_j^\varepsilon((x,y),\tilde{x}_{j-1})-\xi_j\big|<\delta\} \tag{1.5.6}$$
$$\le\exp\big\{-\tfrac{1}{\varepsilon}\big(\sum_{j=1}^N(t_j-t_{j-1})L_b(\tilde{x}_{j-1},\xi_j)-\eta_{b,T}(\varepsilon,\Delta)-C_T(b)(d+\delta)\big)\big\}$$

where $d=|x-x_0|+\max_{0\le j\le N}|x_j-\tilde{x}_j|$, $\eta_{b,T}(\varepsilon,\Delta)$ does not depend on $x,x_j,\tilde{x}_j,\xi_j$ and

$$\lim_{\Delta\to0}\limsup_{\varepsilon\to0}\eta_{b,T}(\varepsilon,\Delta)=0. \tag{1.5.7}$$

In particular, if for each $j=1,...,N$ there exists $\beta_j\in\mathbb{R}^d$ such that

$$L(\tilde{x}_j,\xi_j)=\langle\beta_j,\xi_j\rangle-H(\tilde{x}_j,\beta_j) \tag{1.5.8}$$

and

$$\max_{1\le j\le N}|\beta_j|\le b<\infty \tag{1.5.9}$$

then (1.5.6) holds true with $L(\tilde{x}_j,\xi_j)$ in place of $L_b(\tilde{x}_j,\xi_j)$, $j=1,...,N$.

(ii) For any $b,\lambda,\delta,q > 0$ there exist $\Delta_0 = \Delta_0(b,\lambda,\delta,q) > 0$ and $\varepsilon_0 = \varepsilon_0(b,\lambda,\delta,q,\Delta)$, the latter depending also on $\Delta > 0$, such that if ξ_j and β_j satisfy (1.5.8) and (1.5.9), $\max_{1\leq j\leq N}|\xi_j| \leq q$, $\Delta < \Delta_0$ and $\varepsilon < \varepsilon_0$ then

$$m\{y \in \mathcal{W} : \max_{1\leq j\leq N} \left|\Xi_j^\varepsilon((x,y),\tilde{x}_{j-1}) - \xi_j\right| < \delta\} \geq \exp\left\{ -\tfrac{1}{\varepsilon}\left(\sum_{j=1}^N (t_j - t_{j-1})L(\tilde{x}_{j-1},\xi_j) + \eta_{b,T}(\varepsilon,\Delta) + C_T(b)d + \lambda\right)\right\} \tag{1.5.10}$$

with some $C_T(b) > 0$ depending only on b and T.

PROOF. (i) Assuming that ρ is small and $C \geq 2$ is large so that $C^6\rho$ is still small, we consider for each $x \in \mathcal{X}_T$ and $y \in \mathcal{W}$ closed discs $D_0 \in \mathcal{D}_\varepsilon^u((x,y),\alpha,\rho,C^3)$ and $D \in \mathcal{D}_\varepsilon^u((x,y),\alpha,\rho,C)$ with $D_0 \supset D$. For each small $r \geq 0$ set

$$D(r) = \{v \in D_0 : \inf_{\tilde{v}\in D} d_{D_0}(v,\tilde{v}) \leq r\} \text{ and}$$

$$D(-r) = \{v \in D : \inf_{\tilde{v}\in D_0\setminus D} d_{D_0}(v,\tilde{v}) \geq r\}.$$

Then $D(r) \cap \partial D_0 = \emptyset$ provided $r = r(\rho) < C^2\rho(C-1)$. For any pair of compact sets $\tilde{D} \subset \hat{D} \subset \mathbb{R}^d \times \mathbf{M}$ and $\varrho > 0$ a finite set $G \subset \tilde{D}$ will be called $(s,\varrho,\varepsilon,\tilde{D},\hat{D})$-separated if $v_i, v_j \in G$, $v_i \neq v_j$ implies that $v_i \notin U_{\hat{D}}^\varepsilon(s,v_j,\varrho)$. Choose a maximal $(t_{n-1}\varepsilon^{-1}, C\rho,\varepsilon,\tilde{D},D_0)$-separated set $G_{n-1}(\tilde{D})$ in $\tilde{D} \subset D_0$ (where maximal means that the set cannot be enlarged still remaining $(\cdot,\cdot,\cdot,\cdot,\cdot)$-separated). Then

$$\cup_{v\in G_{n-1}(\tilde{D})} U_{D_0}^\varepsilon(t_{n-1}\varepsilon^{-1},v,C\rho) \supset \tilde{D}$$

and, by Lemma 1.3.2(i) for small ε, $n > 1$, and $v \in \tilde{D}$,

$$U_{D_0}^\varepsilon(t_{n-1}\varepsilon^{-1},v,C\rho) \subset \tilde{D}(\varepsilon).$$

Set

$$\Gamma_{\tilde{D}}^j(r) = \{v \in \tilde{D} : \left|\Xi_j^\varepsilon(v,\tilde{x}_{j-1}) - \xi_j\right| < r\}$$

and

$$G_{n-1}^\Gamma(\tilde{r},r) = \{v \in G_{n-1}(D(\tilde{r})) : U_{D_0}^\varepsilon(t_{n-1}\varepsilon^{-1},v,C\rho) \cap \left(\bigcap_{j=1}^{n-1} \Gamma_{D(\tilde{r})}^j(r)\right) \neq \emptyset\}$$

assuming that $D(\tilde{r}) \subset D_0$. Then for $\tilde{r} \geq 0$, $\tilde{r} < r(\rho) = C^2\rho(C-1)$,

$$\begin{aligned} m_{D_0}\{v \in D(\tilde{r}) : \max_{1\leq j\leq n} \left|\Xi_j^\varepsilon(v,\tilde{x}_{j-1}) - \xi_j\right| < r\} &= m_{D_0}\left(\bigcap_{j=1}^n \Gamma_{D(\tilde{r})}^j(r)\right) \\ &\leq \sum_{v\in G_{n-1}^\Gamma(\tilde{r},r)} m_{D_0}\left(U_{D_0}^\varepsilon(t_{n-1}\varepsilon^{-1},v,C\rho) \cap \Gamma_{D(\tilde{r})}^n(r)\right). \end{aligned} \tag{1.5.11}$$

By Lemma 1.3.2(i) if $n > 1$ and ε is small enough then $d(v',v) \leq \varepsilon$ for any $v' \in U_{D_0}^\varepsilon(t_{n-1}\varepsilon^{-1},v,C\rho)$ and using, in addition, Assumption 1.2.2 and the inequality (1.5.3) we obtain that for any $j \leq n-1$,

$$\left|\Xi_j^\varepsilon(v,x) - \Xi_j^\varepsilon(v',x)\right| \leq K(t_j - t_{j-1})c^{-1}C\rho\int_{t_{j-1}}^{t_j} e^{-\frac{\kappa}{2\varepsilon}(t_{n-1}-s)}ds \leq 4K\Delta^{-1}c^{-1}C\rho\kappa^{-1}\varepsilon. \tag{1.5.12}$$

Hence, if $v \in G_{n-1}^\Gamma(\tilde{r},r)$ then for $C_1 = 8Kc^{-1}C\rho\kappa^{-1}$ and $\tilde{r} < r(\rho) - \varepsilon$,

$$U_{D_0}^\varepsilon(t_{n-1}\varepsilon^{-1},v,C\rho) \subset \bigcap_{j=1}^{n-1} \Gamma_{D(\tilde{r}+\varepsilon)}^j(r + C_1\varepsilon\Delta^{-1}), \tag{1.5.13}$$

provided ε is small enough, and so, by (1.5.3) and (1.5.4),

$$\text{(1.5.14)} \qquad |\pi_1 v_{t_{n-1}\varepsilon^{-1}} - \tilde{x}_{n-1}| \le d_{n-1} = e^{Kt_{n-1}} \sup_{v\in D_0} |\pi_1 v - x_0| \\ +(e^{Kt_{n-1}}Kt_{n-1}+1)\big(\max_{0\le j\le n-1}|x_j-\tilde{x}_j| + K\Delta + (n-1)\Delta(r+C_1\varepsilon\Delta^{-1})\big)$$

where we set $v_s = \Phi^s_\varepsilon v$. Since $H(x',x'',\beta)$ is (Lipschitz) continuous in β there exists $\beta_n^{(a)}(x',x'') \in \mathbb{R}^d$ such that

(1.5.15)

$$|\beta_n^{(a)}(x',x'')| \le a \text{ and } L_a(x',x'',\xi_n) = \langle \beta_n^{(a)}(x',x''),\xi_n\rangle - H(x',x'',\beta_n^{(a)}(x',x'')).$$

Let $v \in G^\Gamma_{n-1}(r)$ and $\beta_n^{(a)} = \beta_n^{(a)}(\pi_1 v_{t_{n-1}\varepsilon^{-1}}, \tilde{x}_{n-1})$. Since $H(x',x'',\beta)$ is Lipschitz continuous (and even C^1) in x' and x'' (see [**16**]) it follows from (1.5.14) that

$$\text{(1.5.16)} \qquad \big|H(\pi_1 v_{t_{n-1}\varepsilon^{-1}}, \tilde{x}_{n-1}, \beta_n^{(a)}) - H(\tilde{x}_{n-1},\beta_n^{(a)})\big| \le C(a)d_{n-1}$$

where $C(a)>0$ depends only on a. Since $U^\varepsilon_{D(\tilde r)}(t_{n-1}\varepsilon^{-1},v,C\rho)\cap\partial D_0 = \emptyset$ provided $v\in G^\Gamma_{n-1}(\tilde r, r)$, $n>1$, $\tilde r < r(\rho)-\varepsilon$ we derive from Lemma 1.3.2(iv), Proposition 1.3.4, and Lemma 1.4.1(i) that for such $v,n,\tilde r,\varepsilon$ and any $a>0$,

$$\text{(1.5.17)} \qquad m_{D_0}\big(U^\varepsilon_{D_0}(t_{n-1}\varepsilon^{-1},v,C\rho)\cap\Gamma^n_{D(\tilde r)}(r)\big) \\ \le m_{D_0}\big(U^\varepsilon_{D_0}(t_{n-1}\varepsilon^{-1},v,C\rho)\big) \\ \times\exp\big(-\tfrac{(t_n-t_{n-1})}{\varepsilon}(L_a(\tilde{x}_{n-1},\xi_n) - \tilde\eta_{a,T}(\varepsilon,\Delta) - C(a)d_{n-1} - ra)\big)$$

where $\tilde\eta_{a,T}(\varepsilon,\Delta)\to 0$ as, first, $\varepsilon\to 0$ and then $\Delta\to 0$.

Since $U^\varepsilon_{D_0}(t_{n-1}\varepsilon^{-1},v,\frac12 C\rho)$ are disjoint for different $v\in G_{n-1}(D(\tilde r))$ we obtain from (1.5.13) and Lemma 1.3.2(iv) that

$$\text{(1.5.18)} \qquad \textstyle\sum_{v\in G^\Gamma_{n-1}(\tilde r,r)} m_{D_0}\big(U^\varepsilon_{D_0}(t_{n-1}\varepsilon^{-1},v,C\rho)\big) \\ \le c^{-1}_{\frac12\rho,T}c^{-1}_{\rho,T}\textstyle\sum_{v\in G^\Gamma_{n-1}(\tilde r,r)} m_{D_0}\big(U^\varepsilon_{D_0}(t_{n-1}\varepsilon^{-1},v,\frac12 C\rho)\big) \\ \le c^{-1}_{\frac12\rho,T}c^{-1}_{\rho,T}m_{D_0}\big(\textstyle\bigcup_{v\in G^\Gamma_{n-1}(\tilde r,r)} U^\varepsilon_{D_0}(t_{n-1}\varepsilon^{-1},v,\frac12 C\rho)\big) \\ \le c^{-1}_{\frac12\rho,T}c^{-1}_{\rho,T}m_{D_0}\big(\textstyle\bigcap_{j=1}^{n-1}\Gamma^j_{D(\tilde r+\varepsilon)}(r+C_1\varepsilon\Delta^{-1})\big).$$

Employing (1.5.11), (1.5.17) and (1.5.18) for $n=N,N-1,...,2$ with $r=\delta+C_1\varepsilon\Delta^{-1},\delta+2C_1\varepsilon\Delta^{-1},...,\delta+(N-1)C_1\varepsilon\Delta^{-1}$ and $\tilde r=\varepsilon,2\varepsilon,...,(N-1)\varepsilon$, respectively, and using only (1.5.17) for $n=1$ we derive that

$$\text{(1.5.19)} \qquad m_{D_0}\{v\in D:\ \max_{1\le j\le N}\big|\Xi^\varepsilon_j(v,\tilde{x}_{j-1})-\xi_j\big|<\delta\} \\ \le \exp\big\{-\tfrac{1}{\varepsilon}\big(\textstyle\sum_{j=1}^N(t_j-t_{j-1})L_a(\tilde{x}_{j-1},\xi_j) - \eta_{a,\rho,T}(\varepsilon,\Delta) - C(a,T)(d+\delta)\big)\big\}$$

provided $\delta+2C_1\varepsilon T\Delta^{-2}\le\rho$ and $\varepsilon T\Delta^{-1}<r(\rho)$ with $\eta_{a,\rho,T}(\varepsilon,\Delta)$ satisfying (1.5.7) and with the same d as in (1.5.6).

Let $D_x(r,w)$ be a ball on $W^u_x(w,\varrho)$ centered at w and having radius Cr, $\rho\le r\le 2\rho<\varrho$ in the interior metric on $W^u_x(w,\varrho)$ (which, recall, is a semi-invariant extension of the family of local unstable manifolds on Λ_x– see Section 3 and [**69**]). Then $D_x(r,w)\in\mathcal{D}^u_\varepsilon((x,w),\alpha,\rho,C)$ if $C\ge 2$. Recall, that if ρ is small enough then the extended local unstable and stable discs $W^u_x(w,r(\rho))$ and $W^s_x(w,r(\rho))$ are defined for all $w\in\mathcal{W}$ and, in fact, by (1.2.4), the compactness arguments and by [**69**] such discs can be defined for all w from a small neighborhood U of $\bar{\mathcal{W}}$ which is still contained in the basin of attraction of each Λ_z. For each $w\in\bar W$ set

$$Q_x(w,\rho)=\bigcup\{D_x(r(\rho),F^r_x\tilde w):\ |r|\le C\rho,\ \tilde w\in W^s_x(w,C\rho)\}$$

and assume that ρ is small enough so that $Q_x(w,\rho)\subset U$. Then (1.5.19) together with the Fubini theorem yield (1.5.6) with the box $Q_x(w,\rho)$ in place of the whole $\mathcal{W}$. Relying on the transversality of unstable and weakly stable submanifolds together with compactness arguments we conclude that there exist an integer n_ρ depending only on ρ such that $\mathcal{W}$ can be covered by n_ρ boxes $Q_x(w_i,\rho)$, $i=1,2,...,n_\rho$ which yields now (1.5.6) in the required form.

(ii) We start proving (1.5.10) by using (1.5.12) in order to conclude similarly to (1.5.13) that if $n>1$, $v\in G^\Gamma_{n-1}(\tilde r-\varepsilon,r-C_1\varepsilon\Delta^{-1})$, and $\tilde v\in U^\varepsilon_{D_0}(t_{n-1}\varepsilon^{-1},v,C\rho)$ then $\tilde v\in\cap_{j=1}^{n-1}\Gamma^j_{D(\tilde r)}(r)$. Hence,

$$
\begin{aligned}
(1.5.20)\qquad & m_{D_0}\{v\in D(\tilde r):\max_{1\le j\le N}|\Xi^\varepsilon_j(v,\tilde x_{j-1})-\xi_j|<r\}\\
&\ge m_{D_0}\Big(\Gamma^n_{D(\tilde r)}(r)\cap\big(\textstyle\bigcup_{v\in G^\Gamma_{n-1}(\tilde r-\varepsilon,r-C_1\varepsilon\Delta^{-1})}U^\varepsilon_{D_0}(t_{n-1}\varepsilon^{-1},v,C\rho)\big)\Big)\\
&\ge\textstyle\sum_{v\in G^\Gamma_{n-1}(\tilde r-\varepsilon,r-C_1\varepsilon\Delta^{-1})}m_{D_0}\big(\Gamma^n_{D(\tilde r)}(r)\cap U^\varepsilon_{D_0}(t_{n-1}\varepsilon^{-1},v,\tfrac12C\rho)\big)
\end{aligned}
$$

where the last inequality holds true since $U^\varepsilon_{D_0}(t_{n-1}\varepsilon^{-1},v,\frac12C\rho)$ are disjoint for different $v\in G^\Gamma_{n-1}(\tilde r,r-C_1\varepsilon\Delta^{-1})$. Using (1.5.16), Lemma 1.3.2(iv), Proposition 1.3.4, and Lemma 1.4.1(iii) we obtain that for any $v\in G^\Gamma_{n-1}(\tilde r-\varepsilon,r-C_1\varepsilon\Delta^{-1})$, $\varsigma\le\delta$, $\sigma>0$ and $b\ge\max_{1\le j\le N}|\beta_j|$,

$$
\begin{aligned}
(1.5.21)\qquad & m_{D_0}\big(\Gamma^n_{D(\tilde r)}(r)\cap U^\varepsilon_{D_0}(t_{n-1}\varepsilon^{-1},v,\tfrac12C\rho)\big)\\
&\ge m_{D_0}\big(U^\varepsilon_{D_0}(t_{n-1}\varepsilon^{-1},v,\tfrac12C\rho)\big)g_{n,b}(\varepsilon,\Delta,\varsigma,\sigma)\exp\big(-\tfrac{(t_n-t_{n-1})}{\varepsilon}L(\tilde x_{n-1},\xi_n)\big)
\end{aligned}
$$

where

$$
\begin{aligned}
g_{n,b}(\varepsilon,\Delta,\varsigma,\sigma)&=\exp\Big(-\tfrac{(t_n-t_{n-1})}{\varepsilon}\big(\tilde\eta_{b,T}(\varepsilon,\Delta)+C_T(b)d_{n-1}+\varsigma b\big)\Big)\\
&\quad\times\Big(1-\exp\big(-\tfrac{(t_n-t_{n-1})}{\varepsilon}(d(b)-\tilde\eta_{b,T}(\varepsilon,\Delta)-\sigma b-\sigma)\big)\Big),
\end{aligned}
$$

$$
d(b)=\min_{1\le j\le N}\tilde L^{\beta_j}_b(\tilde x_{j-1},\mathcal{K}_{\varsigma,C}(\xi_j)),\ \tilde L^\beta_b(x,\mathcal{K})=\inf_{\alpha\in\mathcal{K}}\tilde L^\beta_b(x,\alpha),
$$

$$
\mathcal{K}_{\varsigma,C}(\alpha)=\bar U_C(0)\setminus U_\varsigma(\alpha),\ \tilde L^\beta_b(x,\alpha)=L_b(x,\alpha)-\langle\beta,\alpha\rangle+H(x,\beta),\ C_T(b)>0,
$$

and $\tilde\eta_{b,T}(\varepsilon,\Delta)\to0$ as, first, $\varepsilon\to0$ and then $\Delta\to0$.

By Lemma 1.3.2(iv) and the definitions of Γ^j and G^Γ,

$$
\begin{aligned}
(1.5.22)\qquad & \textstyle\sum_{v\in G^\Gamma_{n-1}(\tilde r-\varepsilon,r-C_1\varepsilon\Delta^{-1})}m_{D_0}\big(U^\varepsilon_{D_0}(t_{n-1}\varepsilon^{-1},v,\tfrac12C\rho)\big)\\
&\ge c_{\frac12\rho,T}c_{\rho,T}\textstyle\sum_{v\in G^\Gamma_{n-1}(\tilde r-\varepsilon,r-C_1\varepsilon\Delta^{-1})}m_{D_0}\big(U^\varepsilon_{D_0}(t_{n-1}\varepsilon^{-1},v,C\rho)\big)\\
&\ge c_{\frac12\rho,T}c_{\rho,T}m_{D_0}\big(\textstyle\bigcap_{j=1}^{n-1}\Gamma^j_{D(\tilde r-\varepsilon)}(r-C_1\varepsilon\Delta^{-1})\big).
\end{aligned}
$$

Employing (1.5.20)–(1.5.22) for $n=N,N-1,...,2$ with $r=\delta$, $\delta-C_1\varepsilon\Delta^{-1}$,...,$\delta-(N-2)C_1\varepsilon\Delta^{-1}$ and $\tilde r=0,-\varepsilon,-2\varepsilon,...,-(N-2)\varepsilon$, respectively, and using only (1.5.21) for $n=1$ we derive that

$$
\begin{aligned}
(1.5.23)\qquad & m_{D_0}\{v\in D:\max_{1\le j\le N}|\Xi^\varepsilon_j(v,\tilde x_{j-1})-\xi_j|<\delta\}\\
&\ge\exp\Big(-\tfrac1\varepsilon\big(\textstyle\sum_{j=1}^N(t_j-t_{j-1})L(\tilde x_{j-1},\xi_j)+C(\rho,\delta)\varepsilon\Delta^{-1}\big)\Big)\\
&\qquad\times\textstyle\prod_{n=1}^N g_{n,b}(\varepsilon,\Delta,\varsigma,\sigma)
\end{aligned}
$$

for some $C(\rho,\delta)>0$ provided, say, $NC_1\varepsilon\Delta^{-1}\le 2TC_1\varepsilon\Delta^{-2}\le\frac{\delta}{2}$ and $T\varepsilon\Delta^{-1}<C\rho/2$. Since $H(x,\beta)$ is differentiable in β (see [**16**]) then

$$\tilde L(\tilde x_j,\alpha)=L(\tilde x_j,\alpha)-\langle\beta_j,\xi_j\rangle+H(\tilde x_j,\beta_j)>0$$

for any $\alpha\ne\xi_j$ (see Theorems 23.5 and 25.1 in [**70**]), and so by the lower semicontinuity of $L(x,\alpha)$ in α (and, in fact, also in x),

$$\tilde L^{\beta_j}(\tilde x_{j-1},\mathcal{K}_{\varsigma,C}(\xi_j))=\inf_{\alpha\in\mathcal{K}_{\varsigma,C}(\xi_j)}\tilde L^{\beta_j}(\tilde x_{j-1},\alpha)>0.$$

This together with Lemma 1.4.2 yield that $d(b)$ appearing in the definition of $g_{n,b}(\varepsilon,\Delta,\varsigma,\sigma)$ is positive provided b is sufficiently large. In fact, it follows from the lower semicontinuity of $L(x,\alpha)$ that $d(b)$ is bounded away from zero by a positive constant independent of $\tilde x_j$ and ξ_j, $j=1,...,N$ if these points vary over fixed compact sets and (1.5.8) together with (1.5.9) hold true. Now, given $\lambda>0$ choose, first, sufficiently large b as needed and then subsequently choosing small σ and ς, then small Δ, and, finally, small enough ε we end up with an estimate of the form

$$g_{n,b}(\varepsilon,\Delta,\varsigma,\sigma)\ge\exp\Big(-\frac{(t_n-t_{n-1})}{\varepsilon}(\eta_{b,\rho,T}(\varepsilon,T)+C_T(b)d+\lambda)\Big)\tag{1.5.24}$$

where $C_T(b)>0$ and $\eta_{b,\rho,T}(\varepsilon,T)$ satisfies (1.5.7). Finally, (1.5.10) follows from (1.5.23), (1.5.24) and the Fubini theorem (similarly to (i)). □

Next, we pass directly to the proof of Theorem 1.2.3 starting with the lower bound. Some of the details below are borrowed from [**79**] but we believe that our exposition and the way of proof are more precise, complete and easier to follow. Assume that $S_{0T}(\gamma)<\infty$, and so that γ is absolutely continuous, since there is nothing to prove otherwise. Then by (1.2.13), $L(\gamma_s,\dot\gamma_s)<\infty$ for Lebesgue almost all $s\in[0,T]$. By (1.2.15) and (1.3.6),

$$H(x,\beta)\le K|\beta|,\tag{1.5.25}$$

and so if $L(\gamma_s,\dot\gamma_s)<\infty$ it follows from (1.2.12) that $|\dot\gamma|\le K$. Suppose that $\mathcal{D}(L_s)=\{\alpha:\,L(\gamma_s,\alpha)<\infty\}\ne\emptyset$ and let $\mathrm{ri}\mathcal{D}(L_s)$ be the interior of $\mathcal{D}(L_s)$ in its affine hull (see [**70**]). Then either $\mathrm{ri}\mathcal{D}(L_s)\ne\emptyset$ or $\mathcal{D}(L_s)$ (by its convexity) consists of one point and recall that $\dot\gamma_s\in\mathcal{D}(L_s)$ for Lebesgue almost all $s\in[0,T]$. By (1.2.10) and (1.5.25),

$$0=H(\gamma_s,0)=\inf_{\alpha\in\mathbb{R}^d}L(\gamma_s,\alpha).\tag{1.5.26}$$

This together with the nonnegativity and lower semi-continuity of $L(\gamma_s,\cdot)$ yield that there exists $\hat\alpha_s$ such that $L(\gamma_s,\hat\alpha_s)=0$ and by a version of the measurable selection (of the implicit function) theorem (see [**15**], Theorem III.38), $\hat\alpha_s$ can be chosen to depend measurably in $s\in[0,T]$. Of course, if $\mathrm{ri}\mathcal{D}(L_s)=\emptyset$ then $\mathcal{D}(L_s)$ contains only $\hat\alpha_s$ and in this case $\hat\alpha_s=\dot\gamma_s$ for Lebesgue almost all $s\in[0,T]$. Taking $\alpha_s=\hat\alpha_s$ and $\beta_s=0$ we obtain

$$L(\gamma_s,\alpha_s)=\langle\beta_s,\alpha_s\rangle-H(\gamma_s,\beta_s).\tag{1.5.27}$$

Observe that $\ell(s,\alpha)=L(\gamma_s,\alpha)$ is measurable as a function of s and α since it is obtained via (1.2.12) as a supremum in one argument of a family of continuous functions, and so this supremum can be taken there over a countable dense set of β's. Hence, the set $A=\{(s,\alpha):\,s\in[0,T],\,\alpha\in\mathcal{D}(L_s)\}=\ell^{-1}[0,\infty)$ is measurable, and so the set $B=A\setminus\{(s,\dot\gamma_s),\,s\in[0,T]\}$ is measurable, as well. Its projection $V=\{s\in[0,T]:\,(s,\alpha)\in B\,\text{for some}\,\alpha\in\mathbb{R}^d\}$ on the first component of the product space is also measurable and V is the set of $s \in [0,T]$ such that $\mathcal{D}(L_s)$ contains more than one point. Employing Theorem III.22 from [**15**] we select $\bar{\alpha}_s \in \mathbb{R}^d$ measurably in $s \in V$ and such that $(s, \bar{\alpha}_s) \in B$. By convexity and lower semicontinuity of $L(\gamma_s, \cdot)$ it follows from Corollary 7.5.1 in [**70**] that

$$L(\gamma_s, \dot{\gamma}_s) = \lim_{p\uparrow\infty} L(\gamma_s, \alpha_s^{(p)}) \text{ where } \alpha_s^{(p)} = (1-p^{-1})\dot{\gamma}_s + p^{-1}\bar{\alpha}_s. \tag{1.5.28}$$

For each $\delta > 0$ set

$$n_\delta(s) = \min\{n \in \mathbb{N} : |L(\gamma_s, \dot{\gamma}_s) - L(\gamma_s, \alpha_s^{(n)})| + |\dot{\gamma}_s - \alpha_s^{(n)}| < \delta\}.$$

Then, clearly, $n_\delta(s)$ is a measurable function of s, and so $\alpha_s = \alpha_s^{(\delta)} = \alpha_s^{(n_\delta(s))}$ and $L(\gamma_s, \alpha_s)$ are measurable in s, as well. By Theorems 23.4 and 23.5 from [**70**] for each $\alpha_s = \alpha_s^{(\delta)}$ there exists $\beta_s = \beta_s^{(\delta)} \in \mathbb{R}^d$ such that (1.5.27) holds true. Given $\delta', \lambda > 0$ take $\delta = \min(\delta', \lambda/3)$ and for $s \in [0,T] \setminus V$ set $\alpha_s = \hat{\alpha}_s$. Then

$$\int_0^T \big|L(\gamma_s, \dot{\gamma}_s) - L(\gamma_s, \alpha_s)\big| ds < \lambda/3 \text{ and } \int_0^T |\dot{\gamma}_s - \alpha_s| ds < \delta'. \tag{1.5.29}$$

For each $b > 0$ set $\alpha_s^b = \alpha_s$ if the corresponding β_s in (1.5.27) satisfies $|\beta_s| \le b$ and $\alpha_s^b = \hat{\alpha}_s$, otherwise. Note, that (1.5.27) remains true with α_s^b in place of α_s with $\beta_s = 0$ if $\alpha_s^b = \hat{\alpha}_s$. As observed above $|\alpha| \le K$ whenever $L(z, \alpha) < \infty$, and so $|\hat{\alpha}_s| \le K$ for Lebesgue almost all $s \in [0,T]$. We recall also that $|\dot{\gamma}_s - \alpha_s| < \delta$ and $\dot{\gamma}_s \le K$ for Lebesgue almost all $s \in [0,T]$. Since $S_{0T}(\gamma) < \infty$, $|L(\gamma_s, \dot{\gamma}_s) - L(\gamma_s, \alpha_s)| < \delta$, and $L(\gamma_s, \alpha_s^b) \uparrow L(\gamma_s, \alpha_s)$ as $b \uparrow \infty$ for Lebesgue almost all $s \in [0,T]$, we conclude from (1.5.29) and the above observations that for b large enough

$$\int_0^T \big|L(\gamma_s, \alpha_s) - L(\gamma_s, \alpha_s^b)\big| ds < \lambda/3 \,\text{and} \int_0^T |\alpha_s - \alpha_s^b| ds < \delta'. \tag{1.5.30}$$

Next, we apply Lemma 1.4.3 to conclude that there exists a sequence $m_j \to \infty$ such that for each $\Delta_j = T/m_j$ and Lebesgue almost all $c \in [0,T)$,

$$\int_0^T \big|L(\gamma_s, \alpha_s^b) - L(\gamma_{q_j(s,c)}, \alpha_{q_j(s,c)}^b)\big| ds < \lambda/3 \,\text{and} \int_0^T |\alpha_s^b - \alpha_{q_j(s,c)}^b| ds < \delta'. \tag{1.5.31}$$

where $q_j(s,c) = [(s+c)\Delta_j^{-1}]\Delta_j - c$, $[\cdot]$ denotes the integral part and we assume $L(\gamma_s, \alpha_s^b) = 0$ and $\alpha_s^b = 0$ if $s < 0$.

Choose $c = c_j \in [\frac{1}{3}\Delta_j, \frac{2}{3}\delta_j]$ and set $\hat{\gamma}_s = x + \int_0^s \alpha_{q_j(u,c)}^b du$, $\psi_s = \gamma_{q_j(s,c)}$ where $\gamma_u = \gamma_0$ if $u < 0$, $x_0 = \tilde{x}_0 = x$, $x_N = \hat{\gamma}_T$, $\tilde{x}_N = \gamma_T$ and $x_k = \hat{\gamma}_{k\Delta_j - c}$, $\tilde{x}_k = \gamma_{k\Delta_j - c}$ for $k = 1, ..., N-1$ and $\xi_k = \alpha_{(k-1)\Delta_j - c}^b$ for $k = 1, 2, ..., N$ where $N = \min\{k : k\Delta_j - c > T\}$. Since $|\dot{\gamma}_s| \le K$ for Lebesgue almost all $s \in [0,T]$ then $\mathbf{r}_{0T}(\gamma, \psi) \le K\Delta_j$ and, in addition, $\mathbf{r}_{0T}(\gamma, \hat{\gamma}) \le 3\delta'$ by (1.5.29)–(1.5.31). This together with (1.5.3) and (1.5.4) yield that for $v = (x, y)$,

$$\begin{aligned} \mathbf{r}_{0T}(Z_v^\varepsilon, \gamma) &\le \mathbf{r}_{0T}(Z_v^\varepsilon, \psi) + K\Delta_j \le (KTe^{KT} + 1)\mathbf{r}_{0T}(Z_v^{\varepsilon,\psi}, \psi) + K\Delta_j \\ &\le (KTe^{KT} + 1)\big(3\delta' + K\Delta_j + (T+1)\max\nolimits_{1 \le k \le N} \big|\Xi_k^\varepsilon(v, \tilde{x}_{k-1}) - \xi_k\big|\big) + K\Delta_j \end{aligned} \tag{1.5.32}$$

provided $\Delta_j \le 1$ where $Z_v^{\varepsilon,\psi}$ and $\Xi_k^\varepsilon(v, x)$ are the same as in Lemma 1.5.1, the latter is defined with $t_k = k\Delta_j - c$, $k = 1, ..., N-1$ and $t_N = T$. Choose δ' so small and m_j so large that

$$(KTe^{KT} + 1)\big(3\delta' + K\Delta_j + (T+1)\delta'\big) + K\Delta_j < \delta$$

then by (1.5.32),

$$(1.5.33)\quad \{y\in\mathcal{W}:\ \mathbf{r}_{0T}(Z^\varepsilon_{x,y},\gamma)<\delta\}\supset\{y\in\mathcal{W}:\ \max_{1\le k\le N}\big|\Xi^\varepsilon_k(v,\tilde x_{k-1})-\xi_k\big|<\delta'\}.$$

By (1.5.29)–(1.5.31),

$$(1.5.34)\qquad \sum_{k=1}^N(t_k-t_{k-1})L(\tilde x_{k-1},\xi_k)\le S_{0T}(\gamma)+\lambda$$

and by the construction above the conditions of the assertion (ii) of Proposition 1.5.2 hold true, so choosing m_j sufficiently large we derive (1.2.16) (with 2λ in place of λ) from (1.5.10), (1.5.33) and (1.5.34) provided ε is small enough.

Next, we pass to the proof of the upper bound (1.2.17). Assume that (1.2.17) is not true, i.e. there exist $a,\lambda,\delta>0$ and $x\in\mathcal{X}_T$ such that for some sequence $\varepsilon_k\to 0$ as $k\to\infty$,

$$(1.5.35)\qquad m\{y\in\mathcal{W}:\ \mathbf{r}_{0T}\big(Z^{\varepsilon_k}_{x,y},\Psi^a_{0T}(x)\big)\ge 3\delta\}>\exp\big(-\frac{1}{\varepsilon_k}(a-\lambda)\big).$$

Since $\|B(x,y)\|\le K$ by (1.2.15) all paths of $Z^\varepsilon_{x,y}(t)$, $t\in[0,T]$ and of $Z^{\varepsilon,\psi}_{v,x}(t)$, $t\in[0,T]$ given by (1.5.1) (the latter for any measurable ψ) belong to a compact set $\tilde{\mathcal{K}}^x\subset C_{0T}$ which consists of curves starting at x and satisfying the Lipschitz condition with the constant K. Let $\tilde U^x_\rho$ denotes the open ρ-neighborhood of the compact set $\Psi^a_{0T}(x)$ and $\mathcal{K}^x_\rho=\tilde{\mathcal{K}}^x\setminus\tilde U^x_\rho$. For any small $\delta'>0$ choose a δ'-net $\gamma_1,...,\gamma_n$ in $\mathcal{K}^x_{2\delta}$ where $n=n(\delta')$. Since

$$\{y\in\mathcal{W}:\ \mathbf{r}_{0T}\big(Z^{\varepsilon_k}_{x,y},\Psi^a_{0T}(x)\big)\ge 3\delta\}\subset\bigcup_{n\ge j\ge 1}\{y\in\mathcal{W}:\ \mathbf{r}_{0T}\big(Z^{\varepsilon_k}_{x,y},\gamma_j\big)\le\delta'\}$$

then there exists j and a subsequence of $\{\varepsilon_k\}$, for which we use the same notation, such that

$$(1.5.36)\qquad m\{y\in\mathcal{W}:\ \mathbf{r}_{0T}(Z^{\varepsilon_k}_{x,y},\gamma_j)\le\delta'\}>n^{-1}\exp\big(-\frac{1}{\varepsilon_k}(a-\lambda)\big).$$

Denote such γ_j by $\gamma^{\delta'}$, choose a sequence $\delta_l\to 0$ and set $\gamma^{(l)}=\gamma^{\delta_l}$. Since $\mathcal{K}^x_{2\delta}$ is compact there exists a subsequence $\gamma^{(l_j)}$ converging in C_{0T} to $\hat\gamma\in\mathcal{K}^x_{2\delta}$ which together with (1.5.36) yield

$$(1.5.37)\qquad \limsup_{\varepsilon\to 0}\varepsilon\ln m\{y\in\mathcal{W}:\ \mathbf{r}_{0T}(Z^\varepsilon_{x,y},\hat\gamma)\le\delta'\}>-a+\lambda$$

for all $\delta'>0$.

We claim that (1.5.37) contradicts (1.5.2) and the assertion (i) of Proposition 1.5.2. Indeed, set

$$S^\psi_{b,0T}(\gamma)=\int_0^T L_b(\psi(s),\dot\gamma(s))ds\ \text{and}\ S_{b,0T}(\gamma)=S^\gamma_{b,0T}(\gamma).$$

By the monotone convergence theorem

$$(1.5.38)\qquad S^\psi_{b,0T}(\gamma)\uparrow S^\psi_{0T}(\gamma)\ \text{and}\ S_{b,0T}(\gamma)\uparrow S_{0T}(\gamma)\ \text{as}\ b\uparrow\infty.$$

Similarly to our remark (before Assumption 1.2.2) in Section 1.2 it follows from the results of Section 9.1 of [**41**] that the functionals $S^\psi_{b,0T}(\gamma),S^\psi_{0T}(\gamma)$ and $S_{b,0T}(\gamma),S_{0T}(\gamma)$ are lower semicontinuous in ψ and γ (see also Section 7.5 in [**31**]).

This together with (1.5.38) enable us to apply Lemma 1.4.2 in order to conclude that

$$\lim_{b\to\infty} S_{b,0T}(\mathcal{K}_\delta^x) = S_{0T}(\mathcal{K}_\delta^x) = \inf_{\gamma\in\mathcal{K}_\delta^x} S_{0T}(\gamma) > a \tag{1.5.39}$$

where $S_{b,0T}(\mathcal{K}_\delta^x) = \inf_{\gamma\in\mathcal{K}_\delta^x} S_{b,0T}(\gamma)$. The last inequality in (1.5.39) follows from the lower semicontinuity of S_{0T}. Thus we can and do choose $b > 0$ such that

$$S_{b,0T}(\mathcal{K}_\delta^x) > a - \lambda/8. \tag{1.5.40}$$

By the lower semicontinuity of $S^\psi_{b,0T}(\gamma)$ in ψ there exists a function $\delta_\lambda(\gamma) > 0$ on $\mathcal{K}_\delta^x$ such that for each $\gamma \in \mathcal{K}_\delta^x$,

$$S^\psi_{b,0T}(\gamma) > a - \lambda/4 \text{ provided } \mathbf{r}_{0T}(\gamma,\psi) < \delta_\lambda(\gamma). \tag{1.5.41}$$

Next, we restrict the set of functions ψ to make it compact. Namely, we allow from now on only functions ψ for which there exists $\gamma \in \mathcal{K}_\delta^x$ such that either $\psi \equiv \gamma$ or $\psi(t) = \gamma(kT/m)$ for $t \in [kT/m, (k+1)T/m)$, $k = 0, 1, ..., m-1$ and $\psi(T) = \gamma(T)$ where m is a positive integer. It is easy to see that the set of such functions ψ is compact with respect to the uniform convergence topology in C_{0T} and it follows that $\delta_\lambda(\gamma)$ in (1.5.41) constructed with such ψ in mind is lower semicontinuous in γ. Hence

$$\delta_\lambda = \inf_{\gamma\in\mathcal{K}_\delta^x} \delta_\lambda(\gamma) > 0. \tag{1.5.42}$$

Now take $\hat\gamma$ satisfying (1.5.37) and for any integer $m \geq 1$ set $\Delta = \Delta_m = T/m$, $x_k = x_k^{(m)} = \hat\gamma(k\Delta)$, $k = 0, 1, ..., m$ and $\xi_k = \xi_k^{(m)} = \Delta^{-1}\big(\hat\gamma(k\Delta) - \hat\gamma((k-1)\Delta)\big)$, $k = 1, ..., m$. Define a piecewise linear χ_m and a piecewise constant ψ_m by

$$\chi_m(t) = x_k + \xi_k\Delta \text{ and } \psi_k(t) = x_k \text{ for } t \in [k\Delta, (k+1)\Delta) \tag{1.5.43}$$

and $k = 0, 1, ..., m-1$ with $\chi_m(T) = \psi_m(T) = \hat\gamma(T)$. Since $\hat\gamma$ is Lipschitz continuous with the constant K then

$$\mathbf{r}_{0T}(\chi_m, \psi_m) \leq K\Delta \text{ and } \mathbf{r}_{0T}(\hat\gamma, \psi_m) \leq K\Delta. \tag{1.5.44}$$

If m is large enough and $\varepsilon > 0$ is sufficiently small then

$$\Delta < K^{-1}\min(\delta/2, \delta_\lambda) \text{ and } \eta_{b,T}(\varepsilon, \Delta) < \lambda/8 \tag{1.5.45}$$

where $\eta_{b,T}(\varepsilon, \Delta)$ is the same as in (1.5.6). Since $\hat\gamma \in \mathcal{K}_{2\delta}^x$ it follows from (1.5.44) and (1.5.45) that $\chi_m \in \mathcal{K}_\delta^x$ and by (1.5.41) and the first inequality in (1.5.45) we obtain that

$$S^{\psi_m}_{b,0T}(\chi_m) = \Delta\sum_{k=0}^{m-1} L_b(x_k, \xi_k) > a - \frac{\lambda}{4}. \tag{1.5.46}$$

Hence, by (1.5.6) and the second inequality in (1.5.45) for all ε small enough,

$$m\{y \in \mathcal{W} : \max_{1\leq k\leq m} |\Xi_k^\varepsilon((x,y), x_{k-1}) - \xi_k| < \rho\} \leq e^{-\frac{1}{\varepsilon}(a-\lambda/2)} \tag{1.5.47}$$

provided $C_T(b)\rho < \lambda/8$ (taking into account that $x_0 = x$). By (1.5.2) and the definition of vectors ξ_k for any $v \in \mathcal{W}$,

$$\begin{aligned}|\Xi_k^\varepsilon(v, x_{k-1}) - \xi_k| &\leq |\Xi_k^\varepsilon(v, x_{k-1}) - \Delta^{-1}\big(Z_v^\varepsilon(k\Delta) - Z_v^\varepsilon((k-1)\Delta)\big)| \\ &+2\Delta^{-1}\mathbf{r}_{0T}(Z_v^\varepsilon, \hat\gamma) \leq (K + 2\Delta^{-1})\mathbf{r}_{0T}(Z_v^\varepsilon, \hat\gamma) + \tfrac{1}{2}K^2\Delta.\end{aligned} \tag{1.5.48}$$

Therefore,

$$\{y \in \mathcal{W} : \mathbf{r}_{0T}(Z^{\varepsilon}_{x,y}, \hat{\gamma}) \leq \delta'\} \subset \{y \in \mathcal{W} : \max_{1\leq k\leq m} |\Xi^{\varepsilon}_{k}((x,y), x_{k-1}) - \xi_k| \leq (K + 2\Delta^{-1})\delta' + \tfrac{1}{2}K^2\Delta\}. \tag{1.5.49}$$

Choosing, first, m large enough so that Δ satisfies (1.5.45) with all sufficiently small ε and also that $8C_T(b)K^2\Delta < \lambda$, and then choosing δ' so small that $16C_T(b)(K + 2\Delta^{-1})\delta' < \delta$, we conclude that (1.5.47) together with (1.5.49) contradicts (1.5.37), and so the upper bound (1.2.17) holds true. Since $S_{0T}(\gamma) = 0$ if and only if $\gamma = \gamma^u$ satisfying (1.2.14) the estimate (1.2.18) follows from (1.2.17) and the lower semicontinuity of the functional S_{0T}, completing the proof of Theorem 1.2.3. □

1.6. Further properties of S-functionals

In this section we study essential properties of the functionals S_{0T} which will be needed in the proofs of Theorems 1.2.5 and 1.2.7 in the next sections. We will start with the following general fact which do not require specific conditions of Theorems 1.2.5 and 1.2.7.

1.6.1. LEMMA. *There exists $r > 0$ such that if $x \in \bar{\mathcal{X}}$ then any μ_{xx} from the space $\mathcal{M}_x$ of F^t_x-invariant probability measures on Λ_x can be included into a weakly continuous in z family $\mu_{xz} \in \mathcal{M}_z$, $|z - x| < r$ (considered in the space of probability measures on $\overline{\mathcal{W}}$) for which $\bar{B}_{\mu_{xz}}(z) = \int B(x,y)d\mu_{xz}(y)$ is C^1 in z and the entropy $h_{\mu_{xz}}(F^1_z)$ is continuous in z as $|z - x| < r$. Furthermore, there exists $C > 0$ such that*

$$|\bar{B}_{\mu_{xz_1}}(z_1) - \bar{B}_{\mu_{xz_2}}(z_2)| < C|z_1 - z_2| \text{ whenever } x, z_1, z_2 \in \bar{\mathcal{X}},\ |z_i - x| < r,\ i = 1, 2 \tag{1.6.1}$$

and for any $\alpha > 0$ there exists $\beta > 0$ such that if $x, z \in \bar{\mathcal{X}}$, $|z - x| < \beta$ then

$$|h_{\mu_{xz}}(F^1_z) - h_{\mu_{xx}}(F^1_x)| < \alpha \text{ and } |I_z(\mu_{xz}) - I_x(\mu_{xx})| < \alpha. \tag{1.6.2}$$

PROOF. The following argument (whose ingredients appear already in [**64**], [**61**], and [**16**]) was indicated to me by A.Katok. If r is small enough the structural stability theorem for Axiom A flows obtained in [**69**] can be applied in order to compare F^1_x and F^1_z but here we will need its more recent form derived in [**64**], [**61**], and [**16**] which yields a homeomorphism $u_{xz} : \Lambda_x \to \Lambda_z$ and a continuous function c_{xz} on Λ_z both with C^1 dependence on z and such that the conjugate flow $\tilde{F}^t_z = u_{xz}F^t_x u^{-1}_{xz}$ satisfies

$$\frac{d\tilde{F}^t_z y}{dt} = c_{xz}(\tilde{F}^t_z y)b(z, \tilde{F}^t_z y)$$

where u_{xx} is the identity map on Λ_x and $c_{xx} \equiv 1$. By the standard direct verification we see that $\mu = u_{xz}\mu_{xx}$ is an $\tilde{F}_z$-invariant probability measure. It is known (see, for instance, [**76**], Theorem 4.2) that then the probability measure μ_{xz} on Λ_z defined by its Radon–Nikodim derivative

$$\frac{d\mu_{xz}}{d\mu}(y) = c_{xz}(y)\Big(\int_{\Lambda_z} c_{xz}d\mu\Big)^{-1}$$

is F_z^t-invariant. In our case this can be seen easily since for any C^1 function q on Λ_z,

$$\frac{d}{dt}\int_{\Lambda_z} q\circ F_z^t d\mu_{xz}\big|_{t=0} = \big(\int_{\Lambda_z} c_{xz}d\mu\big)^{-1}\int_{\Lambda_z} c_{xz}(b(z,\cdot),\nabla q)d\mu$$
$$= \big(\int_{\Lambda_z} c_{xz}d\mu\big)^{-1}\frac{d}{dt}\int_{\Lambda_z} q\circ \tilde F_z^t d\mu\big|_{t=0} = 0$$

where the last equality holds true by $\tilde F_z^t$-invariance of μ.

Now

$$\bar B_{\mu_{xz}}(z) = \int_{\Lambda_z} B(z,y)d\mu_{xz}(y) = \big(\int_{\Lambda_z} c_{xz}d\mu\big)^{-1}\int_{\Lambda_z} B(z,y)c_{xz}(y)d\mu(y)$$
$$\big(\int_{\Lambda_x} c_{xz}(u_{xz}y)d\mu_{xx}(y)\big)^{-1}\int_{\Lambda_x} B(z,u_{xz}y)c_{xz}(u_{xz}y)d\mu_{xx}(y).$$

This together with (1.2.15) yield the differentiability of $\bar B_{\mu_{xz}}(z)$ in z taking into account that c_{xz} and u_{xz} are C^1 in z (see [**16**]) and since the proof of this fact relies on a version of the implicit function theorem (see [**61**]) which provides derivatives in z uniformly in $x\in\mathcal{X}$ whenever $|z-x|<r$ and r is small enough we derive also (1.6.1). Next, clearly, $h_{\mu_{xx}}(F_x^1) = h_\mu(\tilde F_z^1)$. If we knew that μ_{xx} were ergodic then, of course, μ would be ergodic, as well, and it would follow from Theorem 10.1 in [**76**] that

$$h_{\mu_{xz}}(F_z^1) = h_{\mu_{xx}}(F_x^1)\int_{\Lambda_x} c_{xz}(u_{xz}y)d\mu_{xx}(y)$$

which would yield the differentiability of $h_{\mu_{xz}}(F_z^1)$ in z. In the general case we obtain from [**76**] that

$$h_{\mu_{xz}}(F_z^1)\inf_{y\in\Lambda_z} c_{xz}(y) \le h_\mu(\tilde F_z^1) \le h_{\mu_{xz}}(F_z^1)\sup_{y\in\Lambda_z} c_{xz}(y),$$

and so

$$\big|h_{\mu_{xz}}(F_z^1) - h_{\mu_{xx}}(F_x^1)\big| \le h_{\mu_{xz}}(F_z^1)\max\big(|\sup_{y\in\Lambda_z} c_{xz}(y)-1|, |1-\inf_{y\in\Lambda_z} c_{xz}(y)|\big).$$

Since by Ruelle's inequality (see, for instance [**60**]),

$$h_{\mu_{xz}}(F_z^1) \le \sup_{y\in\Lambda_z}|\varphi_z^u(y)|$$

we derive both the continuity of $h_{\mu_{xz}}(F_z^1)$ in z and the first part of (1.6.2). The second part of (1.6.2) follows from its first part in view of (1.2.8) taking into account that the function $\varphi_x^u(y)$ defined by (1.2.5) is Hölder continuous in y and uniformly Lipschitz continuous (even C^1) in x (see [**16**]) and that $B(x,y)$ is Lipschitz continuous in both variables (see (1.2.15)). □

The following result gives, in particular, sufficient conditions for a set to be an S-compact.

1.6.2. LEMMA. *(i) There exists $C>0$ and for each $x\in\mathcal{X}$ where the vector field B is complete there exists $r=r(x)>0$ such that if $|z_1-x|<r$ and $|z_2-x|<r$ then we can construct $\gamma\in C_{0t}$ with $t\le C|z_1-z_2|$ satisfying*

$$\gamma_0 = z_1,\ \gamma_t = z_2 \text{ and } S_{0t}(\gamma)\le C|z_1-z_2|.$$

It follows that $R(\tilde z,z)$ and $R(z,\tilde z)$ are locally Lipschitz continuous in z belonging to the open r-neighborhood of x when $\tilde z$ is fixed.

(ii) Let $\mathcal{O} \subset \mathcal{X}$ be a compact Π^t-invariant set which either contains a dense in $\mathcal{O}$ orbit of Π^t or $R(x,z) = 0$ for any pair $x, z \in \mathcal{O}$. Suppose that B is complete at each point of $\mathcal{O}$. Then $\mathcal{O}$ is an S-compact.

(iii) Assume that for any $\eta > 0$ there exists $T(\eta) > 0$ such that for each $x \in \mathcal{O}$ its orbit $\{\Pi^t x, t \in [0, T(\eta)]\}$ of length $T(\eta)$ forms an η-net in $\mathcal{O}$ or, equivalently, that Π^t is a minimal flow on $\mathcal{O}$. Suppose that B is complete at a point of $\mathcal{O}$. Then $\mathcal{O}$ is an S-compact.

PROOF. (i) Fix some $x \in \mathcal{X}$. In view of the ergodic decomposition (see, for instance, [**60**]) any $\mu \in \mathcal{M}_x$ can be represented as an integral over the space of ergodic measures from $\mathcal{M}_x$. Using the specification (see [**8**] and [**28**]) any ergodic $\mu \in \mathcal{M}_x$ can be approximated (in the weak sense) by F_x^t-invariant measures sitting on its periodic orbits, i.e. by measures of the form $\mu_\varphi = \frac{1}{t_\varphi}\int_0^{t_\varphi} \delta_{F_x^s y} ds$ where $\varphi = \{F_x^s y, 0 \le s \le t_\varphi\}$, $F_x^{t_\varphi} y = y$ is a periodic orbit of F_x^t with a period t_φ. This is done in a standard way by choosing a generic point of an ergodic measure μ, i.e. a point w which satisfies $\lim_{t\to\infty} t^{-1}\int_0^t g(F_x^s w)ds = \int g d\mu$ for any continuous function g on Λ_x, and then approximating the orbit of w by periodic orbits of F_x^t using the specification theorem (see Theorem 3.8 in [**8**]). It is well known (see [**8**]) that there are countably many periodic orbits of F_x^t which together with the above discussion yield that the closed convex hull $\Gamma_x^{(0)}$ of the set $\{\bar{B}_{\mu_\varphi}(x) : \varphi$ is a periodic orbit of $F_x^t\} \subset \mathbb{R}^d$ coincides with $\Gamma_x = \{\bar{B}_\mu(x) : \mu \in \mathcal{M}_x\}$.

Now assume that B is complete at x. Then $\{\alpha\Gamma_x, \alpha \in [0,1]\} = \{\alpha\Gamma_x^{(0)}, \alpha \in [0,1]\}$ contains an open neighborhood of 0 in $\mathbb{R}^d$. But then we can find a simplex Δ_x with vertices in $\Gamma_x^{(0)}$ such that $\{\alpha\Delta_x, \alpha \in [0,1]\}$ contains an open neighborhood of 0 in $\mathbb{R}^d$ and for some periodic orbits $\varphi_1, ..., \varphi_k$ of F_x^t, $k \ge d+1$,

$$\Delta_x = \{\sum_{i=1}^k \lambda_i \bar{B}_{\mu^{(i)}}(x) : \sum_{i=1}^k \lambda_i = 1, \lambda_i \ge 0\, \forall i\}$$

where we denote $\mu^{(i)} = \mu_{\varphi_i}$. By compactness of Δ_x it follows also that

$$\mathrm{dist}(\Delta_x, 0) = d_x > 0.$$

Now, set $\mu_{xx}^{(i)} = \mu^{(i)}$, $i = 1, ..., k$ and include each $\mu_{xx}^{(i)}$ into the weakly continuous in z families $\mu_{xz}^{(i)}$ constructed in Lemma 1.6.1 for z in some neighborhood of x. If $|z - x| \le r(x)$ and $r(x)$ is small enough each simplex

$$\Delta_z = \{\sum_{i=1}^k \lambda_i \bar{B}_{\mu_{xz}^{(i)}}(z) : \sum_{i=1}^k \lambda_i = 1, \lambda_i \ge 0\, \forall i\}$$

intersects and not at 0 with any ray emanating from $0 \in \mathbb{R}^d$ or, in other words, $\{\alpha\Delta_z, \alpha \in [0,1]\}$ contains an open neighborhood of 0 in $\mathbb{R}^d$ and, moreover,

$$\mathrm{dist}(0, \Delta_z) \ge \frac{1}{2} d_x.$$

Since all $\sum_{i=1}^k \lambda_i \mu_{xz}^{(i)}$ are F_z^t-invariant probability measures provided $\sum_{i=1}^k \lambda_i = 1$, $\lambda_i \ge 0$ we conclude that for any z in the $r(x)$-neighborhood of x and any vector ξ there exists an F_z^t-invariant probability measure μ_ξ such that $\bar{B}_{\mu_\xi}(z)$ has the same direction as ξ and

$$K \ge |\bar{B}_{\mu_\xi}(z)| \ge \frac{1}{2} d_x$$

where K is the same as in (1.2.15). It follows that any two points z_1 and z_2 from the open $r(x)$-neighborhood of x can be connected by a curve γ lying on the interval connecting z_1 and z_2 with $K \geq |\dot{\gamma}_s^{(1)}| \geq \frac{1}{2}d_x$, i.e. $\gamma_0 = z_1$, $\gamma_t = z_2$ with some $t \in [K^{-1}|z_1 - z_2|, 2d_x^{-1}|z_1 - z_2|]$ and by (1.2.9),

$$S_{0t}(\gamma) \leq 2d_x^{-1}|z_1 - z_2| \sup_{z\in\bar{\mathcal{X}}, y\in\Lambda_z} |\varphi_z^u(y)|.$$

In view of the triangle inequality for R what we have proved yields the continuity of $R(\tilde{z}, z)$ and $R(z, \tilde{z})$ in z belonging to the open $r(x)$-neighborhood of x when $\tilde{z}$ is fixed. Covering $\bar{\mathcal{X}}$ by $r(x)$-neighborhoods of points $x \in \bar{\mathcal{X}}$ and choosing a finite subcover we obtain (i) with the same constant for all $\bar{\mathcal{X}}$.

Next, we derive the sufficient conditions of (ii) for the S-compactness. First, observe that both assumptions there imply that for any $\eta > 0$ there exist $t_\eta > 0$ and $\gamma^\eta \in C_{0t_\eta}$ such that γ^η form an $\eta/4C$-net in $\mathcal{O}$ and $S_{0t_\eta}(\gamma^\eta) < \eta/4$ where C is the same as in (i). Indeed, if there exists a dense orbit of Π^t in $\mathcal{O}$ then a sufficiently long piece of this orbit will work as such γ with its S-functional equal 0. If $R(x, z) = 0$ for any $x, z \in \mathcal{O}$ then we can choose an $\eta/4C$-net $x_1, ..., x_n$ in $\mathcal{O}$ and then construct curves $\gamma^{(i)}$ such that $\gamma_0^{(i)} = x_i$, $\gamma_{t_i}^{(i)} = x_{i+1}$, $i = 1, ..., n-1$ with $S_{0t_i}(\gamma^{(i)}) < \eta/4n$. Taking $\gamma_t^\eta = \gamma_{t-\sum_{1\leq j\leq i-1} t_j}^{(i)}$ for $t \in [\sum_{1\leq j\leq i-1} t_j, \sum_{1\leq j\leq i} t_j]$ we obtain the required curve. Now, for each $x \in \mathcal{O}$ let U_x be the open $r(x)$-neighborhood of x in $\mathbb{R}^d$ where the construction of the part (i) can be implemented. Since $\mathcal{O}$ is compact we can choose from the cover $\{U_x, x \in \mathcal{O}\}$ of $\mathcal{O}$ a finite subcover $\mathcal{U} = \{U_{x_1}, ..., U_{x_\ell}\}$ of $\mathcal{O}$. For any positive η such that $\eta/4C$ is less than the Lebesgue number (see [**80**]) of $\mathcal{U}$ we construct γ^η as above and then for any $z \in \mathcal{O}$ there is i and $\tilde{z} \in \gamma^\eta$ such that $z, \tilde{z} \in U_{x_i}$, $|z - \tilde{z}| \leq \eta/4C$, and so by the assertion (i) we can connect z and $\tilde{z}$ by a curve $\gamma^z \in C_{0t_z}$ with $t_z \leq \eta/4$ and $S_{0t_z}(\gamma^z) \leq \eta/4$. It follows that any two points $x, z \in \mathcal{O}$ can be connected by a curve $\gamma \in C_{0t}$ with $t \in [0, t_\eta + \eta/2]$ and $S_{0t}(\gamma) \leq 3\eta/4$. Now set $\mathcal{O}_\rho = \{z : \text{dist}(z, \mathcal{O}) \leq \rho\}$ and suppose that $\mathcal{O}_{\rho_0} \subset \cup_{1\leq i\leq \ell} U_{x_i}$. Let $\eta/4C < \rho_0$ be smaller than the Lebesgue number of the cover $\{U_{x_i}, ..., U_{x_\ell}\}$ of $\mathcal{O}_{\rho_0}$ and set $U_\eta = \{z : \text{dist}(z, \mathcal{O}) < \eta/4C\}$. Then for any $z \in U_\eta$ there exists $x \in \mathcal{O}$ with $|z - x| < \eta/4C$, and so $x, z \in U_{x_i}$ for some i. Hence, by (i) there exists a curve $\tilde{\gamma} \in C_{0\tilde{t}}$ connecting x with z and such that $\tilde{t} \leq \eta/4$ and $S_{0\tilde{t}}(\tilde{\gamma}) \leq \eta/4$. By above we can connect any $\tilde{x} \in \mathcal{O}$ with x by a curve $\gamma \in C_{0t}$ with $t \in [0, t_\eta + \eta/2]$ and $S_{0t}(\gamma) \leq 3\eta/4$ and then using $\tilde{\gamma}$ we arrive at a combined curve connecting $\tilde{x}$ with z and satisfying the conditions required to ensure that $\mathcal{O}$ is an S-compact by taking $T_\eta = t_\eta + \eta$.

(iii) Now assume that for any $\eta > 0$ and each $x \in \mathcal{O}$ its piece of the Π^t-orbit of length $T(\eta)$ forms an η-net in $\mathcal{O}$ and suppose that B is complete at $x_0 \in \mathcal{O}$. Set $L = \sup_{-1\leq t\leq 1} \sup_{z\in\bar{V}} \|D_z\Pi^t\|$ where $D_z\Pi^t$ is the differential of Π^t at z. Let $x \in \mathcal{O}$ and $z \in \mathcal{X}$ with $\text{dist}(z, \mathcal{O}) < \eta L^{-T(\eta/3C)}/3C$ where $\eta < Cr(x_0)$. Then for some $\tilde{z} \in \mathcal{O}$, $|z - \tilde{z}| < \eta L^{-T(\eta/3C)}/3C$ and $|\Pi^{-s}\tilde{z} - x_0| < \eta/3C < \frac{1}{3}r(x_0)$, $|\Pi^{-s}\tilde{z} - \Pi^{-s}z| < \eta/3C < \frac{1}{3}r(x_0)$ for some $s \in [0, T(\eta/3C)]$, and so $|\Pi^{-s}z - x_0| < 2\eta/3C < r(x_0)$. In addition, for any $\eta < 3Cr(x_0)$ there exists $t(\eta) > 0$ so that $|\Pi^{t(\eta)}x - x_0| \leq \eta/3C$ with $t(\eta) \in [0, T(\eta/3C)]$. Now, by the assertion (i) we can connect $\Pi^{t(\eta)}x$ with x_0 by a curve $\gamma^{(1)} \in C_{0t_1}$ with $t_1 \leq \eta/3$ and $S_{0t_1}(\gamma^{(1)}) \leq \eta/3$, then connect x_0 with $\Pi^{-s}z$ by a curve $\gamma^{(2)} \in C_{0t_2}$ with $t_2 \leq 2\eta/3$ and $S_{0t_2}(\gamma^{(2)}) \leq 2\eta/3$. Finally, we can connect x with z by the curve $\gamma \in C_{0,t(\eta)+t_1+t_2+s}$ with $S_{0,t(\eta)+t_1+t_2+s}(\gamma) \leq \eta$

and such that $\gamma_t = \Pi^t x$ for $t \in [0, t(\eta)]$, $\gamma_t = \gamma^{(1)}_{t-t(\eta)}$ for $t \in [t(\eta), t(\eta)+t_1]$, $\gamma_t = \gamma^{(2)}_{t-t(\eta)-t_1}$ for $t \in [t(\eta)+t_1, t(\eta)+t_1+t_2]$, and $\gamma_t = \Pi^{t-t(\eta)-t_1-t_2-s}z$ for $t \in [t(\eta)+t_1+t_2, t(\eta)+t_1+t_2+s]$ yielding that $\mathcal{O}$ is an S-compact. □

The following assertion which relies on Lemma 1.6.1 will be also useful in our analysis.

1.6.3. LEMMA. *For any $\eta > 0$ and $T > 0$ there exists $\zeta > 0$ such that if $\gamma \in C_{0T}$, $\gamma \subset \mathcal{X}$, $S_{0T}(\gamma) < \infty$, $\gamma_0 = x_0$, and $|z_0 - x_0| < \zeta$ then we can find $\tilde{\gamma} \in C_{0T}$, $\tilde{\gamma} \subset \mathcal{X}$ with $\tilde{\gamma}_0 = z_0$ satisfying*

$$\mathbf{r}_{0T}(\gamma, \tilde{\gamma}) < \eta \text{ and } |S_{0T}(\tilde{\gamma}) - S_{0T}(\gamma)| < \eta. \tag{1.6.3}$$

PROOF. By (1.2.13) and the lower semicontinuity of the functionals $I_z(\nu)$ there exist measures $\nu_t \in \mathcal{M}_{\gamma_t}$, $t \in [0,T]$ such that $\dot{\gamma}_t = \bar{B}_{\nu_t}(\gamma_t)$ for Lebesgue almost all $t \in [0,T]$ and $I_{\gamma_t}(\nu_t) = L(\gamma_t, \dot{\gamma}_t)$ for Lebesgue almost all $t \in [0,T]$. Recall also that $\dot{\gamma}_t$ is measurable in t. Introduce the (measurable) map $q : [0,T] \times \mathcal{P}(\bar{\mathcal{W}}) \to \mathbb{R} \cup \{\infty\} \times \mathbb{R}^d$ defined by $q(t,\nu) = \big(I_{\gamma_t}(\nu), \bar{B}_\nu(\gamma_t)\big)$. Recall that $\dot{\gamma}_t$ is measurable in t, and so another map $r : [0,T] \to \mathbb{R} \cup \{\infty\} \times \mathbb{R}^d$ defined by $r(t) = \big(L(\gamma_t, \dot{\gamma}_t), \dot{\gamma}_t\big)$ is also measurable in $t \in [0,T]$. Then $q(t, \nu_t) = r(t)$ and it follows from the measurable selection in the implicit function theorem (see [**15**], Theorem III.38) that measures ν_t satisfying this condition can be chosen to depend measurably on $t \in [0,T]$.

Now, given $\eta > 0$ we pick up a small $\zeta > 0$ which will be specified later on and employ Lemma 1.4.3 in the same way as in (1.5.31) together with (1.2.9), (1.2.11), and (1.2.13) in order to conclude that for all $n \in \mathbb{N}$ large enough there exists $t_1^{(n)} \in [0, T/n)$ such that if $t_{j+1}^{(n)} = t_1^{(n)} + jn^{-1}T$, $j = 1, 2, ..., n-1$, $t_{n+1}^{(n)} = T$ then

$$\int_0^{t_1^{(n)}} \big|\bar{B}_{\nu_s}(\gamma_s)\big|ds + \sum_{j=1}^n \int_{t_j^{(n)}}^{t_{j+1}^{(n)}} \big|\bar{B}_{\nu_s}(\gamma_s) - \bar{B}_{\nu_{t_j^{(n)}}}(\gamma_{t_j^{(n)}})\big|ds \tag{1.6.4}$$
$$+ S_{0t_1^{(n)}}(\gamma) + \sum_{j=1}^n \int_{t_j^{(n)}}^{t_{j+1}^{(n)}} \big|I_{\gamma_s}(\nu_s) - I_{\gamma_{t_j^{(n)}}}(\nu_{t_j^{(n)}})\big|ds < \zeta.$$

Set $\dot{\psi}_s^{(n)} = 0$ for $s \in [0, t_1^{(n)})$ and $\dot{\psi}_s^{(n)} = \bar{B}_{\nu_{t_j^{(n)}}}(\gamma_{t_j^{(n)}})$ for $s \in [t_j^{(n)}, t_{j+1}^{(n)})$, $j = 1, ..., n$. Then $\psi_t^{(n)} = \gamma_0 + \int_0^t \dot{\psi}_s^{(n)} ds$, $t \in [0,T]$ defines a polygonal line such that

$$\mathbf{r}_{0T}(\gamma, \psi^{(n)}) < \zeta.$$

Next, set $\tilde{\gamma}_t = z_0$ for all $t \in [0, t_1^{(n)}]$ and continue the construction of $\tilde{\gamma}$ in the following recursive way. Suppose that $\tilde{\gamma}_t$ is already defined for all $t \in [0, t_j^{(n)}]$ and some $j \geq 1$. Denote $x_j = \gamma_{t_j^{(n)}}$, $y_j = \psi^{(n)}_{t_j^{(n)}}$, $z_j = \tilde{\gamma}_{t_j^{(n)}}$ and suppose that $|z_j - x_j| < r - KTn^{-1}$ where K is the same as in (1.2.15) and r comes from Lemma 1.6.1. For $t \in [t_j^{(n)}, t_{j+1}^{(n)}]$ define $\tilde{\gamma}_t$ as the integral curve starting at z_j of the vector field $\tilde{B}(z) = \bar{B}_{\mu_{x_j z}}(z)$, $|z - x_j| < r$ with $\mu_{x_j z} \in \mathcal{M}_z$ obtained in Lemma 1.6.1 for $\mu_{x_j x_j} = \nu_{t_j^{(n)}}$, i.e. $\tilde{\gamma}_t$ is the solution of the equation

$$\tilde{\gamma}_t = z_j + \int_{t_j^{(n)}}^t \tilde{B}(\tilde{\gamma}_s)ds.$$

This definition is legitimate since in view of (1.2.15) and our assumption on z_j the curve $\tilde{\gamma}_t$, $t \in [t_j^{(n)}, t_{j+1}^{(n)}]$ does not exit the r-neighborhood of x_j. By (1.6.1) and the above for all $t \in [t_j^{(n)}, t_{j+1}^{(n)}]$,

$$\big|\tilde{B}(\tilde{\gamma}_t) - \bar{B}_{\nu_{t_j^{(n)}}}(x_j)\big| < C(|z_j - x_j| + KTn^{-1}) < C(|z_j - y_j| + \zeta + KTn^{-1}),$$

and so

$$|z_{j+1} - y_{j+1}| \le \sup_{t\in[t_j^{(n)}, t_{j+1}^{(n)}]} |\tilde{\gamma}_t - \psi_t^{(n)}| \le |z_j - y_j|(1 + CTn^{-1}) + \zeta T/n + CKT^2 n^{-2}.$$

Assuming that $|z_0 - x_0| < \zeta$ with ζ small enough and since $x_0 = y_0$ we obtain successively from here that for all $j = 1, 2, ..., n$,

$$|z_j - y_j| \le (1 + CTn^{-1})^n (2\zeta + KTn^{-1}) \le e^{CT}(2\zeta + KTn^{-1}),$$

which enables us to continue our construction recursively for $j = 1, 2, ..., n$ if ζ and n^{-1} are small enough yielding also that

$$\mathbf{r}_{0T}(\tilde{\gamma}, \psi^{(n)}) \le e^{CT}(2\zeta + KTn^{-1}).$$

Hence, the first part of (1.6.3) follows provided ζ and n^{-1} are sufficiently small.

Next, observe that

$$\sup_{t\in[t_j^{(n)}, t_{j+1}^{(n)}]} |\tilde{\gamma}_t - x_j| \le |z_j - y_j||y_j - x_j| + KTn^{-1} \le KTn^{-1}(1 + e^{CT}) + \zeta(1 + 2e^{CT})$$

and the right hand side here can be made as small as we wish choosing ζ small and n large. Hence, by Lemma 1.6.1 we can make

$$\max_{0\le j\le n} \sup_{t\in[t_j^{(n)}, t_{j+1}^{(n)}]} \big|I_{\tilde{\gamma}_t}(\mu_{x_j\tilde{\gamma}_t}) - I_{x_j}(\nu_{t_j^{(n)}})\big| < \eta/2T$$

which together with (1.6.4) yield the second part of (1.6.3). □

The following result will enable us to control the time which the slow motion can spend away from the ω-limit set of the averaged motion.

1.6.4. LEMMA. *Let $G \subset \mathcal{X}$ be a compact set not containing entirely any forward semi-orbit of the flow Π^t. Then there exist positive constants $a = a_G$ and $T = T_G$ such that for any $x \in G$ and $t \ge 0$,*

$$\inf\big\{S_{0t}(\gamma) : \gamma \in C_{0t} \text{ and } \gamma_s \in G \text{ for all } s \in [0, t]\big\} \ge a[t/T]$$

where $[c]$ denotes the integral part of c.

PROOF. For each $x \in G$ set $\sigma_x = \inf\{t \ge 0 : \Pi^t x \notin G\}$. By the assumption of the lemma $\sigma_x < \infty$ for each $x \in G$ and it follows from continuous dependence of solutions of (1.1.6) on initial conditions that σ_x is upper semicontinuous. Hence, $\tilde{T} + \sup_{x\in G} \sigma_x < \infty$. Set $T = \tilde{T} + 1$ and $\Gamma = \{\gamma \in C_{0T} : \gamma_s \in G \text{ for all } s \in [0, T]\}$. Since no $\gamma \in \Gamma$ can be a solution of the equation (1.2.14) then $S_{0T}(\gamma) > 0$ for any $\gamma \in \Gamma$. The set Γ is closed with respect to the uniform convergence and since the functional S_{0T} is lower semicontinuous we obtain that

$$\inf_{\gamma\in\Gamma} S_{0T}(\gamma) = a > 0.$$

This together with (1.2.13) yield the assertion of Lemma 1.6.4. □

Untill now we have not used specific assumptions of Theorem 1.2.5 but some of them will be needed for the following auxiliary result.

1.6.5. LEMMA. *Let V be a connected open set with a piecewise smooth boundary and assume that (1.2.20) holds true. Then the function $R_\partial(x)$ is upper semicontinuous at any $x_0 \in V$ for which $R_\partial(x_0) < \infty$. Let $\mathcal{O} \subset V$ be an S-compact.*

(i) Then for each $z \in \bar{V}$ the function $R(x,z)$ takes on the same value $R^{\mathcal{O}}(z)$ for all $x \in \mathcal{O}$, and so $R_\partial(x)$ takes on the same value R_∂ for all $x \in \mathcal{O}$ and the set $\partial_{\min}(x) = \{z \in \partial V : R(x,z) = R_\partial\}$ coincides with the same (may be empty) set $\partial_{\min}$ for all $x \in \mathcal{O}$. Furthermore, for each $\delta > 0$ there exists $T(\delta) > 0$ such that for any $x \in \mathcal{O}$ we can construct $\gamma^x \in C_{0t_x}$ with $t_x \in (0, T(\delta)]$ satisfying

$$\gamma_0^x = x,\ \gamma_{t_x}^x \in \partial V \text{ and } S_{0t_x}(\gamma^z) \le R_\partial + \delta. \tag{1.6.5}$$

(ii) Suppose that $R_\partial < \infty$ and $dist(\Pi^t x, \mathcal{O}) \le d(t)$ for some $x \in V$ and $d(t) \to 0$ as $t \to \infty$. Then $R_\partial(x) \le R_\partial$ and for any $\delta > 0$ there exist $T_{\delta,d} > 0$ (depending only on δ and the function d but not on x) and $\hat{\gamma}^x \in C_{0s_x}$ with $s_x \in (0, T_{\delta,d}]$ satisfying

$$\hat{\gamma}_0^x = x,\ \hat{\gamma}_{s_x}^x \in \partial V \text{ and } S_{0s_x}(\hat{\gamma}^x) \le R_\partial + \delta. \tag{1.6.6}$$

In particular, if $R_\partial < \infty$ then $R_\partial(x) < \infty$ and if $\mathcal{O}$ is an S-attractor of the flow Π^t then $R_\partial(x) < \infty$ for all $x \in V$.

(iii) Suppose that for any open set $U \supset \mathcal{O}$ the compact set $\bar{V} \setminus U$ does not contain entirely any forward semi-orbit of the flow Π^t. Then the function $R^{\mathcal{O}}(z)$ is lower semicontinuous in $z \in \bar{V}$, $R^{\mathcal{O}}(z) \to 0$ as $dist(z, \mathcal{O}) \to 0$, and $\partial_{\min}$ is a nonempty compact set.

PROOF. Let $R_\partial(x_0) < \infty$ for some $x_0 \in V$. Then for any $\alpha > 0$ there exist $T > 0$ and $\gamma \in C_{0T}$ such that $\gamma_0 = x_0$, $\gamma_T \in \partial V$ and $S_{0T}(\gamma) \le R_\partial(x_0) + \alpha$. By Lemma 1.6.3 for any $\eta > 0$ we can choose $\zeta > 0$ so that if $z_0 \in V$, $|z_0 - x_0| < \zeta$ then there exists $\tilde{\gamma} \in C_{0T}$ such that $\tilde{\gamma}_0 = z_0$, $\mathbf{r}_{0T}(\tilde{\gamma}, \gamma) < \eta$ and $|S_{0T}(\tilde{\gamma}) - S_{0T}(\gamma)| < \eta$. Let $\mu_{\gamma_T z} \in \mathcal{M}_z$, $|z - \gamma_T| < r$ be measures obtained in Lemma 1.6.1 for $\mu_{\gamma_T \gamma_T} = \nu$ with ν satisfying the second part of (1.2.20). Since the boundary ∂V is piecewise smooth it follows from the continuous dependence of solutions of ordinary differential equations on initial conditions that for all small $\eta > 0$ there exists $t(\eta) \to 0$ as $\eta \to 0$ such that if ψ_t, $t \in [0, t(\eta)]$ is an integral curve of the vector field $\bar{B}_{\mu_{\gamma_T z}}(z)$, $|z - \gamma_T| < r$ with $\psi_0 = \tilde{z}$, $|\tilde{z} - \gamma_T| < \eta$ then $\psi_{t(\eta)} \notin V$. Since $|\tilde{\gamma}_T - \gamma_T| < \eta$ we can define $\tilde{\gamma}_t = \psi_{t-T}$ for $t \in [T, T + t(\eta)]$. Now, $\tilde{\gamma}_{T+t(\eta)} \notin V$ and by (1.2.9),

$$\left| S_{0,T+t(\eta)}(\tilde{\gamma}) - S_{0T}(\tilde{\gamma}) \right| \le t(\eta) \sup_{x \in \mathcal{X}, y \in \Lambda_x} |\varphi_x^u(y)|.$$

Thus we can choose η so small that $R_\partial(z_0) \le R_\partial(x_0) + 2\alpha$ and the upper semicontinuity of $R_\partial(x)$ at x_0 follows.

From now on till the end of the proof of this lemma we assume that $\mathcal{O}$ is an S-compact and prove, first, the assertion (i). It follows from the definition of an S-compact that $R(x_1, x_2) = 0$ for any pair $x_1, x_2 \in \mathcal{O}$, and so $R(x_1, z) = R(x_2, z)$ for any such x_1, x_2 and each $z \in \bar{V}$. It follows that $R_\partial(x)$ takes on the same value R_∂ for all $x \in \mathcal{O}$ and all sets $\partial_{\min}(x), x \in \mathcal{O}$ coincide with some, may be empty, set $\partial_{\min}$. Fix $x_0 \in \mathcal{O}$. Then for each $\delta > 0$ there exists $t_\delta^{(0)} > 0$ and $\gamma^{(0)} \in C_{0t_\delta^{(0)}}$ such that

$$\gamma_0^{(0)} = x_0,\ \gamma_{t_\delta^{(0)}}^{(0)} \in \partial V \text{ and } S_{0t_\delta^{(0)}}(\gamma^{(0)}) \le R_\partial + \delta/2.$$

By the definition of an S-compact there exists $T_{\delta/2} > 0$ such that for any $z \in \mathcal{O}$ we can construct $\gamma^{(z,\delta)} \in C_{0t_\delta^{(z)}}$ with $t_\delta^{(z)} \in [0, T_{\delta/2}]$ satisfying

$$\gamma_0^{(z,\delta)} = z,\ \gamma_{t_\delta^{(z)}}^{(z,\delta)} = x_0 \text{ and } S_{0t_\delta^{(z)}}(\gamma^{(z,\delta)}) \le \delta/2.$$

Defining γ^z by $\gamma_t^z = \gamma_t^{(z,\delta)}$ for $t \in [0, t_\delta^{(z)}]$ and $\gamma_t^z = \gamma_{t-t_\delta^{(z)}}^{(0)}$ for $t \in [t_\delta^{(z)}, t_\delta^{(z)} + t_\delta^{(0)}]$ we obtain a curve satisfying (1.6.5) with $T(\delta) = t_\delta^{(0)} + T_{\delta/2}$.

Next, we prove (ii) assuming that $R_\partial < \infty$ and that $\text{dist}(\Pi^t x, \mathcal{O}) \le d(t)$ for some $x \in V$ with $d(t) \to 0$ as $t \to \infty$. By (i), for any $\eta > 0$ there exists $T_\eta > 0$ such that for any $z \in \mathcal{O}$ we can construct $\gamma^z \in C_{0t_z}$ with $t_z \in (0, T_\eta]$ and $\gamma^z \in C_{0t_z}$ satisfying (1.6.5) with $\delta = \eta$. For such η and T_η choose ζ by Lemma 1.6.3 so that if $|\tilde{x} - z| < \zeta$ and $z \in \mathcal{O}$ then in the same way as at the beginning of the proof of this lemma we can construct $\tilde{\gamma} \in C_{0,t_z+t(\eta)}$ with $t(\eta) \to 0$ as $\eta \to 0$ such that

$$\tilde{\gamma}_0 = \tilde{x},\ \tilde{\gamma}_{t_z+t(\eta)} \in \partial V \text{ and } |S_{0,t_z+t(\eta)}(\tilde{\gamma}) - S_{0t_z}(\gamma^z)| \le \eta + Ct(\eta).$$

Pick up $\tilde{t} = \tilde{t}(d, \zeta)$ so that $d(\tilde{t}) < \zeta$. Then $|\tilde{x} - z| < \zeta$ for $\tilde{x} = \Pi^{\tilde{t}} x$ and some $z \in \mathcal{O}$. Now construct as above $\tilde{\gamma}$ for such z and define $\hat{\gamma}^x \in C_{0s_x}$ with $s_x = \tilde{t} + t_z + t(\eta)$ setting

$$\hat{\gamma}_t^x = \Pi^t x \text{ for } t \in [0, \tilde{t}] \text{ and } \hat{\gamma}_t^x = \tilde{\gamma}(t - \tilde{t}) \text{ for } t \in [\tilde{t}, \tilde{t} + t_z + t(\eta)].$$

Then $S_{0s_x}(\hat{\gamma}^x) \le R_\partial + 2\eta + Ct(\eta)$ and $s_x \in (0, T_\eta + \tilde{t} + t(\eta))]$. Choosing η so small that $2\eta + Ct(\eta) \le \delta$ and then taking $T_{\delta,d} = T_\eta + \tilde{t} + t(\eta)$ we conclude that $\hat{\gamma}^x$ satisfies (1.6.6). Since η is arbitrary we obtain that $R_\partial(x) \le R_\partial$. If $\mathcal{O}$ is an S-attractor whose basin contains $\bar{V}$ then we can choose $d(t) \to 0$ as $t \to 0$ which in view of the continuous dependence of $\Pi^t x$ on x will be the same for all $x \in \bar{V}$ (though for this lemma $d(t)$ as above depending on x would suffice, as well), so our conditions are satisfied now for all $x \in V$. Hence, in this case $R_\partial(x)$ is finite in the whole V, completing the proof of (ii).

Finally, we prove (iii). Recall, that by the definition of an S-compact $\mathcal{O}$ it follows that $R(x, z) = 0$ whenever $x, z \in \mathcal{O}$. For all $\eta > 0$ let $U_\eta \supset \mathcal{O}$ be open sets appearing in the definition of an S-compact. If $x \in \mathcal{O}$ and $z \in U_\eta$ then $R(x, z) \le \eta$. Hence, if $\text{dist}(z_n, \mathcal{O}) \to 0$ as $n \to \infty$ then $R(x, z_n) \to 0$. Now, let $z_0 \in \bar{V} \setminus \mathcal{O}$ and $z_n \to z_0$ as $n \to \infty$. For each $\delta \ge 0$ set $\mathcal{O}_\delta = \{z \in V : \text{dist}(z, \mathcal{O}) \le \delta\}$ and let $\delta(\eta) = \frac{1}{2} \inf\{|x - z| : x \in \mathcal{O},\ z \in \bar{V} \setminus U_\eta\}$. Without loss of generality we will assume that $z_i \notin U_{\eta_0}$ for some $\eta_0 > 0$ and all $i = 0, 1, 2, \ldots$. Fix $x \in \mathcal{O}$. By the definition of the function R for any $\zeta > 0$ we can choose $t_{n,\zeta} > 0$ and $\gamma^{(n,\zeta)} \in C_{0t_{n,\zeta}}$, $n = 0, 1, 2, \ldots$ such that

$$\gamma_0^{(n,\zeta)} = x,\ \gamma_{t_{n,\zeta}}^{(n,\zeta)} = z_n \text{ and } S_{0t_{n,\zeta}}(\gamma^{(n,\zeta)}) \le R(x, z_n) + \zeta. \tag{1.6.7}$$

For each $\eta \le \eta_0$ set

$$s_n = s_{n,\eta,\zeta} = \sup\{t \ge 0 : \gamma_t^{(n,\zeta)} \in \mathcal{O}_{\delta(\eta)}\}.$$

Consider $\tilde{\gamma}^{(n)} \in C_{0,t_{n,\zeta}-s_n}$ defined by $\tilde{\gamma}_t^{(n)} = \gamma_{t+s_n}^{(n,\zeta)}$ for $t \in [0, t_{n,\zeta} - s_n]$ which stays in $\bar{V} \setminus \text{int}\mathcal{O}_{\delta(\eta)}$ (where $\text{int}G$ means the interior of a set G), and so by Lemma 1.6.4 we conclude that

$$t_{n,\zeta} - s_n \le a_\eta^{-1}(R(x, z_n) + 1)$$

provided, say, $\zeta \le 1/2$ where $a_\eta > 0$ depends only on η. In order to verify the lower semicontinuity of $R(x,z)$ at $z = z_0$ we have only to consider the case

$$\liminf_{n\to\infty} R(x, z_n) = A < \infty,$$

and so we can assume that $R(x, z_n) \le 2A$ for all $n = 0, 1, 2, \dots$. Passing to a subsequence and denoting its members by the same letters we can assume also that

$$\lim_{n\to 0} R(x, z_n) = A.$$

The curves $\tilde\gamma^{(n)}$ are Lipschitz continuous with a constant K from (1.2.15), and so this sequence is relatively compact. Hence, we can choose a uniformly converging subsequence and denoting, again, its members by the same letters we obtain now that

$$\tilde\gamma^{(n)} \to \tilde\gamma^{(0)} \text{ as } n \to \infty$$

where $\tilde\gamma^{(0)} \in C_{0t_0}$ with $t_0 \in (0, a_\eta^{-1}(A+1)]$, $\mathrm{dist}(\tilde\gamma_0^{(0)}, \mathcal{O}) = \delta(\eta)$ and $\tilde\gamma_{t_0}^{(0)} = z_0$. Each curve $\tilde\gamma^{(n)}$, $n = 0, 1, 2, \dots$ can be extended to a curve in $C_{0,T}$ with $T = a_\eta^{-1}(2A+1)$ and the same S-functional by adding to one of its ends a piece of the orbit of the flow Π^t. Hence, we can rely on the lower semicontinuity of the functional S_{0T} in order to derive from (1.6.7) that

$$S_{0t_0}(\tilde\gamma^{(0)}) \le A + \zeta.$$

By the definition of an S-compact there exists $\hat\gamma \in C_{0r}$ with $r \in [0, T_{2\eta}]$ such that $\hat\gamma_0 = x$, $\hat\gamma_r = \tilde\gamma_0$ and $S_{0r}(\hat\gamma) \le 2\eta$. It follows that

$$R(x, z_0) \le A + 2\eta + \zeta$$

and since η and ζ can be chosen arbitrarily small we conclude that $R(x, z_0) \le A$ obtaining the lower semicontinuity of $R(x,z)$ at $z = z_0$. Finally, the lower semicontinuity of $R(x,z)$ in $z \in \partial V$ for a fixed $x \in \mathcal{O}$ implies that $\partial_{\min}(x)$ is nonempty and compact and since $\partial_{\min}(x)$ is the same for all $x \in \mathcal{O}$ by (i), the proof of Lemma 1.6.5 is complete. □

1.7. "Very long" time behavior: exits from a domain

In this section we derive Theorems 2.2.5 relying on certain "Markov property type" arguments which are substantial modifications of the corresponding arguments from Sections 4 and 5 of [**49**]. In this and the following section in order to simplify notations we will write $\mathcal{D}_\varepsilon^u(z, \alpha, \rho, C)$ for $\hat{\mathcal{D}}_\varepsilon^u(z, \alpha, \rho, C, L)$ (both introduced in Section 1.3) with some large L so that appropriate discs on (extended) unstable leaves W_x^u and all their Φ_ε^s-iterates belong to this set. We start with the following result which will not only yield Theorem 1.2.5 but also will play an important role in the proof of Theorem 1.2.7 in the next section.

1.7.1. Proposition. *Let V be a connected open set with a piecewise smooth boundary ∂V such that $\bar V = V \cup \partial V \subset \mathcal{X}$. Assume that for each $z \in \partial V$ there exist $\iota = \iota(z) > 0$ and an F^t-invariant probability measure ν on Λ_z so that*

$$z + s\bar B_\nu(z) \in \mathbb{R}^d \setminus \bar V \text{ for all } s \in (0, \iota], \tag{1.7.1}$$

i.e. $\bar B_\nu(z) \ne 0$ and it points out into the exterior of $\bar V$.

(i) Suppose that for some $A_1, T > 0$ and any $z \in \bar V$ there exists $\varphi^z \in C_{0T}$ such that for some $t = t(z) \in (0, T]$,

$$\varphi_0^z = z,\ \varphi_t^z \notin V \text{ and } S_{0t}(\varphi^z) \le A_1. \tag{1.7.2}$$

Then for each $x \in V$,

$$\limsup_{\varepsilon\to 0} \varepsilon \log \int_{\mathcal{W}} \tau^{\varepsilon}_{x,y}(V) dm(y) \le A_1 \tag{1.7.3}$$

and for any $\alpha > 0$ *there exists* $\lambda(\alpha) = \lambda(x,\alpha) > 0$ *such that for all small* $\varepsilon > 0$,

$$m\{y \in \mathcal{W} : \tau^{\varepsilon}_{x,y}(V) \ge e^{(A_1+\alpha)/\varepsilon}\} \le e^{-\lambda(\alpha)/\varepsilon}. \tag{1.7.4}$$

(ii) Assume that there exists an open set G *such that* V *contains its closure* $\bar{G}$ *and the intersection of* $\bar{V} \setminus G$ *with the* ω*-limit set of the flow* Π^t *is empty. Let* Γ *be a compact subset of* ∂V *such that*

$$\inf_{x\in G, z\in\Gamma} R(x,z) \ge A_2 \tag{1.7.5}$$

for some $A_2 > 0$. *Then for some* $T > 0$ *and any* $\beta > 0$ *there exists* $\lambda(\beta) > 0$ *such that for each* $x \in V$ *and any small* $\varepsilon > 0$,

$$\begin{aligned} m\{y \in \mathcal{W} : Z^{\varepsilon}_{x,y}(\tau^{\varepsilon}_{x,y}(V)) \in \Gamma,\ \tau^{\varepsilon}_{x,y}(V) \le e^{(A_2-\beta)/\varepsilon}\} \\ \le m\{y \in \mathcal{W} : Z^{\varepsilon}_{x,y}(\tau^{\varepsilon}_{x,y}(V)) \in \Gamma,\ \tau^{\varepsilon}_{x,y}(V) < T\} + e^{-\lambda(\beta)/\varepsilon}. \end{aligned} \tag{1.7.6}$$

Suppose that for some $x \in V$,

$$a(x) = \inf_{t\ge 0} dist(\Pi^t x, \partial V) > 0. \tag{1.7.7}$$

Then $R_{\partial}(x) > 0$ *and for each* $T > 0$ *there exists* $\hat{\lambda}(T) = \hat{\lambda}(T,x) > 0$ *such that for all small* $\varepsilon > 0$,

$$m\{y \in \mathcal{W} : \tau^{\varepsilon}_{x,y}(V) < T\} \le e^{-\hat{\lambda}(T)/\varepsilon} \tag{1.7.8}$$

and if the set Γ *from (1.7.5) coincides with the whole* ∂V *then*

$$\liminf_{\varepsilon\to 0} \varepsilon \log \int_{\mathcal{W}} \tau^{\varepsilon}_{x,y}(V) dm(y) \ge A_2. \tag{1.7.9}$$

The corresponding to (1.7.3), (1.7.4), (1.7.8) and (1.7.9) assertions hold true also when $\mathcal{W}$ *and* m *in these estimates are replaced by a disc* $D \in \mathcal{D}^u_{\varepsilon}(z,\alpha,\rho,C)$ *with* $\pi_1 z = x$ *and by* m_D, *respectively ((see (1.7.21) and1.7.22), (1.7.34), (1.7.36) and (1.7.37) below).*

PROOF. Observe that applying to (1.5.19) and (1.5.23) the arguments which were used in order to derive Theorem 1.2.3 from Proposition 1.5.2 and the latter from Proposition 1.3.4 and Lemma 1.4.1 we obtain that (1.2.16) and (1.2.17) can be written for any disc $D \in \mathcal{D}^u_{\varepsilon}(z,\alpha,\rho,C)$, $z = (x,y)$ in place of the whole $\mathcal{W}$, namely, for any $\gamma \in C_{0T}$ with $\gamma_0 = x$, $\delta, \lambda, a > 0$, and ε small enough

$$m_D\{v \in D : \mathbf{r}_{0T}(Z^{\varepsilon}_v, \gamma) < \delta\} \ge \exp\left\{-\frac{1}{\varepsilon}(S_{0T}(\gamma) + \lambda)\right\} \tag{1.7.10}$$

and

$$m_D\{v \in D : \mathbf{r}_{0T}(Z^{\varepsilon}_v, \Psi^a_{0T}(x)) \ge \delta\} \le \exp\left\{-\frac{1}{\varepsilon}(a - \lambda)\right\} \tag{1.7.11}$$

which holds true in the same sense as (1.2.16)–(1.2.17) and (1.7.10)–(1.7.11) are uniform in D as above.

In order to prove (i) we observe, first, that the assumption (1.7.1) above together with Lemma 1.6.2(i) and the compactness of ∂V considerations enable us to extend any φ^z, $z \in V$ slightly so that it will exit some fixed neighborhood of V

with only slight increase in its S-functional. Hence, from the beginning we assume that for each $\beta > 0$ there exists $\delta = \delta(\beta) > 0$ such that for any $z \in V$ we can find $T > 0$, $\varphi^z \in C_{0T}$ and $t = t(z) \in (0, T]$ satisfying

$$\varphi_0^z = z,\ \varphi_t^z \notin V_\delta \text{ and } S_{0t}(\varphi^z) \le A_1 + \beta$$

where $V_\delta = \{x : \mathrm{dist}(x, V) \le \delta\}$. It follows that for any $x \in V$, $n \ge 1$, and $D \in \mathcal{D}_\varepsilon^u\big((x, w), \alpha, \rho, C\big)$,

$$\begin{aligned}(1.7.12)\qquad \big\{v \in D :\ \tau_v^\varepsilon(V) > nT\big\} &= \big\{v \in D :\ Z_v^\varepsilon(t) \in V, \forall t \in [0, nT]\big\}\\ &= \big\{v \in D :\ \tau^\varepsilon_{\Phi_\varepsilon^{kT/\varepsilon} v}(V) > T,\ \forall k = 0, 1, ..., n-1\big\} \subset G_{n,\delta}^\varepsilon\\ &\overset{\text{def}}{=} \big\{v \in D :\ \Phi_\varepsilon^{kT/\varepsilon} v \notin \textstyle\bigcup_{z \in V} A_\delta^{\varepsilon,T}(\varphi^z), \forall k = 0, 1, ..., n-1\big\}\end{aligned}$$

where for any subset $H \subset C_{0T}$ and $c > 0$,

$$A_c^{\varepsilon,T}(H) = \{v \in V \times \mathcal{W} :\ \mathbf{r}_{0T}(Z_v^\varepsilon, H) < c\}.$$

For $k = 1, 2, ...$ define

$$Q_{k,\delta}^\varepsilon = \big\{v \in D :\ U_D^\varepsilon(kT/\varepsilon, v, \delta/4) \cap G_{k,\delta}^\varepsilon \ne \emptyset\big\}$$

and

$$R_{k,\delta}^\varepsilon = \bigcup_{v \in Q_{k,\delta}^\varepsilon} U_D^\varepsilon(kT/\varepsilon, v, \delta/4)$$

which are, clearly, compact sets satisfying

$$G_{k,\delta}^\varepsilon \subset Q_{k,\delta}^\varepsilon \subset R_{k,\delta}^\varepsilon \subset G_{k,\delta/2}^\varepsilon.$$

Let E_k be a maximal $\big(kT/\varepsilon, \delta/2, \varepsilon, Q_{k,\delta}^\varepsilon, D\big)$- separated set in $Q_{k,\delta}^\varepsilon$. Then

$$(1.7.13)\qquad \bigcup_{v \in E_k} U_D(kT/\varepsilon, v, \delta/4) \subset R_{k,\delta}^\varepsilon \subset \bigcup_{v \in E_k} U_D(kT/\varepsilon, v, 3\delta/4)$$

and the left hand side of (1.7.13) is a disjoint union. This together with Lemma 1.3.6 give

$$\begin{aligned}(1.7.14)\qquad m_D(R_{k,\delta}^\varepsilon) &\le \textstyle\sum_{v \in E_k} m_D\big(U_D(kT/\varepsilon, v, 3\delta/4)\big)\\ &\le c_{3\delta/4}^{-1} c_{\delta/4}^{-1} \textstyle\sum_{v \in E_k} m_D\big(U_D(kT/\varepsilon, v, \delta/4)\big).\end{aligned}$$

By Lemma 1.3.2(ii),

$$(1.7.15)\qquad D_k(v) = \Phi_\varepsilon^{kT/\varepsilon} U_D(kT/\varepsilon, v, \delta/4) \in \mathcal{D}_\varepsilon^u(\Phi_\varepsilon^{kT/\varepsilon} v, \alpha, \frac{\delta}{4\sqrt{C}}, \sqrt{C}).$$

Clearly, for any $v \in E_k$,

$$\begin{aligned}(1.7.16)\qquad \Gamma_k(v) = \big\{w \in U_D(kT/\varepsilon, v, \delta/4) :&\\ \Phi_\varepsilon^{kT/\varepsilon} w \in \textstyle\bigcup_{z \in V} A_{\delta/2}^{\varepsilon,T}(\varphi^z)\big\} &\subset R_{k,\delta}^\varepsilon \setminus R_{k+1,\delta}^\varepsilon.\end{aligned}$$

In view of (1.7.15) we can apply (1.7.10) which together with the choice of curves φ^x yield that for any $\lambda > 0$ and ε small enough,

$$\begin{aligned}(1.7.17)\qquad m_{D_k(v)}\big(\Phi_\varepsilon^{kT/\varepsilon}\Gamma_k(v)\big) \ge m_{D_k(v)}\big\{w \in D_k(v) :&\\ \mathbf{r}_{0,T}(Z_w^\varepsilon, \varphi^{z_k^\varepsilon(v)}) < \delta/2\big\} &\ge \exp\big(-\tfrac{1}{\varepsilon}(A_1 + \beta + \lambda)\big)\end{aligned}$$

where $z_k^\varepsilon(v) = \pi_1(\Phi_\varepsilon^{kT/\varepsilon} v)$. By Lemma 1.3.6 it follows that

$$(1.7.18)\qquad m_D(\Gamma_k(v)) \ge c(\delta) m_D\big(U_D(kT/\varepsilon, v, \delta/4)\big) \exp\big(-\frac{1}{\varepsilon}(A_1 + \beta + \lambda)\big)$$

for some $c(\delta) > 0$. Since $U_D(kT/\varepsilon, v, \delta/4)$ are disjoint for different $v \in E_k$ we derive from (1.7.14), (1.7.16) and (1.7.18) that

$$(1.7.19)\quad m_D(R^\varepsilon_{k,\delta}) - m_D(R^\varepsilon_{k+1,\delta}) \geq m_D\big(\textstyle\bigcup_{v\in E_k}\Gamma_k(v)\big) = \sum_{v\in E_k} m_D(\Gamma_k(v))$$
$$\geq c(\delta)\exp\big(-\tfrac{1}{\varepsilon}(A_1+\beta+\lambda)\big)\textstyle\sum_{v\in E_k} U_D(kT/\varepsilon, v, \delta/4)$$
$$\geq \tilde c_\delta m_D(R^\varepsilon_{k,\delta})\exp\big(-\tfrac{1}{\varepsilon}(A_1+\beta+\lambda)\big)$$

where $\tilde c_\delta = c(\delta)c_{3\delta/4}c_{\delta/4}$. Applying (1.7.19) for $k = 1, 2, ..., n-1$ we obtain that

$$(1.7.20)\qquad m_D(G^\varepsilon_{n,\delta}) \leq m_D(R^\varepsilon_{n,\delta}) \leq \Big(1 - \tilde c_\delta \exp\big(-\frac{1}{\varepsilon}(A_1+\beta+\lambda)\big)\Big)^n m_D(D).$$

This together with (1.7.12) yield that for any $\beta > 0$ there exists $c(\beta) > 0$ such that for all small $\varepsilon > 0$,

$$(1.7.21)\qquad m_D\{v \in D : \tau^\varepsilon_v(V) > e^{(A_1+\beta)/\varepsilon}\} < e^{-c(\beta)/\varepsilon}.$$

Observe that by (1.7.12) and (1.7.20),

$$(1.7.22)\qquad \textstyle\int_D \tau^\varepsilon_v(V)dm_D(v) \leq \sum_{n=0}^\infty (n+1)T\big(m_D\{v\in D:$$
$$\tau^\varepsilon_v(V) > nT\} - m_D\{v \in D : \tau^\varepsilon_v(V) > (n+1)T\}\big)$$
$$= T\textstyle\sum_{n=0}^\infty m_D\{v\in D : \tau^\varepsilon_v(V) > nT\}$$
$$\leq Tm_D(D)\tilde c_\delta^{-1}\exp\big(\tfrac{1}{\varepsilon}(A_1+\beta+\lambda)\big).$$

In the same way as at the end of the proof of Proposition 1.5.2(i) we fix now an initial point $x = Z^\varepsilon_v(0) \in \mathcal{X}$ and choose discs D to be small balls on the (extended) local unstable manifolds $W^u_x(w, \varrho)$, $w \in \mathcal{W}$ which by means of the Fubini theorem and compactness arguments enable us to extend (1.7.21) and (1.7.22) to the case when m_D is replaced by m and D by $\mathcal{W}$ yielding (1.7.3) and (1.7.4) since β and λ in (1.7.22) can be chosen arbitrarily small as $\varepsilon \to 0$.

Next, we derive the assertion (ii). Let $t > 0$ and n be the integral part of t/T where $T > 0$ will be chosen later. Let, again, $D \in \mathcal{D}^u_\varepsilon\big((x,w),\alpha,\rho,C\big)$ and $x \in V$. Then

$$(1.7.23)\qquad m_D\{v\in D : Z^\varepsilon_v(\tau^\varepsilon_v) \in \Gamma,\ \tau^\varepsilon_v(V) < t\}$$
$$\leq m_D\{v \in D : Z^\varepsilon_v(\tau^\varepsilon_v(V)) \in \Gamma,\ \tau^\varepsilon_v(V) < (n+1)T\}$$
$$= \textstyle\sum_{k=0}^n m_D\{v\in D : Z^\varepsilon_v(\tau^\varepsilon_v(V)) \in \Gamma,\ kT \leq \tau^\varepsilon_v(V) < (k+1)T\}.$$

Let K be the intersection of the ω-limit set of the flow Π^t with $\bar V$. Then K is a compact set and by our assumption $K \subset G$. Hence,

$$\delta = \frac{1}{3}\inf\{|x-z| : x \in K,\ z \in \bar V \setminus G\} > 0$$

and if we set $U_\eta = \{z \in V : \operatorname{dist}(z,K) < \eta\}$ then $U_{3\delta} \subset G$. Now suppose that $kT \leq \tau^\varepsilon_{x,w}(V) < (k+1)T$ for some $k \geq 1$ and $Z^\varepsilon_{x,w}(\tau^\varepsilon_{x,w}(V)) \in \Gamma$ with $x \in V$ and $w \in \mathcal{W}$. Then either there is $t_1 \in [(k-1)T, kT]$ such that $Z^\varepsilon_{x,w}(t) \in \bar V \setminus U_{2\delta}$ for all $t \in [t_1, t_1+T]$ or there exist $t_2, t_3 > 0$ such that $(k-1)T \leq t_2 < t_3 < (k+1)T$ and $Z^\varepsilon_{x,w}(t_2) \in U_{2\delta}$ while $Z^\varepsilon_{x,w}(t_3) \in \Gamma$. Set $\mathcal{T}_z = \{\gamma \in C_{0,2T} : \gamma_0 = z$ and either there is $t_1 \in [0,T]$ so that $\gamma_t \in \bar V \setminus U_{2\delta}$ for all $t \in [t_1, t_1+T]$ or $\gamma_{t_2} \in U_{2\delta}$ and $\gamma_{t_3} \in \Gamma$ for some $0 \leq t_2 < t_3 < 2T\}$. Then for any $k \geq 1$,

$$(1.7.24)\qquad \{v \in D : Z^\varepsilon_v(\tau^\varepsilon_v(V)) \in \Gamma,\ kT \leq \tau^\varepsilon_v(V) < (k+1)T\}$$
$$\subset \big\{v\in D : Z^\varepsilon_v(\tau^\varepsilon_v(V)) \in \Gamma,\ \Phi_\varepsilon^{(k-1)T/\varepsilon}v \in A_0^{\varepsilon,2T}(\mathcal{T}_{z^\varepsilon_{k-1}(v)})\big\}$$

where $z_k^\varepsilon(v)=\pi_1(\Phi_\varepsilon^{kT/\varepsilon}v)$ and $A_0^{\varepsilon,T}(H)=\{w\in V\times\mathcal{W}:\ Z_w^\varepsilon\in H\}$.

Let $D\subset D_0\in\mathcal{D}_\varepsilon^u\big((x,y),\alpha,\rho,C^3\big)$, $x\in V$ (later both discs will be small balls on $W_x^u(y,C^6\rho)$) assuming that ρ is small and $C\geq 2$ is large so that $C^6\rho$ is still small. Choose a maximal $\big((k-1)T/\varepsilon,C\rho,\varepsilon,D_0,D_0\big)$-separated set $\tilde E_{k-1}$ in D_0 and let

$$E_{k-1}=\big\{v\in\tilde E_{k-1}:\ U_{D_0}^\varepsilon((k-1)T/\varepsilon,v,C\rho)\cap D\neq\emptyset\big\}.$$

Then for ε small enough,

$$D_0\supset\bigcup_{v\in E_{k-1}}U_{D_0}^\varepsilon\big((k-1)T/\varepsilon,v,C\rho\big)\supset D \tag{1.7.25}$$

and for any $v,w\in E_{k-1}$, $v\neq w$,

$$U_{D_0}^\varepsilon\big((k-1)T/\varepsilon,v,C\rho/2\big)\cap U_{D_0}^\varepsilon\big((k-1)T/\varepsilon,w,C\rho/2\big)=\emptyset. \tag{1.7.26}$$

If $v\in E_{k-1}$, $w\in U_{D_0}^\varepsilon\big((k-1)T/\varepsilon,v,C\rho\big)$ and $\Phi_\varepsilon^{(k-1)T/\varepsilon}w\in A_0^{\varepsilon,2T}(\mathcal{T}_{z_{k-1}^\varepsilon(w)})$ then by Lemma 1.3.2(iii), $|z_{k-1}^\varepsilon(w)-z_{k-1}^\varepsilon(v)|$ is of order ε, and so for each $\eta>0$ if ε is small enough then $\Phi_\varepsilon^{(k-1)T/\varepsilon}w\in A_\eta^{\varepsilon,2T}(\mathcal{T}_{z_{k-1}^\varepsilon(v)})$. For each $q>0$ set $\mathcal{T}_z^q=\{\gamma\in C_{0,2T}:\ \gamma_0=z$ and $\mathbf{r}_{0,2T}(\gamma,\mathcal{T}_z)\leq q\}$ and suppose that for some $\eta>0$ there is $d_\eta\geq 0$ so that

$$\inf_{z\in V}\ \inf_{\gamma\in\mathcal{T}_z^{2\eta}}S_{0,2T}(\gamma)>d_\eta. \tag{1.7.27}$$

Then $\mathcal{T}_z^{2\eta}\cap\Psi_{0,2T}^{d_\eta}(z)=\emptyset$, where $\Psi_{0,t}^u(z)$ is the same as in Theorem 1.2.3, and so

$$\mathcal{T}_z^\eta\subset\big\{\gamma\in C_{0,2T}:\ \gamma_0=z\text{ and }\mathbf{r}_{0,2T}(\gamma,\Psi_{0,2T}^{d_\eta}(z))\geq\eta\big\}. \tag{1.7.28}$$

Hence,

$$A_\eta^{\varepsilon,2T}(\mathcal{T}_z)\subset\big\{(z,w)\in V\times\mathcal{W}:\ \mathbf{r}_{0,2T}(Z_{(z,w)}^\varepsilon,\Psi_{0,2T}^{d_\eta}(z))\geq\eta\big\}. \tag{1.7.29}$$

By Lemma 1.3.2(ii),

$$D_{k-1}(v)=\Phi_\varepsilon^{(k-1)T/\varepsilon}U_{D_0}^\varepsilon\big((k-1)T/\varepsilon,v,C\rho\big)\in\mathcal{D}_\varepsilon^u\big(\Phi_\varepsilon^{(k-1)T/\varepsilon}v,\alpha,\rho,\sqrt{C}\big),$$

and so applying (1.7.11) to $D_{k-1}(v)$ we obtain from (1.7.27)– (1.7.29) that for any $\beta>0$ and sufficiently small ε uniformly in discs $D_{k-1}(v)$ as above,

$$m_{D_{k-1}(v)}\big(A_\delta^{\varepsilon,2T}(\mathcal{T}_{z_{k-1}^\varepsilon(v)})\big)\leq\exp(-(d_\eta-\beta)/\varepsilon).$$

This together with Lemma 1.3.6 yield that for each $v\in E_{k-1}$,

$$\begin{aligned}m_{D_0}\big\{\tilde v\in U_{D_0}^\varepsilon\big((k-1)T/\varepsilon,v,C\rho\big):\ \Phi_\varepsilon^{(k-1)T/\varepsilon}\tilde v\in A_\eta^{\varepsilon,2T}(\mathcal{T}_{z_{k-1}^\varepsilon(v)})\big\}\\ \leq\tilde Ce^{-(d_\eta-\beta)/\varepsilon}m_{D_0}\big(U_{D_0}^\varepsilon((k-1)T/\varepsilon,v,C\rho)\big)\end{aligned}$$

for some $\tilde C>0$ depending only on $C\rho$. Combining this with (1.7.24)–(1.7.26) and Lemma 1.3.6 we obtain that for any $k\geq 1$,

$$m_D\{v\in D:\ Z_v^\varepsilon(\tau_v^\varepsilon)\in\Gamma,\ kT\leq\tau_v^\varepsilon<(k+1)T\}\leq\hat Ce^{-(d_\eta-\beta)/\varepsilon} \tag{1.7.30}$$

for some $\hat C>0$ depending only on $C\rho$.

Next, we will specify d_η in (1.7.27) choosing $\eta\leq\frac{1}{2}\delta$. For each $z\in V$ we can write

$$\mathcal{T}_z^{2\eta}\subset\tilde{\mathcal{T}}_z^\eta\cup\hat{\mathcal{T}}_z^\eta \tag{1.7.31}$$

where $\tilde{\mathcal{T}}_z^\eta = \{\gamma \in C_{0,2T} : \gamma_0 = z,\, \gamma_{t_2} \in U_{3\delta}$ and $\gamma_{t_3} \in \Gamma_{2\eta}$ for some $0 \le t_2 < t_3 < 2T\}$ with $\Gamma_r = \{z : \mathrm{dist}(z,\Gamma) \le r\}$ and $\hat{\mathcal{T}}_z = \{\gamma \in C_{0,2T} : \gamma_0 = z$ and there is $t_1 \in [0,T]$ so that $\gamma_t \in V_{2\eta} \setminus U_\delta$ for all $t \in [t_1, t_1+T]\}$. By (1.7.5) and the lower semicontinuity of the functional $S_{0,2T}$ it follows that for any $\zeta > 0$ we can choose $\eta > 0$ small enough so that

$$\inf_{z\in V}\inf_{\gamma\in\tilde{\mathcal{T}}_z^\eta} S_{0,2T}(\gamma) > A_2 - \zeta. \tag{1.7.32}$$

Since $\bar V \setminus U_\delta$ is disjoint with the ω-limit set of the flow Π^t and the latter is closed then if η is sufficiently small $V_{2\eta} \setminus U_\delta$ is also disjoint with this ω-limit set and, in particular, it does not contain any forward semi-orbit of Π^t. Hence we can apply Lemma 1.6.4 which in view of (1.2.13) implies that there exists $a > 0$ such that for all small $\eta > 0$,

$$\inf_{z\in V}\inf_{\gamma\in\hat{\mathcal{T}}_z} S_{0,2T}(\gamma) > aT \tag{1.7.33}$$

which is not less than A_2 if we take $T = A_2/a$. Now, (1.7.32) and (1.7.33) produce (1.7.27) with $d = A_2 - \zeta$, and so (1.7.30) follows with such d_η. This together with (1.7.23) yield that for any $\beta > 0$ we can choose sufficiently small $\zeta, \lambda > 0$ and then $\eta > 0$ so that for all ε small enough

$$\begin{aligned} &m_D\{v \in D : Z_v^\varepsilon(\tau_v^\varepsilon) \in \Gamma,\, \tau_v^\varepsilon(V) \le e^{(A_2-\beta)/\varepsilon}\} \\ &\le m_D\{v \in D : Z_v^\varepsilon(\tau_v^\varepsilon) \in \Gamma,\, \tau_v^\varepsilon(V) < T\} + e^{-\lambda/2\varepsilon}. \end{aligned} \tag{1.7.34}$$

Now assume that (1.7.7) holds true for some $x \in V$. Recall, that $S_{0T}(\gamma) = 0$ implies that γ is a piece of an orbit of the flow Π^t. Since no $\gamma \in C_{0T}$ satisfying

$$\gamma_0 = x \text{ and } \inf_{t\in[0,T]} \mathrm{dist}(\gamma_t, \partial V) \le a(x)/2 \tag{1.7.35}$$

can be such piece of an orbit we conclude by the lower semicontinuity of S_{0T} that $S_{0T}(\gamma) > c(x)$ whenever (1.7.35) holds true for some $c(x) > 0$ independent of γ (but depending on x). Hence, by (1.7.11),

$$\begin{aligned} m_D\{v \in D : \tau_v^\varepsilon < T\} &\le m_D\{v \in D : \mathbf{r}_{0T}(Z_v^\varepsilon, \Psi_{0T}^{c(x)}(x)) \\ &\ge a(x)/2\} \le \exp(-c(x)/2\varepsilon) \end{aligned} \tag{1.7.36}$$

provided ε is small enough and (1.7.8) follows. Observe also that any $\gamma \in C_{0t}$ with $\gamma_0 = x \in V$ and $\gamma_t \in \partial V$ should contain a piece which either belongs to some $\tilde{\mathcal{T}}_z^\eta$ or $\hat{\mathcal{T}}_z^\eta$, as above, or to satify (1.7.35). By (1.7.32), (1.7.33), and the above remarks it follows that $S_{0t}(\gamma) \ge q(x)$ for such γ where $q(x) > 0$ depends only on x, and so $R_\partial(x) \ge q(x)$.

Finally, similarly to Proposition 1.5.2 we fix $x = Z_v^\varepsilon(0) \in V$, choose discs D and D_0 to be small balls on the (extended) local unstable manifolds $W_x^u(w,q)$, $w \in \mathcal{W}, q > 0$ and using the Fubini theorem we extend (1.7.34) and (1.7.36) to the case when D and m_D are replaced by $\mathcal{W}$ and m, respectively, yielding (1.7.6). If $\Gamma = \partial V$ then by (1.7.6) and (1.7.8),

$$\begin{aligned} \textstyle\int_{\mathcal{W}} \tau_{x,y}^\varepsilon(V)dm(y) &\ge e^{(A_2-\beta)/\varepsilon} m\{y \in \mathcal{W} : \tau_{x,y}^\varepsilon(V) \ge e^{(A_2-\beta)/\varepsilon}\} \\ &\ge e^{(A_2-\beta)/\varepsilon}(1 - e^{-\lambda(\beta)/\varepsilon} - e^{-\hat\lambda(T)/\varepsilon}) \end{aligned} \tag{1.7.37}$$

and, since $\beta > 0$ is arbitrary, (1.7.9) follows completing the proof of Proposition 1.7.1. □

Now we will derive Theorem 1.2.5 from Proposition 1.7.1. Assume, first, that $R_\partial < \infty$. Then by Lemma 1.6.4, $R_\partial(x)$ is finite and continuous in the whole V. Moreover, since $\mathcal{O}$ is an S-attractor the conditions of Lemma 1.6.5 are satisfied with some $d(t) \to 0$ as $t \to \infty$ the same for all points of V which yields the conditions of Proposition 1.7.1(i) with $A_1 = R_\partial + \delta$ for any $\delta > 0$. Hence, (1.7.3) and (1.7.4) hold true with $A_1 = R_\partial$. Since $\mathcal{O}$ is an S-attractor of the flow Π^t and its basin contains $\bar{V}$ then the intersection of $\bar{V} \setminus \mathcal{O}$ with the ω-limit set of Π^t is empty. By the definition of an S-attractor for any $\eta > 0$ there exists an open set $U_\eta \supset \mathcal{O}$ such that $R(x, z) \leq \eta$ whenever $x \in \mathcal{O}$ and $z \in U_\eta$. Hence, by the triangle inequality for the function R and Lemma 1.6.5 for any set $\Gamma \subset \partial V$,

$$\inf_{z \in U_\zeta, \tilde{z} \in \Gamma} R(z, \tilde{z}) \geq \inf_{\tilde{z} \in \Gamma} R^{\mathcal{O}}(\tilde{z}) - \eta. \tag{1.7.38}$$

If $\Gamma = \partial V$ then by Lemma 1.6.5 the right hand side of (1.7.38) equals $A_2 = R_\partial - \eta$. Assuming that $R_\partial < \infty$ we can apply Proposition 1.7.1(ii) with such A_2 yielding (1.7.6), (1.7.8) and since $\eta > 0$ is arbitrary (1.2.21) and (1.2.22) follow in this case. If $R_\partial = \infty$ then (1.2.22) is trivial and by (1.7.38), $R(z, \tilde{z}) = \infty$ for any $z \in U_\zeta$ and $\tilde{z} \in \partial V$, and so we can apply Proposition 1.7.1(ii) with any A_2 which sais that the left hand side in (1.7.9) equals ∞, and so (1.2.21) holds true in this case, as well.

Next, we establish (1.2.23). For small $\delta, \beta > 0$ and large $T > 0$ which will be specified later on set $\Gamma_1 = \{v \in V_\delta \times \mathcal{W} : Z_v^\varepsilon(T) \in V \setminus U_{\delta/2}(\mathcal{O})\}$, $\Gamma_2 = \{v \in U_{\delta/2}(\mathcal{O}) \times \mathcal{W} : \tau_v^\varepsilon(U_\delta(\mathcal{O})) \leq e^{\beta/\varepsilon}\}$ and $t_\varepsilon = T + e^{\beta/\varepsilon}$. Then

$$\begin{aligned} &\Theta_v^\varepsilon((n+1)t_\varepsilon \wedge \tau_v^\varepsilon(V)) - \Theta_v^\varepsilon(nt_\varepsilon \wedge \tau_v^\varepsilon(V)) \\ &\leq T + t_\varepsilon\big(\mathbb{I}_{\Gamma_1}(\Phi_\varepsilon^{t_\varepsilon n/\varepsilon} v) + \mathbb{I}_{V_\delta \times \mathcal{W} \setminus \Gamma_1}(\Phi_\varepsilon^{t_\varepsilon n/\varepsilon} v)\mathbb{I}_{\Gamma_2}(\Phi_\varepsilon^{t_\varepsilon n/\varepsilon} v)\big). \end{aligned} \tag{1.7.39}$$

If δ is sufficiently small then V_δ is still contained in the basin of $\mathcal{O}$ with respect to the flow Π^t, and so we can choose T (depending only on δ) so that

$$\Pi^T V_\delta \subset U_{\delta/4}(\mathcal{O}).$$

Then for some $a > 0$,

$$\inf\big\{S_{0T}(\gamma) : \gamma \in C_{0T},\ \gamma_0 \in V_\delta,\ \gamma_T \notin U_{\delta/3}(\mathcal{O})\big\} > a,$$

and so if $\gamma_0 \in V_\delta$ and $\gamma_T \notin U_{\delta/2}(\mathcal{O})$ then $\text{dist}(\gamma, \Psi_{0T}^a(z)) \geq \delta/6$ for any $z \in V_\delta$. Relying on (1.7.11) we obtain that for any $\tilde{D} \in \mathcal{D}_\varepsilon^u(z, \alpha, \rho, \sqrt{C})$ with $\tilde{D} \subset V_\rho \times \mathcal{W}$,

$$m_{\tilde{D}}(\Gamma_1 \cap \tilde{D}) \leq e^{-a/2\varepsilon} \tag{1.7.40}$$

provided ε is small enough. Next, the same arguments which yield (1.7.34) and (1.7.36) enable us to conclude that if $\beta > 0$ is small enough then for any $\tilde{D} \in \mathcal{D}_\varepsilon^u(z, \alpha, \rho, \sqrt{C})$ with $\tilde{D} \subset U_{\delta/2}(\mathcal{O}) \times \mathcal{W}$,

$$m_{\tilde{D}}(\Gamma_2 \cap \tilde{D}) \leq e^{-\beta/\varepsilon}. \tag{1.7.41}$$

Now let $D \in \mathcal{D}_\varepsilon^u(z, \alpha, \rho, C)$, $z = (x, y)$, $D \subset V_\delta \times \mathcal{W}$ and $D \subset D_0 \in \mathcal{D}_\varepsilon^u(z, \alpha, \rho, C^3)$ with ρ small and C large so that $C^6\rho$ is still small. Let $E_n^{(1)}$ and $E_n^{(2)}$ be maximal $(nt_\varepsilon\varepsilon^{-1}, C\rho, \varepsilon, D, D_0)-$ and $((nt_\varepsilon + T)\varepsilon^{-1}, C\rho, \varepsilon, D, D_0)-$separated sets, respectively. Then

$$\cup_{v \in E_n^{(1)}} U_{\tilde{D}}^\varepsilon(nt_\varepsilon\varepsilon^{-1}, v, C\rho) \supset D$$

and since the last union is contained in a small neighborhood of D and $U_{\tilde{D}}^\varepsilon(nt_\varepsilon\varepsilon^{-1}, v, C\rho/2)$ are disjoint for different $v \in E_n^{(1)}$ we obtain using Lemma

1.3.6 that

$$\sum_{v\in E_n^{(1)}} m_D\big(U_D^\varepsilon(nt_\varepsilon\varepsilon^{-1},v,C\rho)\big)\le 2c_{C\rho}^{-1}c_{C\rho/2}^{-1}m_D(D).$$

Similarly,

$$\cup_{v\in E_n^{(2)}}U_D^\varepsilon((nt_\varepsilon+T)\varepsilon^{-1},v,C\rho)\supset D$$

and

$$\sum_{v\in E_n^{(2)}} m_D\big(U_D^\varepsilon((nt_\varepsilon+T)\varepsilon^{-1},v,C\rho)\big)\le 2c_{C\rho}^{-1}c_{C\rho/2}^{-1}m_D(D).$$

Since by Lemma 1.3.2(ii),

$$\Phi_\varepsilon^t U_D^\varepsilon(t,v,C\rho)\in\mathcal{D}(\Phi_\varepsilon^t v,\alpha,\rho,\sqrt{C})$$

we can apply (1.7.39)–(1.7.41) together with Lemma 1.3.6 (similarly to the proof of (1.7.30)) in order to conclude that for sufficiently small β and any much smaller ε,

$$\int_D\big(\Theta_v^\varepsilon((n+1)t_\varepsilon\wedge\tau_v^\varepsilon(V))-\Theta_v^\varepsilon(nt_\varepsilon\wedge\tau_v^\varepsilon(V))\big)dm_D(v)\le t_\varepsilon e^{-\beta/\varepsilon}(T+1).$$

Choosing discs D to be small balls on the (extended) local unstable manifolds $W_x^u(w,\varrho)$, $w\in\mathcal{W}$ together with the Fubini theorem we extend this estimate to

$$\text{(1.7.42)}\qquad \int_{\mathcal{W}}\big(\Theta_v^\varepsilon((n+1)t_\varepsilon\wedge\tau_v^\varepsilon(V))-\Theta_v^\varepsilon(nt_\varepsilon\wedge\tau_v^\varepsilon(V))\big)dm(v)\le\tilde{C}t_\varepsilon e^{\beta/\varepsilon}$$

for some $\tilde{C}>0$ depending on δ but independent of n and ε. Finally, (1.2.22) and (1.7.42) together with the Chebyshev inequality yield that for $n(\varepsilon)=[e^{(R_\partial+\beta/4)/\varepsilon}t_\varepsilon^{-1}]$, each $x\in V$, a small $\beta>0$ and any much smaller $\varepsilon>0$,

$$\text{(1.7.43)}\qquad \begin{aligned} &m\{w\in\mathcal{W}:\,\Theta_{x,w}^\varepsilon(\tau_{x,w}^\varepsilon(V))\ge e^{-\beta/4\varepsilon}\tau_{x,w}^\varepsilon(V)\}\\ &\le m\{w\in\mathcal{W}:\,\Theta_{x,w}^\varepsilon((n(\varepsilon)+1)t_\varepsilon)\ge e^{-\beta/4\varepsilon}e^{(R_\partial-\beta/4)/\varepsilon}\}\\ &+m\{w\in\mathcal{W}:\,\tau_{x,w}^\varepsilon(V)<e^{(R_\partial-\beta/4)/\varepsilon}\text{ or }\tau_{x,w}^\varepsilon(V)>e^{(R_\partial+\beta/4)/\varepsilon}\}\\ &\le\tilde{C}e^{-\beta/4\varepsilon}\big(1+e^{-(R_\partial+\beta/4)/\varepsilon}(T+e^{\beta/\varepsilon})\big)+e^{-\lambda(\beta/4)/\varepsilon}.\end{aligned}$$

Since $R_\partial>0$ and we can choose β to be arbitrarily small, (1.7.43) yields (1.2.23).

In order to complete the proof of Theorem 1.2.5 we have to derive (1.2.24). If $\partial_{\min}=\partial V$ then there is nothing to prove, so we assume that $\partial_{\min}$ is a proper subset of ∂V and in this case, clearly, $R_\partial<\infty$. Since $\Gamma=\{z\in\partial V:\,\text{dist}(z,\partial_{\min})\ge\delta\}$ is compact and disjoint with $\partial_{\min}$ which is also compact then by the lower semicontinuity of $R^{\mathcal{O}}(z)$ established in Lemma 1.6.5(iii) it follows that $R^{\mathcal{O}}(z)\ge R_\partial+\beta$ for some $\beta>0$ and all $z\in\Gamma$. Then by (1.7.38), $R(z,\tilde{z})\ge R_\partial+\beta/2$ for any $z\in U_{\beta/2}$ and $\tilde{z}\in\Gamma$. Hence, applying Proposition 1.7.1 we obtain that

$$m\{y\in\mathcal{W}:\,\tau_{x,y}^\varepsilon(V)\ge e^{(R_\partial+\frac{1}{3}\beta)/\varepsilon}\}\le e^{-\lambda/\varepsilon}$$

and

$$m\{y\in\mathcal{W}:\,Z_{x,y}^\varepsilon(\tau_{x,y}^\varepsilon(V))\in\Gamma,\,\tau_{x,y}^\varepsilon(V)\le e^{(R_\partial+\frac{1}{3}\beta)/\varepsilon}\}<e^{-\lambda/\varepsilon}$$

for some $\lambda>0$ and all ε small enough yielding (1.2.24) and completing the proof of Theorem 1.2.5. □

1.8. Adiabatic transitions between basins of attractors

In this section we will prove Theorem 1.2.7 relying, again, on Proposition 1.7.1 together with "Markov property type" arguments and at the end of the proof we will apply even some rough "strong Markov property type" arguments in order to deal with subsequent transitions between basins of attractors. In view of (1.2.27) and Lemma 1.6.2i any curve $\gamma \in C_{0t}$ starting at $\gamma_0 = x \in V_{j_1}$ and ending at $\gamma_t = z \in \cap_{1 \leq i \leq k} \partial V_{j_i}$, $k \leq \ell$ can be extended into each V_{j_i}, $i = 1, ..., k$ with arbitrarily small increase in its S-functional. Hence,

$$R_\partial^{(i)} = \min_{j \neq i} R_{ij} \tag{1.8.1}$$

where $R_\partial^{(i)} = \inf\{R(x,z) : x \in \mathcal{O}_i,\, z \in \partial V_i\}$. Let Q be an open ball of radius at least r_0 centered at the origin of $\mathbb{R}^d$. By Assumption 1.2.6 the slow motion $Z_{x,y}^\varepsilon$ cannot exit Q provided $x \in Q$ and $y \in \mathcal{W}$. Furthermore, it is clear that Q contains the ω-limit set of the averaged flow Π^t. Assumption 1.2.6 enables us to deal only with restricted basins $V_i^Q = V_i \cap Q$ and though the boundaries ∂V^Q of V_i^Q may include now parts of the boundary ∂Q of Q it makes no difference since Z^ε cannot reach ∂Q if it starts in Q. Set $V^{(i)} = Q \setminus \cup_{j \neq i} U_\delta(\mathcal{O}_j)$ where $\delta > 0$ is small enough. We claim that in view of (1.2.27) each V_i satisfies conditions of Proposition 1.7.1(i) for any $\beta > 0$ with $A_1 = R_\partial^{(i)} + \beta$ and some $T = T_\beta$ depending on β. Indeed, set

$$\partial(\eta) = \{v \in Q : \mathrm{dist}(v, \cup_{1 \leq j \leq \ell} \partial V_j) \leq \eta\},\ \eta > 0.$$

In view of (1.2.9) and (1.2.27) there exists $L > 0$ such that if η is small enough and $z \in \partial(\eta)$ we can construct a curve $\varphi^z \in C_{0,L\eta}$ with $S_{0,L\eta}(\varphi^z) \leq \tilde{L}\eta$, $\varphi_0^z = z$, $\varphi_t^z \in V_j \setminus \partial(\eta)$ for some $t \in [0, L\eta]$ and $j = 1, ..., \ell$ where $\tilde{L} = L \sup_{x,y} |\varphi_x^u(y)|$. Since V_j is the basin of $\mathcal{O}_j$ there exists $T = T_{\eta,\delta}$ such that $\Pi^T \varphi_t^z \in U_\delta(\mathcal{O}_j)$ and extending φ^z by the piece of the orbit of Π^t we obtain a curve $\tilde{\varphi}^z \in C_{0,L\eta+T}$ starting at z, entering $U_\delta(\mathcal{O}_j)$ and satisfying $S_{0,L\eta+T}(\tilde{\varphi}^z) \leq \tilde{L}\eta$. Hence, for $z \in \partial(\eta)$ the condition (1.7.2) holds true with $V = V^{(i)}$ and $A_1 = \tilde{L}\eta$. Since the ω-limit set of the flow Π^t is contained in $Q \cap \big(\cup_{1 \leq j \leq \ell}(\partial V_j \cup \mathcal{O}_j)\big)$ it follows from Assumption 1.2.6 and compactness considerations that there exists $\tilde{T} = \tilde{T}_{\eta,\delta}$ such that for any $z \in Q \setminus V_i$ we can find $t_z \in [0, \tilde{T}]$ with $\Pi^{t_z} z \in \partial(\eta) \cup \big(\cup_{j \neq i} U_\delta(\mathcal{O}_j)\big)$. If $\Pi^{t_z} z \in \cup_{j \neq i} U_\delta(\mathcal{O}_j)$ then we take $\varphi_t^z = \Pi^t z$, $t \in [0, \tilde{T}]$ to satisfy (1.7.2) for $V = V^{(i)}$ and $A_1 = 0$. If $\Pi^{t_z} z \in \partial(\eta)$ then we extend the curve $\varphi_t^z = \Pi^t z$, $t \in [0, t_z]$ as in the above argument which yields a curve $\tilde{\varphi}^z$ starting at z, ending in some $U_\delta(\mathcal{O}_j)$, $j \neq i$ and having its S-functional not exceeding $\tilde{L}\eta$. Finally, in the same way as in the proof of Theorem 1.2.5 for any $\beta > 0$ there exists $\hat{T} = \hat{T}_{\eta,\delta,\beta}$ such that whenever $z \in V_i(\eta) = V_i \cap Q \setminus \partial(\eta)$ we can construct $\varphi^z \in C_{0\hat{T}}$ such that (1.7.2) holds true with $V = V_i(\eta)$ and $A_1 = R_\partial^{(i)} + \beta/2$ and, moreover, $\mathrm{dist}(\varphi_t^z, V_j) \leq \eta$ for some $t \leq \hat{T}$ and $j \neq i$ with $R_{ij} = R_\partial^{(i)}$. Then in the same way as above we can extend φ^z to some $\tilde{\varphi}^z \in C_{\hat{T}+\tilde{T}}$ so that $\tilde{\varphi}_t^z \in U_\delta(V_j)$ for some j as above, $t \leq \hat{T} + \tilde{T}$ and $S_{0,\hat{T}+\tilde{T}}(\tilde{\varphi}^z) \leq R_\partial^{(i)} + \beta/2 + \tilde{L}\eta$ which gives (1.7.2) for all $z \in V = V^{(i)}$ with $A_1 = R_\partial^{(i)} + \beta$ provided η is small enough. Hence, Proposition 1.7.1(i) yields the estimates (1.7.3) and (1.7.4) for $\tau_{x,y}^\varepsilon(i)$ in place of $\tau_{x,y}^\varepsilon(V)$ with $A_1 = R_\partial^{(i)}$. In order to obtain the corresponding bounds in the other direction observe that in view of

(1.2.27),

$$R_\partial^{(i)}(\delta) = \inf\{R(x,z) : x \in \mathcal{O}_i,\ z \notin V_i(\eta)\} \to R_\partial^{(i)} \text{ as } \delta \to 0. \tag{1.8.2}$$

Since $\overline{V_i(\eta)}$ is contained in the basin of $\mathcal{O}_i$ we can apply to $V_i(\eta)$ the same estimates as in Theorem 1.2.5 which together with (1.8.2) and the fact that the exit time of Z^ε from $V_i(\eta)$ is smaller than its exit time from V_i provide the remaining bounds yielding (1.2.28) and (1.2.29).

Next, we derive (1.2.30) similarly to (1.2.23) but taking into account that $\cup_{1\le j\le \ell}\partial V_j$ may contain parts of the ω-limit set of the flow Π^t which allows the slow motion Z^ε to stay long time near these boundaries. Still, set

$$\theta_v^\varepsilon = \inf\{t \ge 0 : Z_v^\varepsilon(t) \in \cup_{1\le j\le \ell} U_{\delta/3}(\mathcal{O}_j)\}.$$

Using the same arguments as above we conclude that for any $\eta > 0$ there exists $T = T_{\eta,\delta}$ such that whenever $z \in Q$ we can construct $\varphi^z \in C_{0T}$ with $\varphi_0^z = z$, $\varphi_T^z \in \cup_{1\le j\le \ell} U_\delta(\mathcal{O}_j)$ and $S_{0T}(\varphi^z) \le \eta$. This together with (1.7.21) and Assumption 1.2.6 yield that for any disc $D \in \mathcal{D}_\varepsilon^u(z,\alpha,\rho,C)$, $D \subset Q \times \mathcal{W}$,

$$m_D\{v \in D : \theta_v^\varepsilon \ge e^{2\eta/\varepsilon}\} \le e^{-\lambda(\eta)/\varepsilon}$$

for some $\lambda(\eta) = \lambda(x,\eta) > 0$ and all small ε. Set

$$\Gamma_1 = \{v \in Q \times \mathcal{W} : Z_v^\varepsilon(e^{2\eta/\varepsilon}) \in Q \setminus \cup_{1\le j\le \ell} U_{\delta/2}(\mathcal{O}_j)\},$$

$$\Gamma_2 = \{v \in \cup_{1\le j\le \ell} U_{\delta/2}(\mathcal{O}_j) : \tau_v^\varepsilon(\cup_{1\le j\le \ell} U_\delta(\mathcal{O}_j)) \le e^{\beta/\varepsilon}\}$$

and $t_\varepsilon = e^{2\eta/\varepsilon} + e^{\beta/\varepsilon}$ where η is much smaller than β. Then proceeding similarly to the proof of (1.2.23) as in (1.7.40)–(1.7.43) above we arrive at (1.2.30).

Next, we obtain (1.2.31) relying on additional assumptions specified in the statement of Theorem 1.2.7. Let V_i^Q be the same as above and $\partial_0^{(i)}(x) = \{z \in \partial V_i^Q : R(x,z) = R_\partial^{(i)}\}$. Since $\mathcal{O}_i$ is an S-attractor it follows from Lemma 1.6.5(i) that $R(x,z)$ and $\partial_0^{(i)}(x)$ coincide with the same function $R^{\mathcal{O}_i}(z)$ and the same (in general, may be empty) set $\partial_0^{(i)}$, respectively, for all $x \in \mathcal{O}_i$. By Lemma 1.6.2(i), our assumption that B is complete on ∂V_i implies that $R^{\mathcal{O}_i}(z)$ is continuous in a neighborhood of ∂V_i, and so $\partial_0^{(i)}$ is a nonempty compact set. Since we assume that $\iota(i) \ne i$ is the unique index j for which $R_{ij} = R_{i\iota(i)} = R_\partial^{(i)}$ then by (1.2.27),

$$\min_{j\ne i,\iota(i)} \inf_{z\in\partial_0^{(i)}} \operatorname{dist}(z, \partial V_j) > 0.$$

Observe that if $\tilde{\mathcal{O}} \subset \partial V_i$ is an S-compact then either $\tilde{\mathcal{O}} \subset \partial_0^{(i)}$ or $\tilde{\mathcal{O}} \cap \partial_0^{(i)} = \emptyset$. Denote by L_Π the ω-limit set of the averaged flow Π^t. Since $L_\Pi \cap \partial V_i$ consists of a finite number of S-compacts it follows that

$$\inf\{|z - \tilde z| : z \in L_\Pi \cap \partial_0^{(i)},\ \tilde z \in L_\Pi \setminus \partial_0^{(i)}\} > 0.$$

By the continuity of $R^{\mathcal{O}_i}(z)$ in $z \in \partial V_i$ there exists $a > 0$ such that

$$\inf\{R^{\mathcal{O}_i}(z) : z \in (\cup_{j\ne i,\iota(i)}(\partial V_i \cap \partial V_j)) \cup ((L_\Pi \setminus \partial_0^{(i)}) \cap \partial V_i)\} \ge R_\partial^{(i)} + 9a.$$

These considerations enable us to construct a connected open set G with a piecewise smooth boundary ∂G such that

$$\bar G \subset V_i \cup (V_{\iota(i)} \setminus \mathcal{O}_{\iota(i)}) \cup ((\partial V_i \cap \partial V_{\iota(i)}) \setminus (L_\Pi \setminus \partial_0^{(i)}))$$

and for $\Gamma = \partial G \setminus U_\delta(\mathcal{O}_{\iota(i)})$ and some $a(\delta) > 0$,

$$\inf_{z\in\Gamma} R^{\mathcal{O}_i}(z) \geq R_\partial^{(i)} + 8a \tag{1.8.3}$$

provided $a \leq a(\delta)$. The idea of this construction is that if $Z^\varepsilon_{x,y}(\tau^\varepsilon_{x,y}(i)) \notin V_{\iota(i)}$ then the slow motion should exit G through the part Γ of its boundary. Somewhat similarly to the proof of Proposition 1.7.1(ii) we will show that for "most" initial conditions y this can only occur after the time $\exp\big((R_\partial^{(i)} + 2a)/\varepsilon\big)$ and, on the other hand, we conclude from (1.2.29) that for "most" initial conditions y the exit time $\tau^\varepsilon_{x,y}(i)$ does not exceed $\exp\big((R_\partial^{(i)} + a)/\varepsilon\big)$.

Let U_0 be a sufficiently small open neighborhood of $\partial_0^{(i)}$ so that, in particular,

$$\sup_{z\in U_0} R^{\mathcal{O}_i}(z) \leq R_\partial^{(i)} + a$$

and set

$$\tau^\varepsilon_{x,y}(G) = \inf\{t \geq 0 :\, Z^\varepsilon_{x,y}(\tau^\varepsilon_{x,y}(G)) \notin G\}.$$

For each disc $D \in \mathcal{D}^u_\varepsilon((x,w),\alpha,\rho,C)$ we can write

$$\{v \in D :\, \tau^\varepsilon_v(G) \leq e^{(R_\partial^{(i)}+a)/\varepsilon},\, Z^\varepsilon_v(\tau^\varepsilon_v(G)) \in \Gamma\} \subset \bigcup_{0\leq n\leq n(\varepsilon)+1} \big(A_D^{(1)}(n) \cup \bigcup_{(n-1)t_\varepsilon\leq k\leq(n+1)t_\varepsilon} \big(A_D^{(2)}(k) + A_D^{(3)}(k) + \bigcup_{k-2t_\varepsilon\leq m\leq k-2T} A_D^{(4)}(m) \cap A_D^{(5)}(k)\big)\big) \tag{1.8.4}$$

where $t_\varepsilon = e^{\beta/\varepsilon}$ for some small $\beta > 0$, $n(\varepsilon) = \big[e^{(R_\partial^{(i)}+a-\beta)/\varepsilon}\big]$, $A_D^{(1)}(n) = \{v \in D :\, Z^\varepsilon_v(t) \in G \setminus \big(U_\eta(\mathcal{O}_i) \cup U_\delta(\mathcal{O}_{\iota(i)})\big)$ for all $t \in [(n-1)t_\varepsilon, nt_\varepsilon]\}$ for a sufficiently small $\eta > 0$, $A_D^{(2)}(k) = \{v \in D :\, \exists t_1, t_2 \,\text{with}\, k \leq t_1 < t_2 < k + 3T,\, Z^\varepsilon_v(t_1) \in U_\eta(\mathcal{O}_i),\, Z^\varepsilon_v(t_2) \in \Gamma\}$, $A_D^{(3)}(k) = \{v \in D :\, Z^\varepsilon_v(t) \in G \setminus \big(U_0 \cup U_\eta(\mathcal{O}_i) \cup U_\delta(\mathcal{O}_{\iota(i)})\big)$ for all $t \in [k, k+T]\}$, $A_D^{(4)}(m) = \{v \in D :\, \exists t_1, t_2 \,\text{with}\, m \leq t_1 < t_2 < m + T,\, Z^\varepsilon_v(t_1) \in U_\eta(\mathcal{O}_i),\, Z^\varepsilon_v(t_2) \in U_0\}$, and $A_D^{(5)}(k) = \{v \in D :\, \exists t_3, t_4 \,\text{with}\, k \leq t_3 < t_4 < k + T,\, Z^\varepsilon_v(t_3) \in U_0,\, Z^\varepsilon_v(t_4) \in \Gamma\}$. Observe that $G \setminus \big(U_\eta(\mathcal{O}_i) \cup U_\delta(\mathcal{O}_{\iota(i)})\big)$ satisfies conditions of Proposition 1.7.1(i) with arbitrarily small A_1, so similarly to (1.7.21) (and taking into account Lemma 1.3.6) we can estimate

$$m_D(A_D^{(1)}(n)) \leq \exp(-\frac{1}{2}e^{\beta/\varepsilon}). \tag{1.8.5}$$

Similarly to the proof of Proposition 1.7.1(ii) we obtain also that

$$\max\big(m_D(A_D^{(2)}(k)), m_D(A_D^{(3)}(k))\big) \leq e^{-(R_\partial^{(i)}+3a)/\varepsilon} \tag{1.8.6}$$

where we, first, choose η small and then T large enough.

Next, we estimate $m_D\big(A_D^{(4)}(m) \cap A_D^{(5)}(k)\big)$ for $m \leq k - 2T$ by the following Markov property type argument. Let $D \subset D_0 \in \mathcal{D}^u_\varepsilon((x,y),\alpha,\rho,C^3)$ and choose a maximal $(k/\varepsilon, C\rho, \varepsilon, D, D_0)$-separated set E in D. Let $\tilde{E} = \{v \in E :\, U^\varepsilon_{D_0}(k/\varepsilon, v, C\rho) \cap A_D^{(4)}(m) \neq \emptyset\}$, $\tilde{A} = \cup_{v\in\tilde{E}} U^\varepsilon_{D_0}(k/\varepsilon, v, C\rho)$ and $\mathcal{T}^\varepsilon = \{\gamma \in C_{0T} :\, \exists t_1, t_2 \,\text{with}\, 0 \leq t_1 < t_2 \leq T,\, \gamma_{t_1} \in U_{2\eta}(\mathcal{O}_i) \,\text{and}\, \mathrm{dist}(\gamma_{t_2}, U_0) \leq \eta\}$. Assume that $\mathrm{dist}(z, \mathbb{R}^d \setminus U_0) \leq \eta$ for any $z \in \partial V_i$. By Lemma 1.6.2(i), $R(x,z)$ is continuous in z when z belong to a sufficiently small neighborhood of ∂V_i which together with the definition of S-compacts yields that $S_{0T}(\gamma) \geq R_\partial^{(i)} - 3a/2$ for any $\gamma \in \mathcal{T}^\varepsilon$, provided η is small enough. Observe that $Z^\varepsilon_{\Phi_\varepsilon^{m/\varepsilon} v} \in \mathcal{T}^\varepsilon$ for any $v \in \tilde{A}$, provided

ε is sufficiently small. These together with the arguments similar to the proof of Proposition 1.7.1(ii) yield the estimate

$$m_D(\tilde{A}) \le e^{-(R_\partial^{(i)}-2a)/\varepsilon} \tag{1.8.7}$$

for all ε small enough. Since $U^\varepsilon_{D_0}(k/\varepsilon, v, \frac{1}{2}C\rho)$ are disjoint for different $v \in E$ we obtain by Lemma 1.3.6,

$$m_D(\tilde{A}) \ge \sum_{v\in\tilde{E}} U^\varepsilon_{D_0}(k/\varepsilon, v, \frac{1}{2}C\rho) \ge \tilde{c}\sum_{v\in\tilde{E}} U^\varepsilon_{D_0}(k/\varepsilon, v, C\rho) \ge \tilde{c}m_D(\tilde{A}). \tag{1.8.8}$$

where $\tilde{c} = c_{C\rho}c_{C\rho/2}$. In a similar way we obtain that for each disc $\tilde{D} = \Phi_\varepsilon^{k/\varepsilon}U^\varepsilon_{D_0}(k/\varepsilon, v, C\rho)$,

$$m_{\tilde{D}}\big(\tilde{D}\cap A_D^{(5)}(k)\big) \le e^{-6a/\varepsilon} \tag{1.8.9}$$

provided ε is small enough. By (1.8.7)–(1.8.9) together with Lemma 1.3.6,

$$m_D\big(A_D^{(4)}(m)\cap A_D^{(5)}(k)\big) \le e^{-(R_\partial^{(i)}+3a)/\varepsilon} \tag{1.8.10}$$

provided $m \le k - 2T$ and ε is small enough. Summing in m, k and n we obtain from (1.8.4)–(1.8.6) and (1.8.10) that for a small β and all sufficiently small ε,

$$m_D\{v\in D:\ \tau_v^\varepsilon(G) \le e^{(R_\partial^{(i)}+a)/\varepsilon},\ Z_v^\varepsilon(\tau_v^\varepsilon(G))\in\Gamma\} \le e^{-a/\varepsilon}. \tag{1.8.11}$$

Taking discs D to be small balls on the (extended) local unstable manifolds $W_x^u(w,q)$ and using the Fubini theorem as before we obtain (1.8.11) for m and $\mathcal{W}$ in place of m_D and D, respectively. On the other hand, employing Proposition 1.7.1(i) we derive that

$$m\{v\in\mathcal{W}:\ \tau_v^\varepsilon(G) > e^{(R_\partial^{(i)}+a)/\varepsilon}\} \le e^{-\lambda/\varepsilon}$$

for some $\lambda > 0$ and all ε small enough which together with (1.8.11) considered for m and $\mathcal{W}$ in place of m_D and D yield (1.2.31).

In order to complete the proof of Theorem 1.2.7 it remains to derive (1.2.32) and (1.2.33). Both statements hold true for $n = 1$ in view of (1.2.29) and (1.2.31) but, in fact, we will use them as the induction base with m_D and D in place of m and $\mathcal{W}$ where $D \in \mathcal{D}_\varepsilon^u(z,\alpha,\rho,C)$ and $D \subset (Q\cap V_i)\times\mathcal{W}$ which holds true in view of (1.8.4)–(1.8.11) together with the corresponding form of Proposition 1.7.1. For such D and $n\in\mathbb{N}$ set

$$H_D(n,\alpha) = \{v\in D:\ \Sigma_i^\varepsilon(k,-\alpha) \le \tau_v(i,k) \le \Sigma_i^\varepsilon(k,\alpha)\ \forall k\le n\}$$

and

$$G_D(n) = \{v\in D:\ Z_v^\varepsilon(\tau_v(i,k))\in V_{\iota_k(i)}\ \forall k\le n\}.$$

As the induction hypotesis we assume that for any $\alpha > 0$ there exist $\lambda(\alpha) > 0$ and $\lambda > 0$ such that for all small ε,

(1.8.12)

$$m_D(H_D(n,\alpha)) \ge m_D(D) - ne^{-\lambda(\alpha)/\varepsilon} \text{ and } m_D(G_D(n)) \ge m_D(D) - ne^{-\lambda/\varepsilon}.$$

Set $a = \delta/4K$ where K is the same as in (1.2.15) so that if

$$\Gamma_D(l) = \Gamma_D(l,n,\alpha) = \{v\in D:\ (l-1)a \le \tau_v^\varepsilon(i,n) < la\}\cap G_D(n)\cap H_D(n,\alpha)$$

then

$$3\delta/4 \le \operatorname{dist}\big(Z_v^\varepsilon(t), \mathcal{O}_{\iota_n(i)}\big) \le 5\delta/4 \text{ for all } v\in\Gamma_D(l) \text{ and } t\in[(l-1)a, la]. \tag{1.8.13}$$

Choose also $N = N_a$ so that for any $t \geq (N-1)a$,

$$\Pi^t U_{2\delta}(\mathcal{O}_j) \subset U_{\delta/4}(\mathcal{O}_j) \text{ for each } j = 1, ..., \ell. \tag{1.8.14}$$

Let E_l be a maximal $\big(la/\varepsilon, C\rho, \varepsilon, \overline{\Gamma_D(l)}, D_0\big)$-separated set where $D \subset D_0 \in \mathcal{D}^u_\varepsilon(z, \alpha, \rho, C^3)$ as before. Set

$$\Gamma^U_D(l) = \cup_{v \in E_l} U^\varepsilon_{D_0}(la/\varepsilon, v, C\rho),$$

then for ε small enough,

$$D_0 \supset \Gamma^U_D(l) \supset \Gamma_D(l).$$

We claim that there exists $\beta > 0$ such that if $C\rho \leq \delta/4$ then for all small ε,

$$m_D\big(\Gamma^U_D(l) \cap \bigcup_{j=l+N}^{\infty} \Gamma^U_D(j)\big) \leq e^{-\beta/\varepsilon} m_D(\Gamma^U_D(l)). \tag{1.8.15}$$

Indeed, let $\mathcal{T} = \big\{\gamma \in C_{0,Na} : \gamma_0 \in U_{2\delta}(\mathcal{O}_{\iota_n(i)}),\ \gamma_{t_1} \in U_{3\delta/4}(\mathcal{O}_{\iota_n(i)})$ for some $t_1 \in [0, Na]$ and $\gamma_t \notin U_{\delta/2}(\mathcal{O}_{\iota_n(i)})$ for all $t \in [0, Na]\big\}$. Then by (1.8.14) and the lower semicontinuiti of the functional $S_{0,Na}$ we obtain that

$$\inf\big\{S_{0,Na}(\gamma) : \gamma \in \mathcal{T}\big\} = \eta > 0. \tag{1.8.16}$$

Since by Lemma 1.3.2(ii) and (iii) for any $w \in \tilde{D}(t, v) = \Phi^t_\varepsilon U^\varepsilon_{D_0}(t, v, C\rho)$ the distance $|\pi_1 w - \pi_1 \Phi^t_\varepsilon v|$ has the order of ε we conclude from (1.8.13) and (1.8.16) that for each $v \in \overline{\Gamma_D(l)}$,

$$\mathbf{r}_{0,Na}\big(Z^\varepsilon_{\tilde{v}}, \Psi^\eta_{0,Na}(\tilde{z})\big) \geq \delta/8 \text{ for any } \tilde{v} \in \tilde{D}(la/\varepsilon, v) \cap \Phi^{la/\varepsilon}_\varepsilon \cup_{j=l+N}^{\infty} \Gamma^U_D(j).$$

Hence, by (1.7.11) and Lemma 1.3.6 it follows that for any $v \in \overline{\Gamma_D(l)}$ and all ε small enough,

$$m_{D_0}\big(U^\varepsilon_{D_0}(la/\varepsilon, v, C\rho) \cap \bigcup_{j=l+N}^{\infty} \Gamma^U_D(j)\big) \leq e^{-\eta/2\varepsilon} m_{D_0}\big(U^\varepsilon_{D_0}(la/\varepsilon, v, C\rho)\big)$$

and since $U^\varepsilon_{D_0}(la/\varepsilon, v, C\rho/2)$ are disjoint for different $v \in E_l$ we apply Lemma 1.3.6 once more and obtain (1.8.15).

Set $Q_D(n) = H_D(n, \alpha) \cap G_D(n) \setminus H_D(n+1, \alpha) \cap G_D(n+1)$. Applying (1.8.12) with $n = 1$ to each $\tilde{D} = \tilde{D}(la/\varepsilon, v)$, $v \in E_l$ and using Lemma 1.3.6 we derive also that

$$m_D\big(\Gamma^U_D(l) \cap Q_D(n)\big) \leq e^{-\beta/\varepsilon} m_{D_0}(\Gamma^U_D(l)) \tag{1.8.17}$$

for some $\beta > 0$ and all small ε. By (1.8.17) we can write

$$\begin{aligned} m_D(Q_D(n)) &= \textstyle\sum_{\Sigma^\varepsilon_i(n,-\alpha) \leq l \leq \Sigma^\varepsilon_i(n,\alpha)} m_D\big(\Gamma_D(l) \cap Q_D(n)\big) \leq \\ &\textstyle\sum_{\Sigma^\varepsilon_i(n,-\alpha) \leq l \leq \Sigma^\varepsilon_i(n,\alpha)} m_D\big(\Gamma^U_D(l) \cap Q_D(n)\big) \leq e^{-\beta/\varepsilon} \sum_{1 \leq l \leq \Sigma^\varepsilon_i(n,\alpha)} m_{D_0}\big(\Gamma^U_D(l)\big). \end{aligned} \tag{1.8.18}$$

Observe that for any finite measure μ, measurable sets $A_1, A_2, ...$ and integers $k, N > 0$,

$$\sum_{l=1}^{kN} \mu(A_i) = \sum_{j=1}^{N} \big(\mu(\cup_{i=0}^{k-1} A_{j+ik}) + \sum_{i=0}^{k-2} \mu(A_{j+iN} \cap \bigcup_{r=i+1}^{k-1} A_{j+rN})\big) \tag{1.8.19}$$

which follows applying $\mu(A\cup\tilde A)=\mu(A)+\mu(\tilde A)-\mu(A\cap\tilde A)$ to $A=A_{j+iN}$ and $\tilde A=\cup_{r=i+1}^{k-1}A_{j+rN}$ for $i=0,1,...,k-2$. Applying (1.8.19) for $\mu=m_{D_0}$ and $A_l=\Gamma_D^U(l)$ it follows from (1.8.15) that

$$\sum_{l=1}^{kN} m_{D_0}\big(\Gamma_D^U(l)\big)\le Nm_{D_0}(D_0)+e^{-\beta/\varepsilon}\sum_{l=1}^{kN} m_{D_0}\big(\Gamma_D^U(l)\big),$$

i.e. for any $k\in\mathbb{N}$,

$$\sum_{l=1}^{kN} m_{D_0}\big(\Gamma_D^U(l)\big)\le (1-e^{-\beta/\varepsilon})^{-1}Nm_{D_0}(D_0).$$

This together with (1.8.12) and (1.8.18) complete the induction step and proves (1.2.32) and (1.2.33) for D and m_D in place of $\mathcal{W}$ and m. Finally, as before we complete the proof of Theorem 1.2.7 by choosing discs D to be small balls on the (extended) local unstable manifolds $W_x^u(w,\varrho)$, $w\in\mathcal{W}$ which together with the Fubini theorem enables us to extend the estimates to $\mathcal{W}$ and m as required in (1.2.32) and (1.2.33). □

1.9. Averaging in difference equations

For readers convenience we start this section with the setup and necessary technical results from [**53**] refering there for the corresponding proofs. These results are similar to Section 1.3 and we refer the reader also to [**55**] where more details of proofs can be found than in [**53**] and though [**55**] deals only with the continuous time case the corresponding discrete time proofs can be obtained, essentially, by simplification. We will discuss below mainly the Axiom A case since the corresponding proofs for expanding transformations can be obtained, essentially, by simplification of the same arguments, roughly speaking, by ignoring the stable direction.

As in Section 1.3 we will use the representations $\xi=\xi^{\mathcal{X}}+\xi^{\mathcal{W}}$ of vectors $\xi\in T(\mathbb{R}^d\times\mathbf{M})=\mathbb{R}^d\oplus T\mathbf{M}$, the norms $|||\xi|||$ and the distances $d_{\mathbf{M}}$ and $d(\cdot,\cdot)$ on $\mathbf{M}$ and on $\mathbb{R}^d\times\mathbf{M}$, respectively. It is known (see [**40**]) that the hyperbolic splitting $T_{\Lambda_x}\mathbf{M}=\Gamma_x^s\oplus\Gamma_x^u$ over Λ_x can be continuously extended to the splitting $T_{\mathcal{W}}\mathbf{M}=\Gamma_x^s\oplus\Gamma_x^u$ over $\mathcal{W}$ which is forward invariant with respect to DF_x and satisfies exponential estimates (1.3.1) with a uniform in $x\in\mathcal{X}$ exponent $\kappa>0$. Moreover, by [**71**] (see also [**16**]) we can choose these extensions so that $\Gamma_x^s(w)$ and $\Gamma_x^u(w)$ will be Hölder continuous in w and C^1 in x in the corresponding Grassmann bundle. Actually, since $\mathcal{W}$ is contained in the basin of each attractor Λ_x, any point $w\in\mathcal{W}$ belongs to the stable manifold $W_x^s(v)$ of some point $v\in\Lambda_x$ (see [**13**]), and so we choose naturally $\Gamma_x^s(w)$ to be the tangent space to $W_x^s(v)$ at w. Now each vector $\xi\in T_{x,w}(\mathcal{X}\times\mathcal{W})=T_x\mathcal{X}\oplus T_w\mathcal{W}$ can be represented uniquely in the form $\xi=\xi^{\mathcal{X}}+\xi^s+\xi^u$ with $\xi^{\mathcal{X}}\in T_x\mathcal{X}$, $\xi^s\in\Gamma_x^s(w)$, and $\xi^u\in\Gamma_x^u(w)$. For each small $\varepsilon,\alpha>0$ set $\mathcal{C}^u(\varepsilon,\alpha)=\{\xi\in T(\mathcal{X}\times\mathcal{W}):\ \|\xi^s\|\le\varepsilon\alpha^{-2}\|\xi^u\|\text{ and }\|\xi^{\mathcal{X}}\|\le\varepsilon\alpha^{-1}\|\xi^u\|\}$ and $\mathcal{C}^u_{x,w}(\varepsilon,\alpha)=\mathcal{C}^u(\varepsilon,\alpha)\cap T_{x,w}(\mathcal{X}\times\mathcal{W})$ which are cones around Γ^u and $\Gamma_x^u(w)$, respectively. Similarly, we define $\mathcal{C}^s(\varepsilon,\alpha)=\{\xi\in T(\mathcal{X}\times\mathcal{W}):\ \|\xi^u\|\le\varepsilon\alpha^{-2}\|\xi^s\|\text{ and }\|\xi^{\mathcal{X}}\|\le\varepsilon\alpha^{-1}\|\xi^s\|\}$ and $\mathcal{C}^u_{x,w}(\varepsilon,\alpha)=\mathcal{C}^u(\varepsilon,\alpha)\cap T_{x,w}(\mathcal{X}\times\mathcal{W})$ which are cones around Γ^s and $\Gamma_x^s(w)$, respectively. The corresponding version of Lemma 1.3.1 is proved in [**53**] and the discrete time versions of Lemmas 1.3.2 and 1.3.3 follow in the same way as in [**55**]. Let, again, $\mathcal{D}_\varepsilon^u(z,\alpha,\rho,C)$ be the set of all C^1 embedded n_u-dimensional closed discs $D\subset\mathcal{X}\times\mathcal{W}$ such that $z\in D$, $TD\subset$

$\mathcal{C}^u(\varepsilon,\alpha)$ and if $v\in\partial D$ then $C\rho\le d_D(v,z)\le C^2\rho$. For $D\in\mathcal{D}^u_\varepsilon(z,\alpha,\rho,C)$ and $z=(x,y)\in D\subset\mathcal{X}\times\mathcal{W}$ set $U^\varepsilon_D(n,z,\varrho)=\{\tilde z\in D:\ \max_{0\le k\le n}d_D(\Phi^k_\varepsilon z,\Phi^k_\varepsilon\tilde z)\le\varrho\}$ and let $\pi_1:\mathcal{X}\times\mathcal{W}\to\mathcal{X}$ and $\pi_2:\mathcal{X}\times\mathcal{W}\to\mathcal{W}$ be natural projections on the first and second factors, respectively. The same proof as in [**55**] yields the following discrete time version of Proposition 1.3.4.

1.9.1. PROPOSITION. *For any $\rho,C,b>0$ with C large and $C\rho$ small enough there exists a positive function $\zeta_{b,\rho,T}(\Delta,s,\varepsilon)$ satisfying (1.3.7) such that for any $x,x'\in\mathcal{X}$, $y\in\mathcal{W}$, $n\ge n_1$, $k\le\frac{T}{\varepsilon}-n$, $\beta\in\mathbb{R}^d$, $|\beta|\le b$, $D\in\mathcal{D}^u_\varepsilon((x,y),\alpha,\rho,C)$, $z\in D$ and $V=U^\varepsilon_D(t,z,C\rho)\subset D$ we have*

$$\Big|\tfrac{1}{k}\log\int_V\exp\langle\beta,\textstyle\sum_{j=n}^{n+k-1}B(x',Y^\varepsilon_v(j))\rangle dm_D(v)+\tfrac{1}{k}\log J^u_\varepsilon(n,z)\tag{1.9.1}$$

$$-P_{F_{\pi_1z_n}}(\langle\beta,B(x',\cdot)\rangle+\varphi^u_{\pi_1z_n})\Big|\le\zeta_{b,\rho,T}(\varepsilon k,\min(k,(\log\tfrac{1}{\varepsilon})^\lambda),\varepsilon)$$

where $z_n=\Phi^n_\varepsilon z$ and $\lambda\in(0,1)$.

Next, observe that the results of Section 1.4 above are so general that they work both for the continuous and the discrete time case. Now, we will discuss the discrete time version of Lemma 1.5.1.

1.9.2. LEMMA. *Let $x_i,\tilde x_i\in\mathcal{X}$, $i=0,1,...,N$, $0=t_0<t_1<...<t_{N-1}<t_N=T$, $\Delta=\max_{0\le i\le N-1}(t_{i+1}-t_i)$, $\xi_i=(x_i-x_{i-1})(t_i-t_{i-1})^{-1}$, $n(t)=\max\{j\ge0: t\ge t_j\}$, $\psi(t)=\tilde x_{n(t)}$ and*

$$\Xi^\varepsilon_j(v,x)=([\varepsilon^{-1}t_j]-[\varepsilon^{-1}t_{j-1}])^{-1}\sum_{[\varepsilon^{-1}t_{j-1}]\le k\le[\varepsilon^{-1}t_j]}B(x,Y^\varepsilon_v(k)).\tag{1.9.2}$$

Set for $t=\varepsilon n\in[0,T]$, $n\in\mathbb{N}$,

$$Z^{\varepsilon,\psi}_{v,x}(t)=x+\varepsilon\sum_{0\le k<n}B(\psi(\varepsilon k),Y^\varepsilon_v(k))\tag{1.9.3}$$

and for $t\in[n\varepsilon,(n+1)\varepsilon)$,

$$Z^{\varepsilon,\psi}_{v,x}(t)=(n+1-t/\varepsilon)Z^{\varepsilon,\psi}_{v,x}(\varepsilon n)+(t/\varepsilon-n)Z^{\varepsilon,\psi}_{v,x}(\varepsilon(n+1)).\tag{1.9.4}$$

Then

$$\big|\Xi^\varepsilon_j(v,x_{j-1})-(t_j-t_{j-1})^{-1}(Z^\varepsilon_v(t_j)-Z^\varepsilon_v(t_{j-1}))\big|\le K\big|Z^\varepsilon_v(t_{j-1})-x_{j-1}\big|\tag{1.9.5}$$
$$+\tfrac12K^2(t_j-t_{j-1})+K\varepsilon(4+KT+KT^2+T|\pi_1v|+|x_{j-1}|),$$

$$\sup\nolimits_{0\le s\le t}\big|Z^{\varepsilon,\psi}_{v,x}(s)-\psi(s)\big|\le|x-x_0|+\max\nolimits_{0\le j\le n(t)}|x_j-\tilde x_j|\tag{1.9.6}$$
$$+2\varepsilon(K+\max\nolimits_{1\le i\le N}|\xi_i|)+K\Delta+n(t)\Delta\max\nolimits_{1\le j\le n(t)}\big|\Xi^\varepsilon_j(v,\tilde x_{j-1})-\xi_j\big|$$

and

$$\sup_{0\le s\le t}\big|Z^\varepsilon_v(s)-Z^{\varepsilon,\psi}_{v,x}(s)\big|\le e^{Kt}\big(4K\varepsilon+|\pi_1v-x|+Kt\sup_{0\le s\le t}\big|Z^{\varepsilon,\psi}_{v,x}(s)-\psi(s)\big|\big)\tag{1.9.7}$$

where, recall, $Z^\varepsilon_v(s)=X^\varepsilon_v(s/\varepsilon)$ and $\pi_1v=z\in\mathcal{X}$ if $v=(z,y)\in\mathcal{X}\times\mathbf{M}$.

PROOF. The proof of (1.9.5) and (1.9.6) is strightforward using the definitions (1.9.2)–(1.9.4) in the same way as the proof of (1.5.2) and (1.5.3) only the integrals in the latter case should be replaced by the corresponding sums in the former one. The estimate (1.9.7) follows in the same way as (1.5.4) only the use of the standard

Gronwall inequality in the latter proof should be replaced by the discrete time version of the Gronwall inequality as in Lemma 4.20 of [**26**]. □

Now the proof of the discrete time version of Proposition 1.5.2 and of the remaining part of the proof of large deviations bounds (1.2.16) and (1.2.17) for the discrete time case proceeds almost verbatim as the corresponding continuous time proofs in Section 1.5. Observe that in the discrete time case the functionals S_{0T} are given again by (1.2.13) with $I_x(\nu)$ defined by (1.2.8) where $F_x^1 = F_x$ and $\varphi^u(x)$ is given by (1.2.34). The property of I-functionals described in Lemma 1.6.1 follows directly in the discrete time case via conjugation since we do not have to deal with the time change here. Other auxiliary results of Section 1.6 are derived in the discrete time case exactly in the same way as there. The proof of the discrete time versions of Theorems 1.2.5 and 1.2.7 under the corresponding assumptions goes through exactly in the same way as its continuous time counterpart in Section 1.6 yielding the assertion of Theorem 1.2.10. □

Next, we exhibit computations demonstrating a discrete time version of Theorem 1.2.7 for simple examples. The maps F_x in both examples have the form $F_x y = 3y + x \,(\mathrm{mod}\, 1)$ where $x \in \mathbb{R}^1$ and $y \in [0, 1]$ but by identifying the end points of the unit interval we view F_x as expanding maps of the circle $\mathbb{T}^1$. The function B from (1.1.10) is given in the first example by

$$B(x, y) = x(x^2 - 4)(1 - x^2) + 50 \sin 2\pi y.$$

Hence, we are dealing here with the maps $\Phi_\varepsilon : \mathbb{R}^1 \times \mathbb{T}^1 \to \mathbb{R}^1 \times \mathbb{T}^1$ defined by

$$\Phi_\varepsilon(x, y) = \big(x + \varepsilon(x(x^2 - 4)(1 - x^2) + 50 \sin 2\pi y),\, 3y + x \,(\mathrm{mod}\, 1)\big).$$

All maps F_x preserve the normalized Lebesgue measure Leb on $\mathbb{T}^1$ and it is the SRB measure μ_x^{SRB} for each F_x in this simple case. The averaged equation (1.1.11) for $\bar{Z}(t) = \bar{X}^\varepsilon(t/\varepsilon)$ has here the form

$$\frac{d\bar{Z}(t)}{dt} = \bar{B}(\bar{Z}(t)),$$

where $\bar{B}(x) = x(x^2 - 4)(1 - x^2)$. The one dimensional vector field $\bar{B}(x)$ has three attracting fixed points $\mathcal{O}_1 = 2, \mathcal{O}_2 = 0, \mathcal{O}_3 = -2$ and two repelling fixed points 1 and -1. In order to apply the discrete time version of Theorem 1.2.7 (i.e. Theorem 1.2.10) to this example we have to verify that B is complete at the fixed points $-2, -1, 0, 1, 2$ of the averaged system. Since at these points F_x coincides with the map $y \to 3y(\mathrm{mod}\ 1)$ we can take the periodic orbits $1/8, 3/8$ and $5/8, 7/8$ of the latter and notice that the average of $\sin 2\pi y$ along the former is $1/\sqrt{2}$ and along the latter $-1/\sqrt{2}$ which yields completness of B at zeros of $\bar{B}$.

According to the corresponding part of Theorem 1.2.10 which is a discrete time version of Theorem 1.2.7 the transitions between $\mathcal{O}_1, \mathcal{O}_2$, and $\mathcal{O}_3$ are determined by R_{ij}, $i, j = 1, 2, 3$ which are obtained via the functionals $S_{0t}(\gamma)$ given by (1.2.13). Even here these functionals are not easy to compute though their main ingredients the functionals $I_x(\nu)$ from (1.2.8) are given now by the simple formula

$$I_x(\nu) = \begin{cases} \ln 3 - h_\nu(F_x) & \text{if } \nu \text{ is } F_x\text{-invariant} \\ \infty & \text{otherwise} \end{cases}$$

and the set of F_x-invariant measures can be reasonably described since all F_x's are conjugate to the simple map $y \to 3y \,(\mathrm{mod}\, 1)$. We plot below the histogram of a single orbit of the slow motion $X_{x,y}^\varepsilon(n)$, $n = 0, 1, 2, ..., 10^9$ with $\varepsilon = 10^{-3}$ and

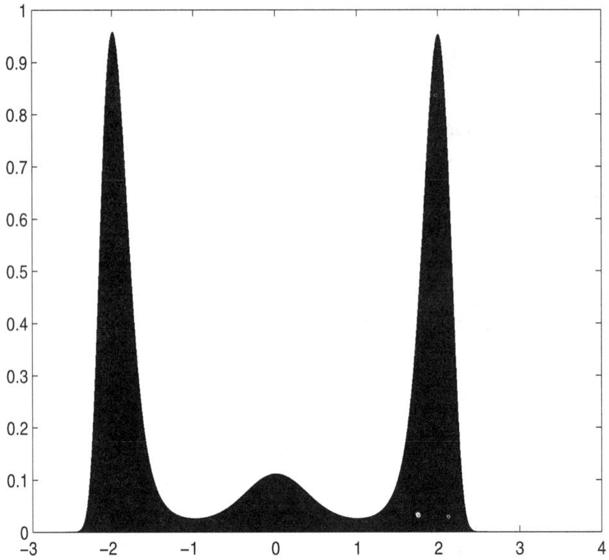

FIGURE 1.9.1. Symmetrical basins case

the initial values $x = 0$, $y = 0.001$. The histogram shows that most of the points of the orbit stay near the attractors $\mathcal{O}_1$, $\mathcal{O}_2$ and $\mathcal{O}_3$ and $X^\varepsilon_{x,y}(n)$ hops between basins of attraction of these points. The form of the histogram indicates (according to Theorem 1.2.7) the equality $R_{21} = R_{23}$ and in this case Theorem 1.2.7 (or its discrete time version) cannot specify whether the slow motion exits from the basin of $\mathcal{O}_2$ to the basin of $\mathcal{O}_1$ or to the basin of $\mathcal{O}_3$. Observe that Theorem 1.2.7 is an asymptotical as $\varepsilon \to 0$ result and it takes an exponential in $1/\varepsilon$ time for a typical orbit to exit from the basin of one attractor and to hop to the basin of another one. Hence, the computations should be done for small ε and exponentially long in $1/\varepsilon$ orbits which is time consuming, so we put a big coefficient in front of sin which makes this exponent smaller. Of course, it is hard to be absolutely sure that ε in our computations is small enough and the number of iterates is large enough to demonstrate faithfully the real situation in this case but we found that our histograms are rather robust, for instance, their shapes have the same form for $\varepsilon = 10^{-3}$ when the number of iterates ranges from 10^8 to, at least, 10^{11} and various initial conditions were checked, as well.

Our second example differs from the first one only in B which is given now by

$$B(x, y) = x(x^2 - 4)(1 - x)(1.5 + x) + 50 \sin 2\pi y.$$

Here the averaged system has the same attracting fixed points $\mathcal{O}_1 = 2, \mathcal{O}_2 = 0, \mathcal{O}_3 = -2$ but one of two repelling fixed points moves from -1 to $-3/2$. This makes the basin of attraction of -2 smaller while the left interval of the basin of attraction of 0 becomes larger. The latter leads to the inequality $R_{23} > R_{21}$ which according to the discrete time version of Theorem 1.2.7 makes it more difficult for the slow motion to exit to the left from the basin of $\mathcal{O}_2$ than to the right. As in the first example in order

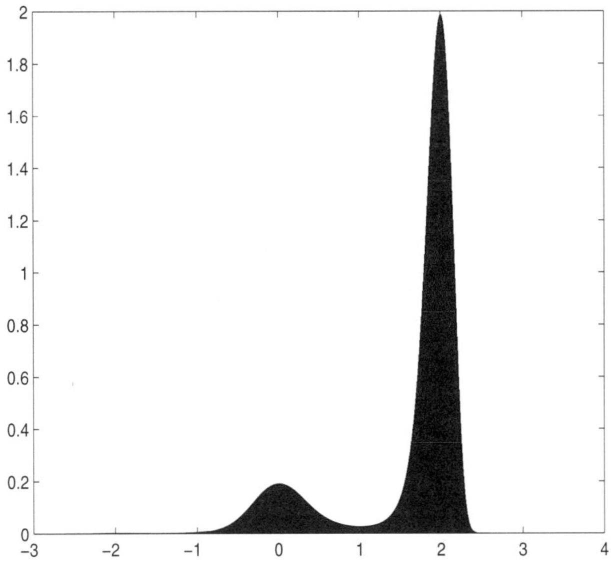

FIGURE 1.9.2. Asymmetrical basins case

to apply the latter result we have to check that B is complete at all zeros of $\bar{B}$ but since we did this already for all integer points it remains to verify completness only for $x = -3/2$ which follows since $\sin 2\pi y$ equals 1 and -1 at two fixed points $1/4$ and $3/4$ of $F_{-3/2}$, respectively. In the histogram here we plot $X^{\varepsilon}_{x,y}(n)$, $n = 0, 1, 2, ..., 10^9$ with $\varepsilon = 10^{-3}$ and the initial values $x = -2$, $y = 0.001$. In compliance with the discrete time version of Theorem 1.2.7 the histogram demonstrates that the slow motion leaves the basin of $\mathcal{O}_3$ and after arriving at the basin of $\mathcal{O}_2$ it exits mostly to the basin of $\mathcal{O}_1$, and so the slow motion hops mostly between basins of $\mathcal{O}_1$ and $\mathcal{O}_2$ staying most of the time in small neighborhoods of these points.

1.10. Extensions: stochastic resonance

The scheme for the stochastic resonance type phenomenon described below is a slight modification of the model suggested by M.Freidlin (cf. [**30**]) and it can be demonstrated in the setup of three scale systems

$$
\begin{aligned}
\frac{dW^{\varepsilon,\delta}(t)}{dt} &= \delta\varepsilon A(W^{\varepsilon,\delta}(t), X^{\varepsilon,\delta}(t), Y^{\varepsilon,\delta}(t)) \\
\frac{dX^{\varepsilon,\delta}(t)}{dt} &= \varepsilon B(W^{\varepsilon,\delta}(t), X^{\varepsilon,\delta}(t), Y^{\varepsilon,\delta}(t)) \qquad (1.10.1)\\
\frac{dY^{\varepsilon,\delta}(t)}{dt} &= b(W^{\varepsilon,\delta}(t), X^{\varepsilon,\delta}(t), Y^{\varepsilon,\delta}(t)),
\end{aligned}
$$

$W^{\varepsilon,\delta} = W^{\varepsilon,\delta}_{w,x,y}$, $X^{\varepsilon,\delta} = X^{\varepsilon,\delta}_{w,x,y}$, $Y^{\varepsilon,\delta} = Y^{\varepsilon,\delta}_{w,x,y}$ with initial conditions $W^{\varepsilon,\delta}(0) = w$, $X^{\varepsilon,\delta}(0) = x$ and $Y^{\varepsilon,\delta}(0) = y$. We assume that $W^{\varepsilon,\delta} \in \mathbb{R}^l$, $X^{\varepsilon,\delta} \in \mathbb{R}^d$ while $Y^{\varepsilon,\delta}$ evolves on a compact $n_{\mathbf{M}}$-dimensional C^2 Riemannian manifold $\mathbf{M}$ and the coefficients A, B, b are bounded smooth vector fields on $\mathbb{R}^l$, $\mathbb{R}^d$ and $\mathbf{M}$, respectively, depending on other variables as parameters. The solution of

(1.10.1) determines the flow of diffeomorphisms $\Phi^t_{\varepsilon,\delta}$ on $\mathbb{R}^l\times\mathbb{R}^d\times\mathbf{M}$ acting by $\Phi^t_{\varepsilon,\delta}(w,x,y)=(W^{\varepsilon,\delta}_{w,x,y}(t),X^{\varepsilon,\delta}_{w,x,y}(t),Y^{\varepsilon,\delta}_{w,x,y}(t))$. Taking $\varepsilon=\delta=0$ we arrive at the (unperturbed) flow $\Phi^t=\Phi^t_{0,0}$ acting by $\Phi^t(w,x,y)=(w,x,F^t_{w,x}y)$ where $F^t_{w,x}$ is another family of flows given by $F^t_{w,x}y=Y_{w,x,y}(t)$ with $Y=Y_{w,x,y}=Y^{0,0}_{w,x,y}$ which are solutions of

$$\frac{dY(t)}{dt}=b(w,x,Y(t)),\ Y(0)=y. \tag{1.10.2}$$

It is natural to view the flow Φ^t as describing an idealized physical system where parameters $w=(w_1,...,w_l)$, $x=(x_1,...,x_d)$ are assumed to be constants of motion while the perturbed flow $\Phi^t_{\varepsilon,\delta}$ is regarded as describing a real system where evolution of these parameters is also taken into consideration but unlike the averaging setup (1.1.1) we have now two sets of parameters moving with very different speeds.

Set $\tilde W^{\varepsilon,\delta}(t)=W^{\varepsilon,\delta}(\frac{t}{\delta\varepsilon})$, $\tilde X^{\varepsilon,\delta}(t)=X^{\varepsilon,\delta}(\frac{t}{\delta\varepsilon})$, $\tilde Y^{\varepsilon,\delta}(t)=Y^{\varepsilon,\delta}(\frac{t}{\delta\varepsilon})$, and pass from (1.10.1) to the equations in the new time

$$\begin{aligned}\frac{d\tilde W^{\varepsilon,\delta}(t)}{dt}&=A(\tilde W^{\varepsilon,\delta}(t),\tilde X^{\varepsilon,\delta}(t),\tilde Y^{\varepsilon,\delta}(t))\\ \frac{d\tilde X^{\varepsilon,\delta}(t)}{dt}&=\delta^{-1}B(\tilde W^{\varepsilon,\delta}(t),\tilde X^{\varepsilon,\delta}(t),\tilde Y^{\varepsilon,\delta}(t))\\ \frac{d\tilde Y^{\varepsilon,\delta}(t)}{dt}&=(\delta\varepsilon)^{-1}b(\tilde W^{\varepsilon,\delta}(t),\tilde X^{\varepsilon,\delta}(t),\tilde Y^{\varepsilon,\delta}(t)).\end{aligned} \tag{1.10.3}$$

Assume that the equation (1.10.1) satisfy the assumptions similar to Assumptions 1.2.1, 1.2.2, 1.2.6 together with other corresponding conditions appearing in the setup of Theorem 1.2.7 (with $\mathbb{R}^l\times\mathbb{R}^d$ in place of $\mathbb{R}^d$), in particular, that $F^t_{w,x}y=Y^{0,0}_{w,x,y}(t)$, $w\in\mathbb{R}^l,x\in\mathbb{R}^d$ form a compact set of flows in the C^2 topology with C^2 dependence on w,x and for all w,x they are Axiom A flows in a neighborhood $\mathcal{W}$ which contains a basic hyperbolic attractor $\Lambda_{w,x}$ for $F^t_{w,x}$ and $\mathcal{W}$ itself is contained in the basin of each $\Lambda_{w,x}$. Set

$$\bar B_w(x)=\bar B(w,x)=\int B(w,x,y)d\mu^{\text{SRB}}_{w,x}(y) \tag{1.10.4}$$

where $\mu^{\text{SRB}}_{w,x}$ is the SRB measure for $F^t_{w,x}$ and let $\bar X^{(w)}$ be the solution of the averaged equation

$$\frac{d\bar X^{(w)}(t)}{dt}=\bar B_w(\bar X^{(w)}(t)). \tag{1.10.5}$$

First, we apply averaging and large deviations estimates in averaging from the previous section to two last equations in (1.10.3) freezing the slowest variable w (i.e. taking for a moment $\delta=0$). Namely, set $\hat X^{\varepsilon}(t)=X^{\varepsilon,0}_{w,x,y}(t/\varepsilon)$ and $\hat Y^{\varepsilon}(t)=Y^{\varepsilon,0}_{w,x,y}(t/\varepsilon)$ so that

$$\begin{aligned}\frac{d\hat X^{\varepsilon}(t)}{dt}&=B(w,\hat X^{\varepsilon}(t),\hat Y^{\varepsilon}(t))\\ \frac{d\hat Y^{\varepsilon}(t)}{dt}&=\varepsilon^{-1}b(w,\hat X^{\varepsilon}(t),\hat Y^{\varepsilon}(t)).\end{aligned} \tag{1.10.6}$$

Suppose for simplicity that $l=d=1$ (i.e. both $W^{\varepsilon,\delta}$ and $X^{\varepsilon,\delta}$ are one dimensional) and that the solution $\bar X^{(w)}(t)$ of (1.10.5) has the limit set consisting of two attracting points $\mathcal{O}_1$ and $\mathcal{O}_2$, which for simplicity we assume to be independent of w, and a repelling fixed point $\mathcal{O}^w_0$ depending on w and separating their basins. As an example of $\bar B$ we may have in mind $\bar B_w(x)=(x-w)(1-x^2),-1<w<1$. Let $S^w_{0T}(\gamma)$ be the large deviations rate functional for the system (1.10.6) defined in (1.2.13) and set for $i, j = 1, 2$,

$$R_{ij}(w) = \inf\{S^w_{0T}(\gamma) : \gamma \in C_{0T},\ \gamma_0 = \mathcal{O}_i,\ \gamma_T = \mathcal{O}_j,\ T \geq 0\} \tag{1.10.7}$$

(cf. with R_{ij} in Theorem 1.2.7). Set

$$\bar{A}_i(w) = \int A(w, \mathcal{O}_i, y) d\mu^{\text{SRB}}_{w,\mathcal{O}_i}(y) \tag{1.10.8}$$

and assume that for all w,

$$\bar{A}_1(w) < 0 \quad \text{and} \quad \bar{A}_2(w) > 0 \tag{1.10.9}$$

which means in view of the averaging principle (see Theorem 1.2.3 and the following it discussion) that $W^{\varepsilon,\delta}_{w,x,y}(t)$ decreases (increases) while $X^{\varepsilon,\delta}_{w,x,y}(t)$ stays close to $\mathcal{O}_1$ (to $\mathcal{O}_2$) for "most" y's with respect to the Riemannian volume on $\mathbf{M}$ restricted to $\mathcal{W}$.

The following statement suggests a "nearly" periodic behavior of the slowest motion.

1.10.1. CONJECTURE. *Suppose that there exist strictly increasing and decreasing functions $w_-(r)$ and $w_+(r)$, respectively, so that*

$$R_{12}(w_-(r)) = R_{21}(w_+(r)) = r$$

and $w_-(\lambda) = w_+(\lambda) = w^*$ for some $\lambda > 0$ while $w_-(r) < w^* < w_+(r)$ for $r < \lambda$. Assume that $\delta \to 0$ and $\varepsilon \to 0$ in such a way that

$$\lim_{\varepsilon,\delta\to 0} \varepsilon \ln(\delta\varepsilon) = -\rho > -\lambda. \tag{1.10.10}$$

Then for any w, x there exists $t_0 > 0$ so that the slowest motion $\tilde{W}^{\varepsilon,\delta}_{w,x,y}(t+t_0)$, $t \geq 0$ converges weakly (as $\varepsilon, \delta \to 0$ so that (1.10.10) holds true) as a random process on the probability space $(\mathcal{W}, m_{\mathcal{W}})$ (where $m_{\mathcal{W}}$ is the normalized Riemannian volume on $\mathcal{W}$) to a periodic function $\psi(t)$, $\psi(t+T) = \psi(t)$ with

$$T = T(\rho) = \int_{w_-(\rho)}^{w_+(\rho)} \frac{dw}{|\bar{A}_1(w)|} + \int_{w_-(\rho)}^{w_+(\rho)} \frac{dw}{|\bar{A}_2(w)|}.$$

The argument supporting this conjecture goes as follows. Set $\check{W}^{\varepsilon,\delta}(t) = W^{\varepsilon,\delta}(t/\varepsilon)$, $\check{X}^{\varepsilon,\delta}(t) = X^{\varepsilon,\delta}(t/\varepsilon)$ and $\check{Y}^{\varepsilon,\delta}(t) = Y^{\varepsilon,\delta}(t/\varepsilon)$ which satisfy

$$\begin{aligned} \frac{d\check{W}^{\varepsilon,\delta}(t)}{dt} &= \delta A(\check{W}^{\varepsilon,\delta}(t), \check{X}^{\varepsilon,\delta}(t), \check{Y}^{\varepsilon,\delta}(t)) \\ \frac{d\check{X}^{\varepsilon,\delta}(t)}{dt} &= B(\check{W}^{\varepsilon,\delta}(t), \check{X}^{\varepsilon,\delta}(t), \check{Y}^{\varepsilon,\delta}(t)) \\ \frac{d\check{Y}^{\varepsilon,\delta}(t)}{dt} &= \varepsilon^{-1} b(\check{W}^{\varepsilon,\delta}(t), \check{X}^{\varepsilon,\delta}(t), \check{Y}^{\varepsilon,\delta}(t)). \end{aligned} \tag{1.10.11}$$

Since $\check{W}^{\varepsilon,\delta}$ moves much slower than $\check{X}^{\varepsilon,\delta}$ we can freeze the former and in place of (1.10.11) we can study (1.10.6). Applying the arguments of Theorem 1.2.7 to the pair $\hat{X}, \hat{Y}$ from (1.10.6) we conclude by (1.2.30) that the intermediate motion $\tilde{X}^{\varepsilon,\delta}$ most of the time stays very close to either $\mathcal{O}_1$ or $\mathcal{O}_2$ before it exits from the corresponding basin, and so in view of an appropriate averaging principle (which follows, for instance, from Theorem 1.2.3) on bounded time intervals the slowest motion $\tilde{W}^{\varepsilon,\delta}$ mostly stays close to the corresponding averaged motion determined by the vector fields $\bar{A}_1$ and $\bar{A}_2$ given by (1.10.4). When $\tilde{X}^{\varepsilon,\delta}$ is close to $\mathcal{O}_1$ the slowest motion $\tilde{W}^{\varepsilon,\delta}$ decreases until $w = w_-(\rho)$ where $R_{12}(w) = \rho$. In view of

(1.2.29) and the scaling (1.10.10) between ε and δ, a moment later $R_{12}(w)$ becomes less than ρ and $\tilde{X}^{\varepsilon,\rho}$ jumps immediately close to $\mathcal{O}_2$. There $\bar{A}_2(w) > 0$, and so $\tilde{W}^{\varepsilon,\delta}$ starts to grow until it reaches $w = w_+(\rho)$ where $R_{21}(w) = \rho$. A moment later $R_{21}(w)$ becomes smaller than ρ and in view of (1.2.29) the intermediate motion $\tilde{X}^{\varepsilon,\delta}$ jumps immediately close to $\mathcal{O}_1$. This leads to a nearly periodic behavior of $\tilde{W}^{\varepsilon,\delta}$. In order to make these arguments precise we have to deal here with an additional difficulty in comparison with the two scale setup considered in previous sections since now the large deviations S-functionals from Theorem 1.2.3 and the R-functions describing adiabatic fluctuations and transitions of Theorems 1.2.5 and 1.2.7 depend on another very slowly changing parameter. Still, the technique of Sections 1.7 and 1.8 above applied on time intervals where changes in the w-variable can be neglected should work here but the details of this approach have not been worked out yet.

On the other hand, when the fast motion $Y^{\varepsilon,\delta}$ does not depend on the slow motions, i.e. when the coefficient b in (1.10.1) depend only on the coordinate y (but not on w and x), then the above arguments can be made precise without much effort. Indeed, we can obtain estimates for transition times $\tau^\varepsilon(1)$ and $\tau^\varepsilon(2)$ of $X^{\varepsilon,\delta}(t/\varepsilon)$ between neighborhoods of $\mathcal{O}_1$ and $\mathcal{O}_2$ as in Theorem 1.2.7 applying the latter to $\hat{X}^\varepsilon$ and $\hat{Y}^\varepsilon$ from (1.10.6) with freezed w-variable. This is possible since the method of Proposition 1.7.1 requires us to make large deviations estimates, essentially, only for probabilities $m\{v \in D : kT \leq \tau_v(i) < (k+1)T\}$, i.e. on bounded time intervals, and then combine them with the Markov property type arguments. During such times the slowest motion $W^{\varepsilon,\delta}$ can move only a distance of order δT. Thus freezing w and using the Gronwall inequality for the equation of X^ε in order to estimate the resulting error we see that the latter is small enough for our purposes. Observe, that it would be much more difficult to justify freezing w in the coefficient b of Y^ε, if we allow the latter to depend on w, since a strightforward application of the Gronwall inequality there would yield an error estimate of an exponential in $1/\varepsilon$ order which is comparable with $1/\delta$. Still, it may be possible to take care about the general case using methods of Sections 1.4 and 1.5 since we produce large deviations estimates there by gluing large deviations estimates on smaller time intervals where the x-variable (and so, of course, w-variable) can be freezed. Next, set

$$W^{\varepsilon,\delta,i}_{w,y}(t) = w + \delta\varepsilon \int_0^t A(W^{\delta,\varepsilon,i}_w(s), \mathcal{O}_i, Y(s))ds$$

where now Y does not depend on ε and δ. Then by (1.10.1) together with the Gronwall inequality we obtain that

$$|W^{\varepsilon,\delta}_{w,x,y}(t) - W^{\varepsilon,\delta,i}_{w,y}(t)| \leq L\delta\varepsilon e^{\delta\varepsilon Lt} \int_0^t |X^{\varepsilon,\delta}_{w,x,y}(s) - \mathcal{O}_i| ds$$

where L is the Lipschitz constant of A. If x belongs to the basin $\mathcal{O}_i$ then according to Theorem 1.2.7 $\hat{X}^\varepsilon$, and so also $X^{\varepsilon,\delta}$, stays most of the time near $\mathcal{O}_i$ up to its exit from the basin of the latter which yields according to the above inequality that $W^{\varepsilon,\delta}$ stays close to $W^{\varepsilon,\delta,i}$ during this time. But now we can employ the averaging principle for the pair $W^{\varepsilon,\delta,i}(t), Y(t)$ which sais that $W^{\varepsilon,\delta,i}(t)$ stays close on the time intervals of order $1/\delta\varepsilon$ to the averaged motion $\bar{W}_w^{\varepsilon,\delta,i}(t)$ defined by

$$\bar{W}_w^{\varepsilon,\delta,i}(t) = w + \int_0^t \bar{A}_i(\bar{W}_w^{\varepsilon,\delta,i}(s))ds$$

and in view of (1.10.9), $\bar{W}_w^{\varepsilon,\delta,1}(t)$ decreases while $\bar{W}_w^{\varepsilon,\delta,2}(t)$ increases which leads to the behavior described in Conjecture 1.10.1.

A similar conjecture can be made under the corresponding conditions for the discrete time case determined by a three scale difference system of equations of the form

$$\begin{aligned} &W^{\varepsilon,\delta}(n+1) - W^{\varepsilon,\delta}(n) = \varepsilon\delta A(W^{\varepsilon,\delta}(n), X^{\varepsilon,\delta}(n), Y^{\varepsilon,\delta}(n)),\ \ W^{\varepsilon,\delta}(0) = w, \\ &X^{\varepsilon,\delta}(n+1) - X^{\varepsilon,\delta}(n) = \varepsilon B(W^{\varepsilon,\delta}(n), X^{\varepsilon,\delta}(n), Y^{\varepsilon,\delta}(n)),\ \ X^{\varepsilon,\delta}(0) = x, \\ &Y^{\varepsilon,\delta}(n+1) = F_{W^{\varepsilon,\delta}(n),X^{\varepsilon,\delta}(n)} Y^{\varepsilon,\delta}(n),\ \ Y^{\varepsilon,\delta}(0) = y \end{aligned} \tag{1.10.12}$$

where A and B are smooth vector functions and $F_{w,x} : \mathbf{M} \to \mathbf{M}$ is a smooth map (a diffeomorphism or an endomorphism). We obtain an example where discrete time versions of conditions of Conjecture 1.10.1 hold true setting, for instance, $A(w,x,y) = x\cos 2\pi w + \sin 2\pi y$, $B(w,x,y) = (x-w)(1-x^2) + \sin 2\pi y$ and $F_{w,x}y = 3y + x + w \pmod 1$.

1.11. Young measures approach to averaging

This section deals with the averaging principle and a bit with the corresponding large deviations in the sense of convergence of Young measures and I thank K.Gelfert for asking me about Young measures applications in averaging and for indicating to me the paper [**3**].

Let μ belongs to the space $\mathcal{P}(\mathbb{R}^d \times \mathbf{M})$ of probability measures on $\mathbb{R}^d \times \mathbf{M}$ and consider the Young measure (which is a map from a measure space to a space of measures, see [**3**]) ζ^ε from $([0,T] \times \mathbb{R}^d \times \mathbf{M}, \ell_T \times \mu)$ to $\mathcal{P}(\mathbb{R}^d \times \mathbf{M})$ defined by

$$\zeta^\varepsilon(t,x,y) = \delta_{X_{x,y}^\varepsilon(t/\varepsilon), Y_{x,y}^\varepsilon(t/\varepsilon)}$$

where ℓ_T is the Lebesgue measure on $[0,T]$, δ_w is the unit mass at w, and X^ε, Y^ε are solutions of (1.1.1) on the product $\mathcal{P}(\mathbb{R}^d \times \mathbf{M})$. We assume that for all $x, z \in \mathbb{R}^d$ and $y, v \in \mathbf{M}$ the coefficients b and B satisfy

$$\begin{aligned} &|B(x,y)| + |b(x,y)| \le K \text{ and} \\ &|B(x,y) - B(z,v)| + |b(x,y) - b(z,v)| \le K\big(|x-z| + d_{\mathbf{M}}(y,v)\big) \end{aligned} \tag{1.11.1}$$

for some $L > 0$ independent of x, y, z, v. Of course, we could require the Lipschitz continuity and the boundedness conditions (1.11.1) only in some open domain as in Section 1.2 but we can always extend these vector fields to the whole $\mathbb{R}^d$ keeping these properties intact.

Suppose that $\mu \in \mathcal{P}(\mathbb{R}^d \times \mathbf{M})$ has a disintegration

$$d\mu(x,y) = d\mu_x(y)d\lambda(x),\ \ \lambda \in \mathcal{P}(\mathbb{R}^d) \tag{1.11.2}$$

such that for each Lipschitz continuous function g on $\mathbf{M}$ and any $x, z \in \mathbb{R}^d$,

$$|\int g d\mu_x - \int g d\mu_z| \le K_{L(g)}|x - z| \tag{1.11.3}$$

for some $K_L > 0$ depending only on L where $L(g)$ is both a Lipschitz constant of g and it also bounds $|g|$. Set

$$\bar{B}(x) = \int B(x, y) d\mu_x(y) \tag{1.11.4}$$

then by (1.11.1) and (1.11.3), $\bar{B}$ is bounded and Lipschitz continuous, and so there exists a unique solution $\bar{X}^\varepsilon(t) = \bar{X}_x^\varepsilon(t)$ of (1.1.3). For any bounded continuous function g on $\mathbb{R}^d \times \mathbf{M}$ define

$$\mathcal{E}_\varepsilon^g(t, \delta) = \{(x, y) \in \mathbb{R}^d \times \mathbf{M} : \big|\frac{1}{t}\int_0^t g(x, Y_{x,y}^\varepsilon(u))du - \bar{g}(x)\big| > \delta\}$$

where $\bar{g}(x) = \int g(x, y) d\mu_x(y)$.

By the definition (see [**3**]), the Young measures ζ^ε converge as $\varepsilon \to 0$ to the Young measure ζ^0 defined by

$$\zeta^0(t, x, y) = \delta_{\bar{Z}_x(t)} \times \mu_{\bar{Z}_x(t)} \in \mathcal{P}(\mathbb{R}^d \times \mathbf{M}),$$

$\bar{Z}_x(t) = \bar{X}_x^\varepsilon(t/\varepsilon)$, if for any bounded continuous function f on $[0, T] \times \mathbb{R}^d \times \mathbf{M}$,

$$\int d\mu(x, y) \int_0^T f(s, \Phi_\varepsilon^{s/\varepsilon}(x, y))ds \to \int d\lambda(x) \int_0^T \bar{f}(s, \bar{Z}_x(s))ds \text{ as } \varepsilon \to 0.$$

The following result provides a verifiable (in some interesting cases) criterion for even stronger convergence.

1.11.1. THEOREM. *Let $\mu \in \mathcal{P}(\mathbb{R}^d \times \mathbf{M})$ has the disintegration (1.11.2) satisfying (1.11.3). Then*

$$\lim_{\varepsilon \to 0} \int_{\mathbb{R}^d} \int_{\mathbf{M}} \sup_{0 \le t \le T} \big| \int_0^t \big(f(s, \Phi^{s/\varepsilon}(x, y)) - \bar{f}(s, \bar{Z}_x(s))\big)ds \big| d\mu_x(y) d\lambda(x) = 0 \tag{1.11.5}$$

for any bounded continuous function $f = f(t, x, y)$ on $[0, T] \times \mathbb{R}^d \times \mathbf{M}$ where $\bar{f}(t, x) = \int f(t, x, y) d\mu_x(y)$ if and only if for each $N \in \mathbb{N}$ and any finite collection $g_1, ..., g_N$ of bounded Lipschitz continuous functions on $\mathbb{R}^d \times \mathbf{M}$ there exists an integer valued function $n = n(\varepsilon) \to \infty$ as $\varepsilon \to 0$ such that for any $\delta > 0$ and $l = 1, ..., N$,

$$\lim_{\varepsilon \to 0} \max_{0 \le j < n(\varepsilon)} \mu\{\Phi_\varepsilon^{-jt(\varepsilon)} \mathcal{E}_\varepsilon^{g_l}(t(\varepsilon), \delta)\} = 0, \tag{1.11.6}$$

where $t(\varepsilon) = \frac{T}{\varepsilon n(\varepsilon)}$ and, recall, $\Phi_\varepsilon^t(x, y) = (X_{x,y}^\varepsilon(t), Y_{x,y}^\varepsilon(t))$.

PROOF. First, we prove that (1.11.5) implies (1.11.6). Let $g_1, ..., g_N$ be bounded Lipschitz continuous functions on $\mathbb{R}^d \times \mathbf{M}$ and set

$$\rho_{x,y}^{\varepsilon,l}(t) = \varepsilon \int_0^t \big(g_l(X_{x,y}^\varepsilon(s), Y_{x,y}^\varepsilon(s)) - \bar{g}_l(\bar{X}_x^\varepsilon(s))\big)ds. \tag{1.11.7}$$

If

$$\rho_{x,y}^{\varepsilon,l} = \sup_{0 \le t \le T/\varepsilon} |\rho_{x,y}^{\varepsilon,l}(t)|$$

then by (1.11.5) for each $l = 1, ..., N$,

$$\rho_l^\varepsilon = \int_{\mathbb{R}^d} \int_{\mathbf{M}} \rho_{x,y}^{\varepsilon,l} d\mu(x, y) \to 0 \text{ as } \varepsilon \to 0. \tag{1.11.8}$$

Choose an integer valued function $n(\varepsilon) \to \infty$ as $\varepsilon \to 0$ so that

$$n(\varepsilon) \max_{1 \le l \le N} \rho_l^\varepsilon \to 0 \text{ as } \varepsilon \to 0 \tag{1.11.9}$$

and let $t(\varepsilon) = T/\varepsilon n(\varepsilon)$. Set $x_k^\varepsilon = X_{x,y}^\varepsilon(kt(\varepsilon))$, $y_k^\varepsilon = Y_{x,y}^\varepsilon(kt(\varepsilon))$ and $\bar{x}_k^\varepsilon = \bar{X}_x^\varepsilon(kt(\varepsilon))$, $k = 0, 1, \ldots$. Then by (1.11.7),

(1.11.10)

$$\rho_{x,y}^{\varepsilon,l}((j+1)t(\varepsilon)) - \rho_{x,y}^{\varepsilon,l}(jt(\varepsilon)) = \varepsilon \int_0^{t(\varepsilon)} \big(g_l(X_{x_j^\varepsilon,y_j^\varepsilon}^\varepsilon(u), Y_{x_j^\varepsilon,y_j^\varepsilon}^\varepsilon(u)) - \bar{g}_l(\bar{X}_{\bar{x}_j^\varepsilon}^\varepsilon(u))\big)du.$$

By (1.11.1),

$$\begin{aligned}(1.11.11)\qquad &\varepsilon\big|\textstyle\int_0^{t(\varepsilon)} \big(g_l(X_{x_j^\varepsilon,y_j^\varepsilon}^\varepsilon(u), Y_{x_j^\varepsilon,y_j^\varepsilon}^\varepsilon(u)) - g_l(x_j^\varepsilon, Y_{x_j^\varepsilon,y_j^\varepsilon}^\varepsilon(u))\big)du\big| \\ &\le \varepsilon L_l \textstyle\int_0^{t(\varepsilon)} |X_{x_j^\varepsilon,y_j^\varepsilon}^\varepsilon(u) - x_j^\varepsilon|du \le L_l L(\varepsilon n(\varepsilon))^2\end{aligned}$$

where L_l is the Lipschitz constant of g_l. Similarly, by (1.11.1) and (1.11.3),

$$(1.11.12)\qquad \varepsilon\Big|\int_0^{t(\varepsilon)} \big(\bar{g}_l(\bar{X}_{\bar{x}_j^\varepsilon}^\varepsilon(u)) - \bar{g}_l(\bar{x}_j^\varepsilon)\big)du\Big| \le (L_l + K_{L_l})K(\varepsilon t(\varepsilon))^2$$

and

$$(1.11.13)\qquad |\bar{g}_l(\bar{x}_j^\varepsilon) - \bar{g}_l(x_j^\varepsilon)| \le (L_l + K_{L_l})|\bar{x}_j^\varepsilon - x_j^\varepsilon| \le (L_l + K_{L_l})\rho_{x,y}^\varepsilon.$$

It follows from (1.11.10)–(1.11.13) that

$$\begin{aligned}(1.11.14)\qquad &\big|\tfrac{1}{t(\varepsilon)} \textstyle\int_0^{t(\varepsilon)} g_l(x_j^\varepsilon, Y_{x_j^\varepsilon,y_j^\varepsilon}^\varepsilon(u))du - \bar{g}_l(x_j^\varepsilon)\big| \\ &\le TK(2L_l + K_{L_l})/n(\varepsilon) + (L_l + K_{L_l} + 2T^{-1}n(\varepsilon))\rho_{x,y}^\varepsilon.\end{aligned}$$

Given $\delta > 0$ choose $\varepsilon_\delta > 0$ such that for all $\varepsilon \le \varepsilon_\delta$ and $l = 1, \ldots, N$,

$$TK(2L_l + K_{L_l})/n(\varepsilon) \le \delta/2.$$

Then by (1.11.14),

$$\Phi_\varepsilon^{-jt(\varepsilon)}\mathcal{E}_\varepsilon^{g_l}(t(\varepsilon), \delta) \subset A_\varepsilon(\delta) = \big\{(x, y) \in \mathbb{R}^d \times \mathbf{M} : (L_l + K_{L_l} + 2T^{-1}n(\varepsilon))\rho_{x,y}^{\varepsilon,l} > \delta/2\big\}.$$

By Chebyshev's inequality

$$(1.11.15)\qquad \mu(A_\varepsilon(\delta)) \le \frac{2}{\delta}(L_l + K_{L_l} + 2T^{-1}n(\varepsilon))\rho_l^\varepsilon.$$

By (1.11.9) the right hand side of (1.11.15) tends to 0 as $\varepsilon \to 0$ yielding (1.11.6).

Next, we derive (1.11.5) from (1.11.6). Since f in (1.11.5) is a bounded function and λ is a probability measure it is easy to see that it suffices to prove (1.11.5) when the integration in x there is restricted to compact subsets of $\mathbb{R}^d$. But if we integrate in (1.11.5) in x running over a compact set $G \subset \mathbb{R}^d$ then by (1.1.1) and (1.11.1),

$$(1.11.16)\qquad \sup_{0\le s\le T} \operatorname{dist}\big(X_{x,y}^\varepsilon(s/\varepsilon), G\big) \le KT,$$

i.e. both $Z_{x,y}^\varepsilon(s) = X_{x,y}^\varepsilon(s/\varepsilon)$ and $\bar{Z}_x(s)$ belong to the $KT-$neighborhood G_{KT} of the set G when $x \in G$ and $s \in [0, T]$. On $[0, T] \times G_{KT} \times \mathbf{M}$ we can approximate f uniformly by Lipschitz continuous functions. Thus, in place of (1.11.5) it suffices to show that for any compact set $G \subset \mathbb{R}^d$ and a bounded Lipschitz continuous function f on $[0, T] \times G_{KT} \times \mathbf{M}$ with a Lipschitz constant $L = L(f)$ in all variables,

(1.11.17)

$$\lim_{\varepsilon\to 0} \varepsilon \int_G \int_{\mathbf{M}} \sup_{0\le t\le T/\varepsilon} \Big|\int_0^t \big(f(\varepsilon s, X_{x,y}^\varepsilon(s), Y_{x,y}^\varepsilon(s)) - \bar{f}(\varepsilon s, \bar{X}_x^\varepsilon(s))\big)ds\Big| d\mu_x(y)d\lambda(x) = 0.$$

By (1.11.2), (1.11.3) and (1.11.4),

$$\begin{aligned}(1.11.18)\qquad &\varepsilon\big|\int_0^t\big(f(\varepsilon s,X^\varepsilon_{x,y}(s),Y^\varepsilon_{x,y}(s))-\bar f(\varepsilon s,\bar X^\varepsilon_x(s))\big)ds\big|\\ &\le\varepsilon\big|\int_0^t\big(f(\varepsilon s,X^\varepsilon_{x,y}(s),Y^\varepsilon_{x,y}(s))-\bar f(\varepsilon s,X^\varepsilon_x(s))\big)ds\big|\\ &+(L+K_L)T\sup_{0\le s\le T/\varepsilon}|X^\varepsilon_x(s)-\bar X^\varepsilon_x(s)|.\end{aligned}$$

Since (1.11.6) holds true also for $g=B$, it follows from Theorem 2.1 of [**55**] that

$$(1.11.19)\qquad \lim_{\varepsilon\to 0}\int_G\int_{\mathbf M}\sup_{0\le s\le T/\varepsilon}|X^\varepsilon_x(s)-\bar X^\varepsilon_x(s)|d\mu(x,y)=0,$$

and so we have only to deal with the first absolute value in the right hand side of (1.11.18). As before set $x^\varepsilon_k=X^\varepsilon_{x,y}(kt(\varepsilon))$, $y^\varepsilon_k=Y^\varepsilon_{x,y}(kt(\varepsilon))$, $\bar x^\varepsilon_k=\bar X^\varepsilon_x(kt(\varepsilon))$ and fix a large $N\in\mathbb N$. Let $l=l(j)=[\varepsilon jt(\varepsilon)N/T]=[jN/n(\varepsilon)]$ then by (1.11.1), (1.11.2) and (1.11.3),

$$\begin{aligned}(1.11.20)\qquad &\varepsilon\big|\int_0^{t(\varepsilon)}\big(f(\varepsilon jt(\varepsilon)+\varepsilon u,X^\varepsilon_{x^\varepsilon_j,y^\varepsilon_j}(u),Y^\varepsilon_{x^\varepsilon_j,y^\varepsilon_j}(u))\\ &-f(lT/N,x^\varepsilon_j,Y^\varepsilon_{x^\varepsilon_j,y^\varepsilon_j}(u))\big)ds\big|\le LT^2/Nn(\varepsilon)\\ &+L\varepsilon\int_0^{t(\varepsilon)}|X^\varepsilon_{x^\varepsilon_j,y^\varepsilon_j}(u)-x^\varepsilon_j|du\le LT^2/Nn(\varepsilon)+LT^2(1+K)(n(\varepsilon))^{-2}\end{aligned}$$

and

$$\begin{aligned}(1.11.21)\qquad &\varepsilon\big|\int_0^{t(\varepsilon)}\big(\bar f(\varepsilon jt(\varepsilon)+\varepsilon u,X^\varepsilon_{x^\varepsilon_j,y^\varepsilon_j}(u))-\bar f(lT/N,x^\varepsilon_j)\big)du\big|\\ &\le LT^2/Nn(\varepsilon)+T^2(L+KL+KK_L)(n(\varepsilon))^{-2}.\end{aligned}$$

Now using (1.11.20), (1.11.21) and assuming that $|f|\le\hat L_f$ for some constant $\hat L_f>0$ we obtain

$$\begin{aligned}(1.11.22)\qquad &\varepsilon\sup_{0\le t\le T/\varepsilon}\big|\int_0^t\big(f(\varepsilon s,X^\varepsilon_{x,y}(s),Y^\varepsilon_{x,y}(s))-\bar f(\varepsilon s,X^\varepsilon_{x,y}(s))\big)ds\big|\\ &\le 2\hat L_f\varepsilon t(\varepsilon)+\varepsilon\sum_{j=0}^{n(\varepsilon)-1}\big|\int_{jt(\varepsilon)}^{(j+1)t(\varepsilon)}\big(f(\varepsilon s,X^\varepsilon_{x,y}(s),Y^\varepsilon_{x,y}(s))-\bar f(\varepsilon s,\bar X^\varepsilon_x(s))\big)ds\big|\\ &\le 2\hat L_f\varepsilon t(\varepsilon)+\varepsilon\sum_{j=0}^{n(\varepsilon)-1}\big|\int_0^{t(\varepsilon)}\big(f(\varepsilon jt(\varepsilon)+\varepsilon s,X^\varepsilon_{x^\varepsilon_j,y^\varepsilon_j}(s),Y^\varepsilon_{x^\varepsilon_j,y^\varepsilon_j}(s))\\ &-\bar f(\varepsilon jt(\varepsilon)+\varepsilon s,X^\varepsilon_{x^\varepsilon_j,y^\varepsilon_j}(s))\big)ds\big|\\ &\le 2LT^2/N+2\big(\hat L_fT+T^2(L+KL+KK_L)\big)/n(\varepsilon)\\ &+\varepsilon t(\varepsilon)\sum_{l=0}^{N-1}\sum_{ln(\varepsilon)/N\le j<(l+1)n(\varepsilon)/N,j\le n(\varepsilon)}\big|\tfrac{1}{t(\varepsilon)}\int_0^{t(\varepsilon)}f(lT/N,x^\varepsilon_j,Y^\varepsilon_{x^\varepsilon_j,y^\varepsilon_j}(s))ds\\ &-\bar f(lT/N,x^\varepsilon_j)\big|\le 2LT^2/N+2\big(\hat L_fT+T^2(L+KL+KK_L)\big)/n(\varepsilon)\\ &+\varepsilon t(\varepsilon)n(\varepsilon)\delta+2\hat L_f\varepsilon t(\varepsilon)\sum_{l=0}^{N-1}\sum_{ln(\varepsilon)/N\le j<(l+1)n(\varepsilon)/N,j\le n(\varepsilon)}\mathbb I_{\mathcal E^{f_l}_\varepsilon(t(\varepsilon),\delta)}(x^\varepsilon_j,y^\varepsilon_j)\end{aligned}$$

where $f_l(z,v)=f(lT/N,z,v)$. Integrating against μ both parts of (1.11.22) over $G\times\mathbf M$ we obtain

$$\begin{aligned}(1.11.23)\qquad &\varepsilon\int_G\int_{\mathbf M}\sup_{0\le t\le T/\varepsilon}\big|\int_0^t\big(f(\varepsilon s,X^\varepsilon_{x,y}(s),Y^\varepsilon_{x,y}(s))\\ &-\bar f(\varepsilon s,X^\varepsilon_{x,y}(s))\big)ds\big|d\mu(x,y)\le 2\big(\hat L_fT+T^2(L+KL+KK_L)\big)/n(\varepsilon)\\ &+2LT^2/N+T\delta+2\hat L_f\max_{0\le l\le N-1}\eta_l(\varepsilon,\delta)\end{aligned}$$

where

$$\eta_l(\varepsilon,\delta)=\max_{0\le j\le n(\varepsilon)-1}\mu\big\{(G\times\mathbf M)\cap\Phi_\varepsilon^{-jt(\varepsilon)}\mathcal E^{f_l}_\varepsilon(t(\varepsilon),\delta)\big\}.$$

By the assumption there exists an integer valued function $n(\varepsilon) \to \infty$ as $\varepsilon \to 0$ such that (1.11.6) holds true for all $g = f_0, f_1, ..., f_{N-1}$ and then $\max_{0 \le l \le N-1} \eta_l(\varepsilon, \delta) \to 0$ as $\varepsilon \to 0$. Hence, letting first $\varepsilon \to 0$, then $\delta \to 0$ and, finally, $N \to 0$ we obtain (1.11.17) in view of (1.11.18) and (1.11.19), completing the proof of Theorem 1.11.1. □

Observe that (1.11.5) holding true for all bounded continuous functions is, in principle, stronger than the averaging principle in the form (1.11.19) since the latter is equivalent to (1.11.5) with $f = B$. In fact, if we require (1.11.6) only for one function $g = B$ then in the same way as in the proof of Theorem 1.11.1 above we conclude that (1.11.6) is equivalent to (1.11.19) if we consider the latter over all compacts $G \subset \mathbb{R}^d$ (which was proved earlier in Theorem 2.1 of [**55**]). Still, the main interesting classes of systems, we are aware of, for which (1.11.5) holds true are the same for which (1.11.19) is satisfied though it is easy to construct examples of (somewhat degenerate) right hand sides b and B in (1.1.1) for which (1.11.19) holds true but (1.11.5) fails (since in the latter we require convergence for all functions f and in the former only for $f = B$).

Set

$$\mathcal{E}_0^g(t, \delta) = \{(x, y) \in \mathbb{R}^d \times \mathbf{M} : \left|\frac{1}{t}\int_0^t g(x, F_x^u y)du - \bar{g}(x)\right| > \delta\}$$

where, recall, $F_x^u y = Y_{x,y}(u)$ and $Y(u)$ satisfies (1.1.2). In the same way as Corollary 3.1 in [**55**] we obtain

1.11.2. Corollary. *Suppose that there exists an integer valued function $n = n(\varepsilon) \to \infty$ as $\varepsilon \to 0$ such that $t(\varepsilon) = T(\varepsilon n(\varepsilon))^{-1} = o(\log(1/\varepsilon))$ and for any $\delta > 0$ and each bounded Lipschitz continuous function $g = g(x, y)$ on $\mathbb{R}^d \times \mathbf{M}$,*

$$\lim_{\varepsilon \to 0} \max_{0 \le j < n(\varepsilon)} \mu\{\Phi_\varepsilon^{-jt(\varepsilon)} \mathcal{E}_0^g(t(\varepsilon), \delta)\} = 0. \tag{1.11.24}$$

Then (1.11.6) is also satisfied, and so (1.11.5) holds true.

In the same way as in [**55**] we obtain that (1.11.24) holds true in the Anosov theorem setup when μ_x is an F_x^t-invariant measure which is ergodic for λ-almost all x, where λ is the normalized Lebesgue measure on a large compact in $\mathbb{R}^d$, and $\mu_x(U) = \int_U q(x, y)dm(y)$ with $q(x, y) > 0$ differentiable in x and y. Furthermore, in the same way as in Theorem 2.4 of [**55**] or similarly to Theorem 2.4 of [**57**] we conclude that (1.11.6) and (1.11.24) hold true under Assumptions 1.2.1 and 1.2.2. Moreover, employing the method of [**57**] the result can be extended to some partially hyperbolic systems.

Observe that under Assumptions 1.2.1 and 1.2.2 we can obtain also large deviations bounds in the form (1.2.16) and (1.2.17) for

$$W_{x,y}^\varepsilon(t) = \int_0^t f(s, X_{x,y}^\varepsilon(s/\varepsilon), Y_{x,y}^\varepsilon(s/\varepsilon))ds$$

with the functional

$$\tilde{S}_{0T}(\tilde{\gamma}) = \inf\{S_{0T}(\gamma) : S_{0T}(\gamma) = \textstyle\int_0^T I_{\gamma_t}(\nu_t)dt,$$
$$\dot{\gamma}_t = \bar{B}_{\nu_t}(\gamma_t), \tilde{\gamma}_t = \textstyle\int_0^t \bar{f}_{\nu_s}(s, \gamma_s)ds \ \forall t \in [0, T]\}, \ \bar{f}_\nu(s, x) = \int f(s, x, y)d\nu(y),$$

where f is a bounded Lipschitz continuous vector function. This assertion, actually, follows directly from Theorem 1.2.3 (together with the contraction principle, see

[**25**]) applying it to the slow motion (Z^ε, τ^ε, W^ε) given by the equations

$$\frac{dZ^\varepsilon(t)}{dt} = B(Z^\varepsilon(t), Y^\varepsilon(t/\varepsilon)), \ \frac{\tau^\varepsilon(t)}{dt} = 1, \ \frac{dW^\varepsilon(t)}{dt} = f(\tau^\varepsilon(t), Z^\varepsilon(t), Y^\varepsilon(t/\varepsilon))$$

combined with the time changed second equation in (1.1.1) for the fast motion Y^ε. Of course, analogous results can be obtained in the discrete time setup of difference equations (1.1.10).

Part 2

Markov Fast Motions

2.1. Introduction

Many real systems can be viewed as a combination of slow and fast motions which leads to complicated double scale equations. Already in the 19th century in applications to celestial mechanics it was well understood (though without rigorous justification) that a good approximation of the slow motion can be obtained by averaging its parameters in fast variables. Later, averaging methods were applied in signal processing and, rather recently, to model climate–weather interactions (see [**36**], [**18**], [**37**] and [**52**]). The classical setup of averaging justified rigorously in [**12**] presumes that the fast motion does not depend on the slow one and most of the work on averaging treats this case only. On the other hand, in real systems both slow and fast motions depend on each other which leads to the more difficult fully coupled case which we study here. This setup emerges, in particular, in perturbations of Hamiltonian systems which leads to fast motions on manifolds of constant energy and slow motions across them.

It is natural to view double scale models as describing physical systems considered as perturbations of an idealized one which depends on parameters $x = (x_1, ..., x_d) \in \mathbb{R}^d$ assumed to be constants (integrals) of motion. In Part 2 we suppose that the evolution of this idealized sistem is described by certain family of Markov processes $Y_x(t) = Y_{x,y}(t) = Y_{x,y}(t, \omega)$, $Y_{x,y}(0) = y$ on a separable metric space $\mathbf{M}$. In the perturbed system parameters start changing slowly in time and we assume that the corresponding slow motion $X^\varepsilon(t) = X^\varepsilon_{x,y}(t) = X^\varepsilon_{x,y}(t, \omega)$ is described by an ordinary differential equations in $\mathbb{R}^d$ having the form

$$\frac{dX^\varepsilon(t)}{dt} = \varepsilon B(X^\varepsilon(t), Y^\varepsilon(t)),\ X^\varepsilon(0) = x,\ Y^\varepsilon(0) = y \tag{2.1.1}$$

where $B : \mathbb{R}^d \times \mathbf{M} \to \mathbb{R}^d$ is Lipschitz continuous and the fast motion $Y^\varepsilon(t) = Y^\varepsilon_{x,y}(t)$ evolves on $\mathbf{M}$, it depends, in general, on the slow one and tends to $Y_{x,y}(t)$ as $\varepsilon \to 0$. Usually, $Y^\varepsilon(t)$ is determined by certain equations, in general, coupled with (2.1.1) which means that their coefficients depend on the slow motion $X^\varepsilon(t)$.

Assume that a nonrandom limit

$$\bar{B}(x) = \lim_{T\to\infty} T^{-1} \int_0^T B(x, Y_{x,y}(t))dt \tag{2.1.2}$$

exists in some sense, it "essentially" does not depend on y and it depends Lipschitz continuously on x. Then there exists a unique solution $\bar{X}^\varepsilon = \bar{X}^\varepsilon_x$ of the averaged equation

$$\frac{d\bar{X}^\varepsilon(t)}{dt} = \varepsilon \bar{B}(\bar{X}^\varepsilon(t)), \quad \bar{X}^\varepsilon(0) = x. \tag{2.1.3}$$

The averaging principle suggests that often

$$\lim_{\varepsilon\to 0} \sup_{0\le t\le T/\varepsilon} |X^\varepsilon_{x,y}(t) - \bar{X}^\varepsilon_x(t)| = 0 \tag{2.1.4}$$

in some sense. If unperturbed motions $Y^\varepsilon_{x,y} = Y_y$ do not depend on the slow variables x and $Y^\varepsilon_{x,y} = Y_y$ then the averaged principle holds true under quite general circumstances but when the fast motion depends on the slow one (coupled case) the situation becomes more complicated and approximation of $X^\varepsilon_{x,y}$ by X^ε_x in the weak or the average sense was justified under some conditions in[**45**] and [**77**]. An extension of the averaging principle in the sense of convergence of Young measures is discussed in Section 2.10 below.

In this work we are interested in large deviations bounds for probabilities that the time changed slow motion $Z^\varepsilon(t) = X^\varepsilon(t/\varepsilon)$ belongs to various sets of curves which leads, in particular, to exponential bounds of the form

$$P\{\sup_{0\le t\le T} |Z^\varepsilon_{x,y}(t) - \bar Z^\varepsilon_x(t)| > \delta\} \le e^{-\kappa/\varepsilon},\ \kappa,\delta > 0 \tag{2.1.5}$$

where $\bar Z_x(t) = \bar X^\varepsilon_x(t/\varepsilon)$ satisfies

$$\frac{d\bar Z^\varepsilon_x(t)}{dt} = \varepsilon \bar B(\bar Z^\varepsilon_x(t)),\quad \bar Z^\varepsilon_x(0) = x. \tag{2.1.6}$$

When the fast motion do not depend on the slow one such results were obtained in [**29**] and [**31**] but the coupled case was dealt with much later in [**78**] though (as we indicated this to the author) the proof there contained a vicious circle and substantial gaps which, essentially, were fixed recently in [**79**]. Still, [**79**] is rather difficult to follow and we find it useful to provide a precise and consistent exposition of this important result which also deals with a more general case including fast motions being random evolutions whose extreme partial cases are diffusions and finite Markov chains with continuous time. Moreover, we go beyond bounded time large deviations and describe the adiabatic behaviour of the slow motion Z^ε on exponentially large in $1/\varepsilon$ time intervals such as its exits from a domain of attraction and transitions between attractors of the averaged system (2.1.6). We observe that essentially the same proof yields the same results for a bit more general case when both B in 2.1.1 and the coefficients of the random evolutions in the next section depend also Lipschitz continuously on ε.

We consider also the discrete time case where (2.1.1) is replaced by a difference equation of the form

$$X^\varepsilon(n+1) - X^\varepsilon(n) = \varepsilon B(X^\varepsilon(n), Y^\varepsilon(n)),\ X^\varepsilon(0) = X^\varepsilon_x = x \tag{2.1.7}$$

where $B(x,y)$ is the same as in (2.1.1) and the fast motion $Y^\varepsilon(n) = Y^\varepsilon_{x,y}(n)$, $n = 0,1,...,$, $Y^\varepsilon(0) = y$ is a perturbation of a family $Y_{x,y}(n), n \ge 0$ of Markov chains parametrized by $x \in \mathbb{R}^d$. For somewhat less general discrete time situation large diviations bounds were obtained in [**35**] by a simpler approach but in our more general situation we can rely only on methods similar to the continuous time case. Moreover, unlike [**35**] we go farther and study also very long time "adiabatic" behaviour of the slow motion similar to the continuous time case and illustrate some of the results by computer simulations for simple models.

The strategy and many of arguments in Part 2 are rather similar to Part 1 where deterministic chaotic fast motions such as Anosov and Axiom A systems were considered. Still, in view of the heavy dynamical systems background and machinery Part 1 is hardly accessible for most of probabilists. By this reason we give full proofs here refering to Part 1 only for proofs of some general results on large deviations, rate functionals and some others which do not rely on the specific dynamical systems setup.

2.2. Preliminaries and main results

We will assume that right hand side of (2.1.1) is bounded and Lipschitz continuous, i.e. for some $K > 0$,

$$\sup_{x,y} |B(x,y)| \le K \text{ and } |B(x,y) - B(z,v)| \le K\big(|x-z| + d_{\mathbf{M}}(y,v)\big) \tag{2.2.1}$$

where $d_{\mathbf{M}}$ is the metric on $\mathbf{M}$. Our large deviations estimates will be derived under the following general assumption on the fast motion which is satisfied, as we explain it below, for random evolutions which are Markov processes with switching at random times between a finite number of diffusion processes.

2.2.1. ASSUMPTION. *There exist a convex differentiable in β and Lipschitz continuous in other variables function $H(x,x',\beta)$ defined for all $\beta \in \mathbb{R}^d$ and for x,x' from the closure $\bar{\mathcal{X}}$ of a relatively compact open connected set $\mathcal{X} \subset \mathbb{R}^d$ and a positive function $\zeta_{b,T}(\Delta,s,\varepsilon)$ satisfying*

$$\limsup_{\Delta\to 0}\limsup_{\varepsilon\to 0}\limsup_{s\to\infty}\zeta_{b,T}(\Delta,s,\varepsilon)=0 \tag{2.2.2}$$

such that for all $t>0$, $x,x'\in\bar{\mathcal{X}}, y\in\mathbf{M}$ and $|\beta|\le b$,

$$\begin{aligned}\big|\tfrac{1}{t}\log E\exp\langle\beta,\textstyle\int_0^t B\big(x',Y^\varepsilon_{x,y}(s)\big)ds\rangle&\\ -H(x,x',\beta)\big|\le\zeta_{b,T}\big(\varepsilon t,\min(t,(\log 1/\varepsilon)^\lambda),\varepsilon\big)&\end{aligned} \tag{2.2.3}$$

where $\lambda\in(0,1)$ and $\langle\cdot,\cdot\rangle$ is the inner product.

Set

$$L(x,x',\alpha)=\sup_{\beta\in\mathbb{R}^d}\big(<\alpha,\beta>-H(x,x',\beta)\big), \tag{2.2.4}$$

$H(x,\beta) = H(x,x,\beta)$ and $L(x,\alpha) = L(x,x,\alpha)$. Since $H(x,h',0) = 0$ then $L(x,x',\alpha)\ge 0$. In view of Assumption 2.2.1 and standard convex analysis duality results (see [**2**] and [**70**]) $L(x,x',\alpha)$ is (strictly) convex, lower semicontinuous and we have also that

$$H(x,x',\beta)=\sup_{\alpha\in\mathbb{R}^d}\big(<\alpha,\beta>-L(x,x',\alpha)\big). \tag{2.2.5}$$

It follows also from Assumption 2.2.1 that

$$|H(x,x',\beta)|\le K|\beta|. \tag{2.2.6}$$

Since $H(x,h',0) = 0$ by (2.2.6) and $L(x,x',\alpha)$ is lower semicontinuous then it follows from (2.2.5) that there exists a unique $\alpha_{x,x'}\in\mathbb{R}^d$ such that

$$L(x,x',\alpha_{x,x'})=0. \tag{2.2.7}$$

Set $\alpha_x=\alpha_{x,x}$. If $\alpha_x=\alpha(x)$ depends Lipschitz continuously in x then we can define the averaged motion $\bar{X}^\varepsilon=\bar{X}^\varepsilon_x$ in this general setup as the solution of the ordinary differential equation

$$\frac{d\bar{X}^\varepsilon(t)}{dt}=\alpha(\bar{X}^\varepsilon(t))\ X^\varepsilon(0)=x. \tag{2.2.8}$$

Denote by C_{0T} the space of continuous curves $\gamma_t=\gamma(t)$, $t\in[0,T]$ in $\mathcal{X}$ which is the space of continuous maps of $[0,T]$ into $\mathcal{X}$. For each absolutely continuous $\gamma\in C_{0T}$ its velocity $\dot\gamma_t$ can be obtained as the almost everywhere limit of continuous functions $n(\gamma_{t+n^{-1}}-\gamma_t)$ when $n\to\infty$. Hence $\dot\gamma_t$ is measurable in t, and so we can set

$$S_{0T}(\gamma)=\int_0^T L(\gamma_t,\dot\gamma_t)dt \tag{2.2.9}$$

Define the uniform metric on C_{0T} by

$$\mathbf{r}_{0T}(\gamma,\eta)=\sup_{0\le t\le T}|\gamma_t-\eta_t|$$

for any $\gamma, \eta \in C_{0T}$. Set $\Psi^a_{0T}(x) = \{\gamma \in C_{0T} : \gamma_0 = x,\ S_{0T}(\gamma) \le a\}$. Since $L(x,\alpha)$ is lower semicontinuous and convex in α and, in addition, $L(x,\alpha) = \infty$ if $|\alpha| > K$ it follows that the conditions of Theorem 3 in Ch.9 of [**41**] are satisfied as we can choose a fast growing minorant of $L(x,\alpha)$ required there to be zero in a sufficiently large ball and to be equal, say, $|\alpha|^2$ outside of it. As a result we conclude that S_{0T} is lower semicontinuous functional on C_{0T} with respect to the metric $\mathbf{r}_{0T}$, and so $\Psi^a_{0T}(x)$ is a closed set which plays a crucial role in the large deviations arguments below. Set $\mathcal{X}_t = \{x \in \mathcal{X} : \inf_{z\in\partial\mathcal{X}} |x - z| \ge 2Kt\}$.

2.2.2. THEOREM. *Suppose that (2.2.1) and Assumption 2.2.1 hold true. Set* $Z^\varepsilon_{x,y}(t) = X^\varepsilon_{x,y}(t/\varepsilon)$ *and let* $x \in \mathcal{X}_T$. *Then for any* $a, \delta, \lambda > 0$ *and every* $\gamma \in C_{0T}$, $\gamma_0 = x$ *there exists* $\varepsilon_0 = \varepsilon_0(x,\gamma,a,\delta,\lambda) > 0$ *such that for* $\varepsilon < \varepsilon_0$ *uniformly in* $y \in \mathbf{M}$,

$$P\left\{\mathbf{r}_{0T}(Z^\varepsilon_{x,y},\gamma) < \delta\right\} \ge \exp\left\{-\frac{1}{\varepsilon}(S_{0T}(\gamma) + \lambda)\right\} \tag{2.2.10}$$

and

$$P\left\{\mathbf{r}_{0T}(Z^\varepsilon_{x,y},\Psi^a_{0T}(x)) \ge \delta\right\} \le \exp\left\{-\frac{1}{\varepsilon}(a - \lambda)\right\}. \tag{2.2.11}$$

Next, let $V \subset \mathcal{X}$ be a connected open set and put $\tau^\varepsilon_{x,y}(V) = \inf\{t \ge 0 : Z^\varepsilon_{x,y}(t) \notin V\}$ where we take $\tau^\varepsilon_{x,y}(V) = \infty$ if $X^\varepsilon_{x,y}(t) \in V$ for all $t \ge 0$. The following result follows directly from Theorem 2.2.2.

2.2.3. COROLLARY. *Under the conditions of Theorem 2.2.2 for any* $T > 0$ *and* $x \in V$,

$$\begin{aligned}&\lim_{\varepsilon\to 0} \varepsilon \log P\left\{\tau^\varepsilon_{x,y}(V) < T\right\}\\ &= -\inf\left\{S_{0t}(\gamma) : \gamma \in C_{0T},\ t \in [0,T],\ \gamma_0 = x,\ \gamma_t \notin V\right\}.\end{aligned}$$

The main class of Markov processes satisfying our conditions which we have in mind consists of random evolutions on $\mathbf{M} = M \times \{1, ..., N\}$ where M is a compact n-dimensional C^2 Riemannian manifold and the unperturbed parametric family of Markov processes $Y_{x,y}(t)$ is the pair $Y_{x,v,k}(t) = (\hat{Y}_{x,v,k}(t), \nu_{x,v,k}(t))$ governed by the stochastic differential equations

$$d\hat{Y}_{x,v,k}(t) = \sigma_{\nu_{x,v,k}(t)}\big(x, \hat{Y}_{x,v,k}(t)\big)dw_t + b_{\nu_{x,v,k}(t)}\big(x, \hat{Y}_{x,v,k}(t)\big)dt \tag{2.2.12}$$

where $\hat{Y}_{x,v,k}(0) = v$, $\nu_{x,v,k}(0) = k$ and for all $1 \le i,j \le N$, $i \ne j$,

$$P\big\{\nu_{x,v,k}(t+\Delta) = j \big| \nu_{x,v,k}(t) = i, \hat{Y}_{x,v,k}(t) = w\big\} = q_{ij}(x,w)\Delta + o(\Delta) \text{ as } \Delta \downarrow 0. \tag{2.2.13}$$

We assume that $q_{kl}(x,w)$, $k,l = 1,...,N$ are bounded positive C^1 functions, $\sigma_k(x,v)\sigma^*_k(x,v) = a_k(x,v) = \big(a^{ij}_k(x,v),\ i,j = 1,...,n\big)$ is a C^1 field of positively definite symmetric matrices on M, $b_k(x,v) = \big(b^1_k(x,v), ..., b^n_k(x,v)\big)$ is a C^1 vector field and all functions are defined and satisfy the above properties for $v \in M$ and x belonging to an open neighborhood of $\bar{\mathcal{X}}$. Here w_t is the Brownian motion and the equation (2.2.12) is written in local coordinats. Observe that the existence and some properties of such Markov processes are discussed in [**72**]. The generator $\mathcal{L}^x$ of the Markov process $(\hat{Y}_x(t), \nu_x(t))$ is the operator acting on C^2 vector functions

$f=(f_1,...,f_N)$ on M by the formula

$$(\mathcal{L}^x f)_k(y)=\mathcal{L}^x_k f_k(y)+\sum_{l=1}^N q_{kl}(x,y)\big(f_l(y)-f_k(y)\big) \tag{2.2.14}$$

where $\mathcal{L}^x_k$ is the elliptic second order differential operator

$$\mathcal{L}^x_k=\frac{1}{2}\langle a_k(x,\cdot)\nabla,\nabla\rangle+\langle b_k(x,\cdot),\nabla\rangle. \tag{2.2.15}$$

Now, the perturbed fast motion $Y^\varepsilon=(\hat{Y}^\varepsilon,\nu^\varepsilon)$ satisfies

(2.2.16)

$$d\hat{Y}^\varepsilon_{x,v,k}(t)=\sigma_{\nu^\varepsilon_{x,v,k}(t)}\big(X^\varepsilon_{x,v,k}(t),\hat{Y}^\varepsilon_{x,v,k}(t)\big)dw_t+b_{\nu^\varepsilon_{x,v,k}(t)}\big(X^\varepsilon_{x,v,k}(t),\hat{Y}^\varepsilon_{x,v,k}(t)\big)dt,$$

$X^\varepsilon_{x,v,k}(0)=x$, $Y^\varepsilon_{x,v,k}(0)=v$, $\nu^\varepsilon_{x,v,k}(0)=k$ and

$$\begin{aligned}P\{\nu^\varepsilon_{x,v,k}(t+\Delta)=j|\nu^\varepsilon_{x,v,k}(t)=i,X^\varepsilon_{x,v,k}(t)=z,\hat{Y}^\varepsilon_{x,v,k}(t)=w\}\\ =q_{ij}(x,w)\Delta+o(\Delta)\text{ as }\Delta\downarrow 0\text{ for all }1\le i,j\le N,\ i\ne j\end{aligned} \tag{2.2.17}$$

where X^ε is given by (2.1.1) with $B(x,y)=B(x,v,k)=B_k(x,v)$, $y=(v,k)$ smoothly depending on x and v, so that the triple $(X^\varepsilon(t),Y^\varepsilon(t),\nu^\varepsilon(t))$ is a Markov processes. The following result which will be proved in Section 2.4 claims, in particular, that random evolutions above satisfy Assumption 2.2.1

2.2.4. PROPOSITION. *For the process $Y_x(t)=(\hat{Y}_x(t),\nu_x(t))$ defined by (2.2.12) and (2.2.13) the limit*

$$H(x,x',\beta)=\lim_{t\to\infty}\frac{1}{t}\log E\exp\langle\beta,\int_0^t B_{\nu_{x,v,k}(s)}(x',\hat{Y}_{x,v,k}(s))ds\rangle \tag{2.2.18}$$

exists uniformly in $x,x'\in\mathcal{X}$, $y\in\mathbf{M}$ and $|\beta|\le b$, it is strictly convex and differentiable in β and Lipschitz continuous in other variables, and it does not depend under our conditions on v and k. In this circumstances the function $L(x,x',\alpha)$ given by (2.2.4) can be represented in the explicit form

$$L(x,x',\alpha)=\inf\big\{I_x(\mu):\sum_{k=1}^N\int_M B_k(x,v)d\mu_k(v)=\alpha\big\} \tag{2.2.19}$$

where

$$I_x(\mu)=-\inf_{u>0}\sum_{k=1}^N\int_M\frac{(\mathcal{L}^x u)_k}{u_k}d\mu_k \tag{2.2.20}$$

and the first infinum is taken over the set $\mathcal{P}(\mathbf{M})$ of probability measures on $\mathbf{M}$, i.e. over the vector measures $\mu=(\mu_1,...,\mu_N)$ with $\sum_{k=1}^N\mu_k(M)=1$, and the second one is taken over positive vector functions u on M belonging to the domain of the operator $\mathcal{L}^x$. Clearly, $I_x(\mu)\ge 0$ and, furthermore, $I_x(\mu)=0$ if and only if μ is the invariant measure $\mu^x=(\mu_1^x,...,\mu_N^x)$ of the Markov process Y_x which is unique in our circumstances since the Doeblin condition (see **[19]***) holds true here. The vector field $\bar{B}(x)=\int_{\mathbf{M}}B(x,y)d\mu^x(y)=\sum_{k=1}^N\int_M B_k(x,v)d\mu_k(v)$ is C^1 in x, and so we can define the averaged motion $\bar{X}^\varepsilon=\bar{X}^\varepsilon_x$ by*

$$\frac{d\bar{X}^\varepsilon(t)}{dt}=\varepsilon\bar{B}(\bar{X}^\varepsilon(t)),\ X^\varepsilon(0)=x. \tag{2.2.21}$$

Hence, $S_{0T}(\gamma) = 0$ if and only if $\gamma_t = \bar{Z}(t) = \bar{X}^\varepsilon(t/\varepsilon)$ for all $t \in [0,T]$. The processes Y^ε given by (2.2.16) and (2.2.17) together with the function $H(x,x',\beta)$ satisfy Assumption 2.2.1.

Clearly, if $N = 1$ above then Y^ε becomes a diffusion process and if all operators $\mathcal{L}_k^x$ are just zero then we arrive to the case of continuous time Markov chains as fast motions which also satify all our assumptions. We observe also that both Proposition 2.2.4 and the results below can be extended to the case when $\mathcal{L}_k^x$ are hypoelliptic operators satisfying natural conditions so that we could rely, in particular, on results of Section 6.3 from [**22**].

Suppose that the coefficients σ_k, b_k and q_{ij} in (2.2.12) and (2.2.13) do not depend on x. Then $Y_{x,y}^\varepsilon(t) = Y_{x,y}(t) = Y_y(t)$ is an ergodic Markov process with the unique invariant measure μ and for any y almost surely

$$\lim_{T\to\infty} T^{-1}\int_0^T B(x,Y_y(t))dt = \bar{B}(x) = \int B(x,y)d\mu(y)$$

and by standard general results on the uncoupled averaging (see [**73**]) it follows that for any y almost surely

$$\sup_{0\le t\le T} |X_{x,y}^\varepsilon(t/\varepsilon) - \bar{X}_x^\varepsilon(t/\varepsilon)| \to 0 \text{ as } \varepsilon \to 0. \tag{2.2.22}$$

In the fully coupled case (i.e. when a_k, b_k, q_{ij} depend on x) Theorem 2.2.2 implies in the case of fast motions given by (2.2.16) and (2.2.17) that for each $\delta > 0$ there is $\alpha(\delta) > 0$ such that for all small ε,

$$P\{\sup_{0\le t\le T} |X_{x,y}^\varepsilon(t/\varepsilon) - \bar{X}_x^\varepsilon(t/\varepsilon)| > \delta\} \le e^{-\alpha(\delta)/\varepsilon} \tag{2.2.23}$$

which means, in particular, that in this case we have in (2.2.22) convergence in probability. Examples from [**11**] show that, in general, in the fully coupled setup we do not have convergence in (2.2.22) with probability one though in some cases such convergence can be derived from (2.2.23) if the derivatives of X^ε and Y^ε in ε grow subexponentially in $1/\varepsilon$ on time intervals of order $1/\varepsilon$ (see Remark 2.3.6).

In the following assertions we assume always that the fast motions are obtained by means of (2.2.12) and (2.2.13) so that we could rely on (2.2.18)–(2.2.20) though, in principle, it is possible to impose some general conditions on functions $L(x,\alpha)$ which would enable us to proceed with our arguments.

Precise large deviations bounds such as (2.2.10) and (2.2.11) of Theorem 2.2.2 are crucial in our study in Sections 2.6 and 2.7 of the "very long", i.e. exponential in $1/\varepsilon$, time "adiabatic" behaviour of the slow motion. Namely, we will describe such long time behavior of Z^ε in terms of the function

$$R(x,z) = \inf_{t\ge 0, \gamma\in C_{0t}} \{S_{0t}(\gamma) : \gamma_0 = x, \gamma_t = z\}$$

under various assumptions on the averaged motion $\bar{Z}$. Observe that R satisfies the triangle inequality $R(x_1,x_2) + R(x_2,x_3) \ge R(x_1,x_3)$ for any $x_1,x_2,x_3 \in \mathcal{X}$ and it determines a semi metric on $\mathcal{X}$ which measures "the difficulty"' for the slow motion to move from point to point in terms of the functional S.

Introduce the averaged flow Π^t on $\mathcal{X}_t$ by

$$\frac{d\Pi^t x}{dt} = \bar{B}(\Pi^t x),\ x \in \mathcal{X}_t \tag{2.2.24}$$

where $\bar{B}(z)$ is the same as in (2.2.21) and set $\bar{B}_\mu(z) = \int_{\mathbf{M}} B(z,y)d\mu(y) = \sum_{k=1}^{N} \int_M B_k(x,v)d\mu_k(v)$ for any probability measure $\mu = (\mu_1,...,\mu_N)$ on $\mathbf{M} = M \times \{1,...,N\}$. Call a Π^t-invariant compact set $\mathcal{O} \subset \mathcal{X}$ an S-compact if for any $\eta > 0$ there exist $T_\eta \geq 0$ and an open set $U_\eta \supset \mathcal{O}$ such that whenever $x \in \mathcal{O}$ and $z \in U_\eta$ we can pick up $t \in [0, T_\eta]$ and $\gamma \in C_{0t}$ satisfying

$$\gamma_0 = x,\ \gamma_t = z \text{ and } S_{0t}(\gamma) < \eta.$$

It is clear from this definition that $R(x,z) = 0$ for any pair points x, z of an S-compact $\mathcal{O}$ and by the above triangle inequality for R we see that $R(x,z)$ takes on the same value when $z \in \mathcal{X}$ is fixed and x runs over $\mathcal{O}$. We say that the vector field B on $\mathcal{X} \times \mathbf{M}$ is complete at $x \in \mathcal{X}$ if the convex set of vectors $\{\beta\bar{B}_\mu(x) : \beta \in [0,1], \mu \in \mathcal{P}(\mathbf{M}),\ I_x(\mu) < \infty\}$ contains an open neigborhood of the origin in $\mathbb{R}^d$. It follows by Lemma 1.6.2 in Part 1 that if $\mathcal{O} \subset \mathcal{X}$ is a compact Π^t-invariant set such that B is complete at each $x \in \mathcal{O}$ and either $\mathcal{O}$ contains a dense orbit of the flow Π^t (i.e. Π^t is topologically transitive on $\mathcal{O}$) or $R(x,z) = 0$ for any $x, z \in \mathcal{O}$ then $\mathcal{O}$ is an S-compact. Moreover, to ensure that $\mathcal{O}$ is an S-compact it suffices to assume that B is complete already at some point of $\mathcal{O}$ and the flow Π^t on $\mathcal{O}$ is minimal, i.e. the Π^t-orbits of all points are dense in $\mathcal{O}$ or, equivalently, for any $\eta > 0$ there exists $T(\eta) > 0$ such that the orbit $\{\Pi^t x,\ t \in [0, T(\eta)]\}$ of length $T(\eta)$ of each point $x \in \mathcal{O}$ forms an η-net in $\mathcal{O}$ which is equivalent to minimality of the flow Π^t on $\mathcal{O}$ (see [**80**]). The latter condition obviously holds true when $\mathcal{O}$ is a fixed point or a periodic orbit of Π^t but also, more generally, when Π^t on $\mathcal{O}$ is uniquely ergodic (see [**80**]).

A compact Π^t-invariant set $\mathcal{O} \subset \mathcal{X}$ is called an attractor (for the flow Π^t) if there is an open set $U \supset \mathcal{O}$ and $t_U > 0$ such that

$$\Pi^{t_U}\bar{U} \subset U \text{ and } \lim_{t\to\infty} \operatorname{dist}(\Pi^t z, \mathcal{O}) = 0 \text{ for all } z \in U.$$

For an attractor $\mathcal{O}$ the set $V = \{z \in \mathcal{X} : \lim_{t\to\infty} \operatorname{dist}(\Pi^t z, \mathcal{O}) = 0\}$, which is clearly open, is called the basin (domain of attraction) of $\mathcal{O}$. An attractor which is also an S-compact will be called an S-attractor .

In what follows we will speak about connected open sets V with piecewise smooth boundaries ∂V. The latter can be introduced in various ways but it will be convenient here to adopt the definition from [**17**] saying that ∂V is the closure of a finite union of disjoint, connected, codimension one, extendible C^1 (open or closed) submanifolds of $\mathbb{R}^d$ which are called faces of the boundary. The extendibility condition means that the closure of each face is a part of a larger submanifold of the same dimension which coincides with the face itself if the latter is a compact submanifold. This enables us to extend fields of normal vectors to the boundary of faces and to speak about minimal angles between adjacent faces which we assume to be uniformly bounded away from zero or, in other words, angles between exterior normals to adjacent faces at a point of intersection of their closures are uniformly bounded away from π and $-\pi$. The following result which will be proved in Section 2.6 describes exits of the slow motion from neighborhoods of attractors of the averaged motion.

2.2.5. THEOREM. *Let $\mathcal{O} \subset \mathcal{X}$ be an S-attractor of the flow Π^t whose basin contains the closure $\bar{V}$ of a connected open set V with a piecewise smooth boundary ∂V such that $\bar{V} \subset \mathcal{X}$ and assume that for each $z \in \partial V$ there exists $\varpi = \varpi(z) > 0$*

and a probability measure $\eta = \eta_z$ *with* $I_z(\eta_z) < \infty$ *such that*

$$z + s\bar{B}(z) \in V \text{ but } z + s\bar{B}_\eta(z) \in \mathbb{R}^d \setminus \bar{V} \text{ for all } s \in (0, \varpi], \tag{2.2.25}$$

i.e. $\bar{B}(z) \neq 0$, $\bar{B}_\eta(z) \neq 0$ *and the former vector points out into the interior while the latter into the exterior of* V. *Set* $R_\partial(z) = \inf\{R(z, \tilde{z}) : \tilde{z} \in \partial V\}$ *and* $\partial_{\min}(z) = \{\tilde{z} \in \partial V : R(z, \tilde{z}) = R_\partial(z)\}$. *Then* $R_\partial(z)$ *takes on the same value* R_∂ *and* $\partial_{\min}(z)$ *coincides with the same compact nonempty set* $\partial_{\min}$ *for all* $z \in \mathcal{O}$ *while* $R_\partial(x) \leq R_\partial$ *for all* $x \in V$. *Furthermore, for any* $x \in V$ *uniformly in* $y \in \mathbf{M}$,

$$\lim_{\varepsilon \to 0} \varepsilon \log E\tau^\varepsilon_{x,y}(V) = R_\partial > 0 \tag{2.2.26}$$

and for each $\alpha > 0$ *there exists* $\lambda(\alpha) = \lambda(x, \alpha) > 0$ *such that uniformly in* $y \in \mathbf{M}$ *for all small* $\varepsilon > 0$,

$$P\{e^{(R_\partial - \alpha)/\varepsilon} > \tau^\varepsilon_{x,y}(V) \text{ or } \tau^\varepsilon_{x,y}(V) > e^{(R_\partial + \alpha)/\varepsilon}\} \leq e^{-\lambda(\alpha)/\varepsilon}. \tag{2.2.27}$$

Next, set

$$\Theta^\varepsilon_v(t) = \Theta^{\varepsilon,\delta}_v(t) = \int_0^t \mathbb{I}_{V \setminus U_\delta(\mathcal{O})}(Z^\varepsilon_v(s))ds$$

where $U_\delta(\mathcal{O}) = \{z \in \mathcal{X} : dist(z, \mathcal{O}) < \delta\}$ *and* $\mathbb{I}_\Gamma(z) = 1$ *if* $z \in \Gamma$ *and* $= 0$, *otherwise. Then for any* $x \in V$ *and* $\delta > 0$ *there exists* $\lambda(\delta) = \lambda(x, \delta) > 0$ *such that uniformly in* $y \in \mathbf{M}$ *for all small* $\varepsilon > 0$,

$$P\{\Theta^\varepsilon_{x,y}(\tau^\varepsilon_{x,y}(V)) \geq e^{-\lambda(\delta)/\varepsilon}\tau^\varepsilon_{x,y}(V)\} \leq e^{-\lambda(\delta)/\varepsilon}. \tag{2.2.28}$$

Finally, for every $x \in V$ *and* $\delta > 0$,

$$\lim_{\varepsilon \to 0} P\{dist(Z^\varepsilon_{x,y}(\tau^\varepsilon_{x,y}(V)), \partial_{\min}) \geq \delta\} = 0 \tag{2.2.29}$$

provided $R_\partial < \infty$ *and the latter holds true if and only if for some* $T > 0$ *there exists* $\gamma \in C_{0T}$, $\gamma_0 \in \mathcal{O}$, $\gamma_T \in \partial V$ *such that* $\dot{\gamma}_t = \bar{B}_{\nu_t}(\gamma_t)$ *for Lebesgue almost all* $t \in [0, T]$ *with* $\nu_t \in \mathcal{M}_{\gamma_t}$ *then* $R_\partial < \infty$.

Theorem 2.2.5 asserts, in particular, that typically the slow motion Z^ε performs rare (adiabatic) fluctuations in the vicinity of an S-attractor $\mathcal{O}$ since it exists from any domain $U \supset \mathcal{O}$ with $\bar{U} \subset V$ for the time much smaller than $\tau^\varepsilon(V)$ (as the corresponding number $R_\partial = R_{\partial U}$ will be smaller) and by (2.2.28) it can spend in $V \setminus U_\delta(\mathcal{O})$ only small proportion of time which implies that Z^ε exits from U and returns to $U_\delta(\mathcal{O})$ (exponentially in $1/\varepsilon$) many times before it finally exits V. We observe that in the much simpler uncoupled setup corresponding results in the case of $\mathcal{O}$ being an attracting point were obtained for a continuous time Markov chain as a fast motion in [**29**] but the proofs there rely on the lower semicontinuity of the function R which does not hold true in general, and so extra conditions like S-compactness of $\mathcal{O}$ or, more specifically, the completness of B at $\mathcal{O}$ should be assumed there, as well. It is important to observe that the intuition based on diffusion type small random perturbations of dynamical systems should be applied with caution to problems of large deviations in averaging since the S-functional of Theorem 2.2.2 describing them is more complex and have rather different properties than the corresponding functional emerging in diffusion type random perturbations of dynamical systems (see [**31**]). The reason for this is the deterministic nature of the slow motion Z^ε which unlike a diffusion can move only with a bounded speed and, moreover, even in order to ensure its "diffusive like" local behaviour (i.e. to let it go in many directions) some extra nondegeneracy type conditions on the vector field B are required.

Our next result describes rare (adiabatic) transitions of the slow motion Z^ε between basins of attractors of the averaged flow Π^t which we consider now in the whole $\mathbb{R}^d$ and impose certain conditions on the structure of its ω-limit set.

2.2.6. Assumption. *Assumption 2.2.1 holds true for* $\mathcal{X}=\mathbb{R}^d$, *the families* $a_k^{ij}(x,\cdot), i,j=1,...,n)$ *and* $b_k(x,\cdot)=\big(b_k^1(x,\cdot),...,b_k^n(x,\cdot)\big)$ *of matrix and vector fields are compact sets in the* C^1 *topology,*

$$\|B(x,y)\|_{C^2(\mathbb{R}^d\times\mathbf{M})}\le K \tag{2.2.30}$$

for some $K>0$ *independent of* x,y *and there exists* $r_0>0$ *such that*

$$\big(x,B(x,y)\big)\le -K^{-1}\ \textit{for any}\ y\in\mathbf{M}\ \textit{and}\ |x|\ge r_0. \tag{2.2.31}$$

The condition (2.2.31) means that outside of some ball all vectors $B(x,y)$ have a bounded away from zero projection on the radial direction which points out to the origin. This condition can be weakened, for instance, it suffices that

$$\lim_{d\to\infty}\inf\{R(x,z):\ \mathrm{dist}(x,z)\ge d\}=\infty$$

but, anyway, we have to make some assumption which ensure that the slow motion stays in a compact region where really interesting dynamics takes place.

Next, suppose that the ω-limit set of the averaged flow Π^t is compact and it consists of two parts, so that the first part is a finite number of S-attractors $\mathcal{O}_1,...,\mathcal{O}_\ell$ whose basins $V_1,...,V_\ell$ have piecewise smooth boundaries $\partial V_1,...,\partial V_\ell$ and the remaining part of the ω-limit set is contained in $\cup_{1\le j\le\ell}\partial V_j$. We assume also that for any $z\in\cap_{1\le i\le k}\partial V_{j_i}$, $k\le\ell$ there exist $\varpi=\varpi(z)>0$ and probability measures $\eta_1,...,\eta_k$ such that $I(\eta_i)<\infty$, $i=1,...,k$ and

$$z+s\bar{B}_{\eta_i}(z)\in V_{j_i}\ \text{for all}\ s\in(0,\varpi]\ \text{and}\ i=1,...,k, \tag{2.2.32}$$

i.e. $\bar{B}_{\eta_i}(z)\ne 0$ and it points out into the interior of V_{j_i} which means that from any boundary point it is possible to go to any adjacent basin along a curve with an arbitrarily small S-functional. Let $\delta>0$ be so small that the δ-neighborhood $U_\delta(\mathcal{O}_i)=\{z\in\mathcal{X}:\ \mathrm{dist}(z,\mathcal{O}_i)<\delta\}$ of each $\mathcal{O}_i$ is contained with its closure in the corresponding basin V_i. For any $x\in V_i$ set

$$\tau^\varepsilon_{x,y}(i)=\tau^{\varepsilon,\delta}_{x,y}(i)=\inf\big\{t\ge 0:\ Z^\varepsilon_{x,y}(t)\in\cup_{j\ne i}U_\delta(\mathcal{O}_j)\big\}.$$

In Section 2.7 we will derive the following result.

2.2.7. Theorem. *The function* $R_{ij}(x)=\inf_{z\in V_j}R(x,z)$ *takes on the same value* R_{ij} *for all* $x\in\mathcal{O}_i$, $i\ne j$. *Let* $R_i=\min_{j\ne i,j\le\ell}R_{ij}$. *Then for any* $x\in V_i$ *uniformly in* $y\in\mathbf{M}$,

$$\lim_{\varepsilon\to 0}\varepsilon\log E\tau^\varepsilon_{x,y}(i)=R_i>0 \tag{2.2.33}$$

and for any $\alpha>0$ *there exists* $\lambda(\alpha)=\lambda(x,\alpha)>0$ *such that for all small* $\varepsilon>0$,

$$P\big\{e^{(R_i-\alpha)/\varepsilon}>\tau^\varepsilon_{x,y}(i)\ \textit{or}\ \tau^\varepsilon_{x,y}(i)>e^{(R_i+\alpha)/\varepsilon}\big\}\le e^{-\lambda(\alpha)/\varepsilon}. \tag{2.2.34}$$

Next, set

$$\Theta^{\varepsilon,i}_v(t)=\Theta^{\varepsilon,i,\delta}_v(t)=\int_0^t\mathbb{I}_{V_i\setminus U_\delta(\mathcal{O}_i)}(Z^\varepsilon_v(s))ds.$$

Then for any $x\in V_i$ *and* $\delta>0$ *there exists* $\lambda(\delta)=\lambda(x,\delta)>0$ *such that uniformly in* $y\in\mathbf{M}$ *for all small* $\varepsilon>0$,

$$P\big\{\Theta^{\varepsilon,i}_{x,y}(\tau^\varepsilon_{x,y}(i))\ge e^{-\lambda(\delta)/\varepsilon}\tau^\varepsilon_{x,y}(i)\big\}\le e^{-\lambda(\delta)/\varepsilon}. \tag{2.2.35}$$

Now, suppose that the vector field B is complete on ∂V_i for some $i \leq \ell$ (which strengthens (2.2.32) there) and the restriction of the ω-limit set of Π^t to ∂V_i consists of a finite number of S-compacts. Assume also that there is a unique index $\iota(i) \leq \ell$, $\iota(i) \neq i$ such that $R_i = R_{i\iota(i)}$. Then for any $x \in V_i$ there exists $\lambda = \lambda(x) > 0$ such that uniformly in $y \in \mathbf{M}$ for all small $\varepsilon > 0$,

$$P\{Z^\varepsilon_{x,y}(\tau^\varepsilon_{x,y}(i)) \notin V_{\iota(i)}\} \leq e^{-\lambda/\varepsilon}. \tag{2.2.36}$$

Finally, suppose that the above conditions hold true for all $i = 1, ..., \ell$. Define $\iota_0(i) = i$, $\tau^\varepsilon_v(i,1) = \tau^\varepsilon_v(i)$ and recursively,

$$\iota_k(i) = \iota(\iota_{k-1}(i)) \text{ and } \tau^\varepsilon_v(i,k) = \tau^\varepsilon_v(i,k-1) + \tau^\varepsilon_{v_\varepsilon(k-1)}\big(j(v_\varepsilon(k-1))\big),$$

where $v_\varepsilon(k) = \Phi_\varepsilon^{\varepsilon^{-1}\tau^\varepsilon_v(i,k)} v$, $j((x,y)) = j$ if $x \in V_j$, and set $\Sigma^\varepsilon_i(k,a) = \sum_{l=1}^k \exp\big((R_{\iota_{l-1}(i),\iota_l(i)} + a)/\varepsilon\big)$. Then for any $x \in V_i$ and $\alpha > 0$ there exists $\lambda(\alpha) = \lambda(x,\alpha) > 0$ such that uniformly in $y \in \mathbf{M}$ for all $n \in \mathbb{N}$ and sufficiently small $\varepsilon > 0$,

$$P\{\Sigma^\varepsilon_i(k,-\alpha) > \tau^\varepsilon_{x,y}(i,k) \text{ or } \tau^\varepsilon_{x,y}(i,k) > \Sigma^\varepsilon_i(k,\alpha) \text{ for some } k \leq n\} \leq ne^{-\lambda(\alpha)/\varepsilon} \tag{2.2.37}$$

and for some $\lambda = \lambda(x) > 0$,

$$P\{Z^\varepsilon_{x,y}(\tau^\varepsilon_{x,y}(i,k)) \notin V_{\iota_k(i)} \text{ for some } k \leq n\} \leq ne^{-\lambda/\varepsilon}. \tag{2.2.38}$$

Generically there exists only one index $\iota(i)$ such that $R_i = R_{i\iota(i)}$ and in this case Theorem 2.2.7 asserts that $Z^\varepsilon_{x,y}$, $x \in V_i$ arrives (for "most" $y \in \mathcal{W}$) at $V_{\iota(i)}$ after it leaves V_i. If $\mathcal{I}(i) = \{j : R_i = R_{ij}\}$ contains more than one index then the method of the proof of Theorem 2.2.7 enables us to conclude that in this case $Z^\varepsilon_{x,y}$, $x \in V_i$ arrives (for "most" $y \in \mathcal{W}$) at $\cup_{j\in\mathcal{I}(i)} V_j$ after leaving V_i but now we cannot specify the unique basin of attraction of one of $\mathcal{O}_j$'s where $Z^\varepsilon_{x,y}$ exits from V_i. If the succession function ι is uniquely defined then it determines an order of transitions of the slow motion Z^ε between basins of attractors of $\bar{Z}$ and because of their finite number Z^ε passes them in certain cyclic order going around such cycle exponentially many in $1/\varepsilon$ times while spending the total time in a basin V_i which is approximately proportional to $e^{R_i/\varepsilon}$. If there exist several cycles of indices $i_0, i_1, ..., i_{k-1}, i_k = i_0$ where $i_j \leq \ell$ and $i_{j+1} = \iota(i_j)$ then transitions between different cycles may also be possible. In the uncoupled case with fast motions being continuous time Markov chains a description of such transitions via certain hierarchy of cycles appeared without a detailed proof in [**29**] and [**31**]. In our fully coupled setup the corresponding description does not seem to be different from the uncoupled situation since its justification relies only on the Markov property arguments and estimates of probabilities of transitions of Z^ε from $U_\delta(\mathcal{O}_i)$ to $U_\delta(\mathcal{O}_j)$.

Set $\mathcal{I} = \{1, ..., \ell\}$. Following [**31**] we call a graph consisting of arrows $(k \to l)$ $(k \neq i,\ k,l \in \mathcal{I},\ k \neq l)$ an i-graph if every point $l \neq i$ is the origin of exactly one arrow and the graph has no circles. Let $G(i)$ be the set of all i-graphs. Next, choose $\delta > 0$ so small that $\overline{U_{2\delta}(\mathcal{O}_j)} \subset V_j$, $j = 1, ..., \ell$ and define stopping times $\sigma^{\varepsilon,\delta}_{x,y}(0) = 0$ and by induction for $k \geq 1$,

$$\hat{\sigma}^{\varepsilon,\delta}_{x,y}(k) = \inf\{t \geq \sigma^{\varepsilon,\delta}_{x,y}(k-1) : Z^\varepsilon_{x,y}(t) \notin \cup_{1\leq j\leq\ell} U_{2\delta}(\mathcal{O}_j)\},$$

$$\sigma^{\varepsilon,\delta}_{x,y}(k) = \inf\{t \geq \hat{\sigma}^{\varepsilon,\delta}_{x,y}(k) : Z^\varepsilon_{x,y}(t) \in \cup_{1\leq j\leq\ell} U_\delta(\mathcal{O}_j)\}.$$

Define the Markov chain

$$W^\varepsilon_{x,y}(k) = \big(Z^\varepsilon_{x,y}(\sigma^{\varepsilon,\delta}_{x,y}(k)),\, Y^\varepsilon_{x,y}(\varepsilon^{-1}\sigma^{\varepsilon,\delta}_{x,y}(k))\big)$$

which evolves on the phase space $\cup_{1\le j\le \ell}\Gamma_j$ where $\Gamma_j = \partial U_\delta(\mathcal{O}_j)\times \mathbf{M}$.

2.2.8. THEOREM. *Let $P(\cdot,\cdot)$ be the transition probability of the Markov chain W^ε. Then for any $\beta > 0$ there exist $\delta_0, \varepsilon_0 > 0$ such that if $\delta < \delta_0$ and $\varepsilon < \varepsilon_0$ then*

$$e^{(-R_{ij}-\beta)/\varepsilon} \le P\big((x,y),\Gamma_j\big) \le e^{(-R_{ij}+\beta)/\varepsilon} \tag{2.2.39}$$

whenever $(x,y)\in\Gamma_i$. Furthermore, if μ^ε_W is an invariant measure of W^ε on $\cup_{1\le j\le\ell}\Gamma_j$ then

$$e^{-2\beta(\ell-1)/\varepsilon}\frac{Q_j}{Q_1+\cdots+Q_\ell} \le \mu^\varepsilon_W(\Gamma_j) \le e^{2\beta(\ell-1)/\varepsilon}\frac{Q_j}{Q_1+\cdots+Q_\ell} \tag{2.2.40}$$

where

$$Q_i = \sum_{g\in G(i)} \exp\big(-\varepsilon^{-1}\sum_{(k\to l)\in g} R_{kl}\big). \tag{2.2.41}$$

Since total times spent by a Markov process in different sets are asymptotically proportional to masses given to these sets by corresponding invariant measures then Theorem 2.2.8 (together with Theorem 2.2.7) yields actually that the slow motion $Z^\varepsilon(t)$ spends in a basin V_j of the attractor $\mathcal{O}_j$ a percentage of total time approximately proportional to Q_j which will be illustrated by computational examples in Section 2.8. In fact, this description is effective only if there is a unique i_0 and a graph $g\in G(i_0)$ such that $\sum_{(k\to l)\in g} R_{kl}$ is minimal possible among all such sums over all j-graphs. In this case the slow motion spends in V_{i_0} a proportion of time close to one.

Next, we formulate our results for the discrete time case of difference equations (2.1.7).

2.2.9. ASSUMPTION. *There exist a convex differentiable in β and Lipschitz continuous in other variables function $H(x,x',\beta)$ defined for all $\beta\in\mathbb{R}^d$ and for x,x' from the closure $\bar{\mathcal{X}}$ of a relatively compact open connected set $\mathcal{X}\subset\mathbb{R}^d$ and a positive function $\zeta_{b,T}(\Delta,s,\varepsilon)$ satisfying (2.2.2) such that for all $k>0$, $x,x'\in\bar{\mathcal{X}}, y\in\mathbf{M}$ and $|\beta|\le b$,*

$$\begin{aligned}\big|\tfrac{1}{k}\log E\exp\langle\beta,\textstyle\sum_{j=1}^k B\big(x',Y^\varepsilon_{x,y}(j)\big)\rangle\\ -H(x,x',\beta)\big| \le \zeta_{b,T}\big(\varepsilon k,\min(k,(\log 1/\varepsilon)^\lambda),\varepsilon\big)\end{aligned} \tag{2.2.42}$$

where $\lambda\in(0,1)$ and $Y^\varepsilon_{x,y}(n), n=0,1,2,\ldots$ appears in (2.1.7).

2.2.10. THEOREM. *Suppose that (2.2.1) and Assumption 2.2.9 are satisfied and that $X^\varepsilon(n)=X^\varepsilon_{x,y}(n), n=0,1,2,\ldots$ are given by (2.1.7). For $t\in[n,n+1]$ define $X^\varepsilon(t)=(t-n)X^\varepsilon(n+1)+(n+1-t)X^\varepsilon(n)$ and set $Z^\varepsilon(t)=Z^\varepsilon_{x,y}(t)=X^\varepsilon_{x,y}(t/\varepsilon)$. Then Theorem 2.2.2 and Corollary 2.2.3 hold true with the corresponding functionals S_{0T}.*

In a bit more restricted situation Theorem 2.2.10 was proved by a simpler method in [**35**].

The main model of Markov chains serving as fast motions $Y^\varepsilon(n), n\ge 0$, we have in mind, is obtained in the following way. We start with a parametric family of Markov chains $Y_{x,y}(n)$, $n\ge 0$, $Y_{x,y}(0)=y$ on a compact C^2 Riemannian manifold M with transition probabilities $P^x(y,\Gamma)=P^x_y\{Y_{x,y}(1)\in\Gamma\}$ having positive densities $p^x(y,z) = P^x(y,dz)/m(dz)$ with respect to the Riemannian volume m, so that $p^x(y,z)$ is C^1 in x and continuous in other variables. Next, we define $X^\varepsilon(n)$ and $Y^\varepsilon(n)$ adding to (2.1.7) another equation

$$P\{Y^\varepsilon(n+1) \in \Gamma \big| X^\varepsilon(n) = x,\, Y^\varepsilon(n) = y\} = P^x(y,\Gamma). \tag{2.2.43}$$

2.2.11. PROPOSITION. *Let $Y_{x,y}(n)$ be as above. Then the limit*

$$H(x,x',\beta) = \lim_{k\to\infty} \frac{1}{k} \log E \exp\langle \beta, \sum_{j=1}^{k} B(x', Y_{x,y}(j))\rangle \tag{2.2.44}$$

exists uniformly in x,x' running over a compact set and in $y \in \mathbf{M}$ and it satisfies conditions of Assumption 2.2.9. In this circumstances the functionals S_{0T} appearing in the large deviations estimates (2.2.10) and (2.2.11) again have the form (2.2.9) with $L(x,\alpha)$ given by (2.2.19) where now

$$I_x(\mu) = -\inf_{u>0} \int_{\mathbf{M}} \log \frac{\int_{\mathbf{M}} p^x(y,v)u(v)dm(v)}{u(y)} d\mu(y). \tag{2.2.45}$$

Clearly, $I_x(\mu) \geq 0$ and, furthermore, $I_x(\mu) = 0$ if and only if μ is the invariant measure μ^x of the Markov chain Y_x which is unique since the Doeblin condition (see [**19**]*) holds true here. The vector field $\bar{B}(x) = \int_{\mathbf{M}} B(x,y)d\mu^x(y)$ is C^1 in x, and so we can define uniquely the averaged motion $\bar{X}^\varepsilon = \bar{X}^\varepsilon_x$ by (2.2.21) and, again, $S_{0T}(\gamma) = 0$ if and only if $\gamma_t = \bar{Z}(t) = \bar{X}^\varepsilon(t/\varepsilon)$ for all $t \in [0,T]$. Furthermore, $Y^\varepsilon(n), n \geq 0$ given by (2.2.43) satisfies (2.2.42).*

The existence of the limit (2.2.44) and its properties in our circumstances are well known (see [**23**], [**24**], [**47**], [**43**], [**39**]) and the fact that (2.2.42) holds true here will be explained at the beginning of Section 2.8.

2.2.12. THEOREM. *Let the fast motion $Y^\varepsilon(n) = Y^\varepsilon_{x,y}(n)$ be constructed as above via (2.2.43) then with the corresponding definitions of S-compacts and under similar conditions the conclusions of Theorems 2.2.5, 2.2.7 and 2.2.8 remain true for the corresponding slow motion $Z^\varepsilon(t)$ defined in Theorem 2.2.10.*

Observe, that we can easily produce a wide class of systems satisfying the conditions of Theorems 2.2.5, 2.2.7, and 2.2.8 or Theorem 2.2.12 by setting $B(x,y) = \tilde{B}(x) + \hat{B}(x,y)$ so that $\int \hat{B}(x,y)d\mu^x(y) = 0$ where μ^x is the unique invariant measure of Y_x and the vector field $\tilde{B}$, which becomes now the averaged vector field $\bar{B}$, has an ω-limit set satisfying conditions of the above theorems. Simple examples of this construction will be exhibited in Section 2.8 for which we also compute historgrams indicating proportions of time the slow motion spends near different attracting points of the averaged motion. We observe that the functional S_{0T}, which plays a crucial role in the above theorems, seems to be quite difficult to compute since this leads to difficult nonclassical variational problems.

2.3. Large deviations

We will need the following version of general large deviations bounds when usual assumptions hold true with errors. The proof is a strightforward modification of the standard one (cf. [**47**]) and its details can be found in Part 1, Lemma 1.4.1.

2.3.1. LEMMA. *Let $H = H(\beta)$, $\eta = \eta(\beta)$ be uniformly bounded on compact sets functions on $\mathbb{R}^d$ and $\{\Xi_\tau, \tau \geq 1\}$ be a family of $\mathbb{R}^d$-valued random vectors on a* *probability space* $(\Omega, \mathcal{F}, P)$ *such that* $|\Xi_\tau| \le C < \infty$ *with probability one for some constant* C *and all* $\tau \ge 1$. *For any* $a > 0$ *and* $\alpha, \beta_0 \in \mathbb{R}^d$ *set*

$$(2.3.1)\quad L_a^{\beta_0}(\alpha) = \sup_{\beta \in \mathbb{R}^d, |\beta+\beta_0| \le a} (\langle \beta, \alpha \rangle - H(\beta)),\ L_a(\alpha) = L_a^0(\alpha),\ L(\alpha) = L_\infty(\alpha).$$

(i) For any $\lambda, a > 0$ *there exists* $\tau_0 = \tau(\lambda, a, C)$ *such that whenever for some* $\tau \ge \tau_0$, $\beta_0 \in \mathbb{R}^d$ *and each* $\beta \in \mathbb{R}^d$ *with* $|\beta + \beta_0| \le a$,

$$(2.3.2)\qquad H_\tau(\beta) = \tau^{-1} \log E e^{\tau \langle \beta, \Xi_\tau \rangle} \le H(\beta) + \eta(\beta)$$

then for any compact set $\mathcal{K} \subset \mathbb{R}^d$,

$$(2.3.3)\qquad P\{\Xi_\tau \in \mathcal{K}\} \le \exp\left(-\tau(L_a^{\beta_0}(\mathcal{K}) - \eta_a^{\beta_0} - \lambda|\beta_0| - \lambda)\right)$$

where

$$(2.3.4)\qquad \eta_a^{\beta_0} = \sup\{\eta(\beta) : |\beta + \beta_0| \le a\} \text{ and } L_a^{\beta_0}(\mathcal{K}) = \inf_{\alpha \in \mathcal{K}} L_a^{\beta_0}(\alpha).$$

(ii) Suppose that $\alpha_0 \in \mathbb{R}^d$, $0 < a \le \infty$ *and there exists* $\beta_0 \in \mathbb{R}^d$ *such that* $|\beta_0| \le a$ *and*

$$(2.3.5)\qquad H(\beta_0) = \langle \beta_0, \alpha_0 \rangle - L_a(\alpha_0).$$

If (2.3.2) holds true then for any $\delta > 0$,

$$(2.3.6)\qquad P\{|\Xi_\tau - \alpha_0| \le \delta\} \le \exp\left(-\tau(L_a(\alpha_0) - \eta(\beta_0) - \delta|\beta_0|)\right).$$

(iii) Assume that $\alpha_0, \beta_0 \in \mathbb{R}^d$ *satisfy (2.3.5). For any* $\lambda, a > 0$ *there exists* $\tau_0 = \tau(\lambda, a, C)$ *such that whenever for some* $\tau \ge \tau_0$ *and each* $\beta \in \mathbb{R}^d$ *with* $|\beta| \le a$ *the inequality (2.3.2) holds true together with*

$$(2.3.7)\qquad \tau^{-1} \log E e^{\tau \langle \beta, \Xi_\tau \rangle} \ge H(\beta) - \eta(\beta)$$

then for any $\gamma, \delta > 0$, $\gamma \le \delta$,

$$(2.3.8)\qquad P\{|\Xi_\tau - \alpha_0| < \delta\} \ge \exp\left(-\tau(L(\alpha_0) + \eta(\beta_0) + \gamma|\beta_0|)\right) \times \Big(1 - \exp\big(-\tau(\tilde{L}_a^{\beta_0}(\mathcal{K}_{\gamma,C}(\alpha_0)) - \eta_a - \eta(\beta_0) - \lambda|\beta_0| - \lambda)\big)\Big)$$

where

$$\tilde{L}_a^{\beta_0}(\alpha) = L_a(\alpha) - \langle \beta_0, \alpha \rangle + H(\beta_0),$$

$\tilde{L}_a^{\beta_0}(\mathcal{K}) = \inf_{\alpha \in \mathcal{K}} \tilde{L}_a^{\beta_0}(\alpha)$, $\eta_a = \eta_a^0$, $\mathcal{K}_{\gamma,C}(\alpha_0) = \overline{U_C(0)} \setminus U_\gamma(\alpha_0)$, $U_\gamma(\alpha) = \{\tilde{\alpha} : |\tilde{\alpha} - \alpha| < \gamma\}$ *and* $\bar{U}$ *denotes the closure of* U.

The proof of the following result is also standard and can be found in Part 1, Lemma 1.4.2.

2.3.2. LEMMA. *Let* S_n, $n = 1, 2, \ldots$ *be a nondecreasing sequence of lower semicontinuous functions on a metric space* M *and let* $S = \lim_{n\to\infty} S_n$. *Assume that* S *is also lower semicontinuous and for any compact set* $\mathcal{K} \subset M$ *denote*

$$S_n(\mathcal{K}) = \inf_{\gamma \in \mathcal{K}} S_n(\gamma) \text{ and } S(\mathcal{K}) = \inf_{\gamma \in \mathcal{K}} S(\gamma).$$

Then

$$(2.3.9)\qquad \lim_{n\to\infty} S_n(\mathcal{K}) = S(\mathcal{K}).$$

We will need also the following general result which will enable us to subdivide time into small intervals freezing the slow variable on each of them so that the estimate (2.2.3) of Assumption 2.2.1 becomes sufficiently precise and, on the other hand, we will not change much the corresponding functionals S_{0T} appearing in required large deviations estimates. This result is certainly not new, it is cited in [**79**] as a folklore fact and a version of it can be found in [**59**], p.67 while for a complete proof we refer the reader to Part 1, Lemma 1.4.3.

2.3.3. LEMMA. *Let $f = f(t)$ be a measurable function on $\mathbb{R}^1$ equal zero outside of $[0,T]$ and such that $\int_0^T |f(t)|dt < \infty$. For each positive integer m and $c \in [0,T]$ define $f_m(t,c) = f([(t+c)\Delta^{-1}]\Delta - c)$ where $\Delta = T/m$ and $[\cdot]$ denotes the integral part. Then there exists a sequence $m_i \to \infty$ such that for Lebesgue almost all $c \in [0,T]$,*

$$\lim_{i\to\infty} \int_0^T |f(t) - f_{m_i}(t,c)|dt = 0. \tag{2.3.10}$$

Next we will need the following simple estimates whose proof uses the Gronwall inequality and can be found in Part 1, Lemma 1.5.1.

2.3.4. LEMMA. *Let $x_i, \tilde{x}_i \in \mathcal{X}$, $i = 0,1,...,N$, $0 = t_0 < t_1 < ... < t_{N-1} < t_N = T$, $\Delta = \max_{0\le i\le N-1}(t_{i+1} - t_i)$, $\xi_i = (x_i - x_{i-1})(t_i - t_{i-1})^{-1}$, $n(t) = \max\{j \ge 0 : t \ge t_j\}$, $\psi(t) = \tilde{x}_{n(t)}$, $v \in \mathcal{X} \times \mathbf{M}$,*

$$\Xi_j^\varepsilon(v,x) = (t_j - t_{j-1})^{-1} \int_{t_{j-1}}^{t_j} B(x, Y_v^\varepsilon(s/\varepsilon))ds,$$

and for $t \in [0,T]$,

$$Z_{v,x}^{\varepsilon,\psi}(t) = x + \int_0^t B(\psi(s), Y_v^\varepsilon(s/\varepsilon))ds. \tag{2.3.11}$$

Then

$$\begin{aligned} &\left|\Xi_j^\varepsilon(v, x_{j-1}) - (t_j - t_{j-1})^{-1}(Z_v^\varepsilon(t_j) - Z_v^\varepsilon(t_{j-1}))\right| \\ &\quad \le K\left|Z_v^\varepsilon(t_{j-1}) - x_{j-1}\right| + \tfrac{1}{2}K^2(t_j - t_{j-1}), \end{aligned} \tag{2.3.12}$$

$$\begin{aligned} \sup\nolimits_{0\le s\le t} \left|Z_{v,x}^{\varepsilon,\psi}(s) - \psi(s)\right| &\le |x - x_0| + \max\nolimits_{0\le j\le n(t)} |x_j - \tilde{x}_j| \\ &+ K\Delta + n(t)\Delta \max\nolimits_{1\le j\le n(t)} \left|\Xi_j^\varepsilon(v, \tilde{x}_{j-1}) - \xi_j\right| \end{aligned} \tag{2.3.13}$$

and

$$\sup_{0\le s\le t} \left|Z_v^\varepsilon(s) - Z_{v,x}^{\varepsilon,\psi}(s)\right| \le e^{Kt}\left(|\pi_1 v - x| + Kt \sup_{0\le s\le t} \left|Z_{v,x}^{\varepsilon,\psi}(s) - \psi(s)\right|\right) \tag{2.3.14}$$

where, recall, $Z_v^\varepsilon(s) = X_v^\varepsilon(s/\varepsilon)$ and $\pi_1 v = z \in \mathcal{X}$ if $v = (z,y) \in \mathcal{X} \times \mathbf{M}$.

For any $x', x'' \in \mathcal{X}$ and $\beta, \xi \in \mathbb{R}^d$ set

$$L_b(x', x'', \xi) = \sup_{\beta\in\mathbb{R}^d, |\beta|\le b} \left(\langle\beta, \xi\rangle - H(x', x'', \beta)\right),$$

and $L_b(x,\xi) = L_b(x,x,\xi)$ with $H(x',x'',\beta)$ given by Assumption 2.2.1. The following result is the crucial step in the proof of Theorem 2.2.2.

2.3.5. PROPOSITION. *Let $x_j, t_j, \xi_j, N, \Delta, T$ and Ξ_j^ε be the same as in Lemma 2.3.4 and assume that*

$$\hat{\Delta} = \min_{0\le i\le N-1}(t_{i+1}-t_i) \ge \Delta/3. \tag{2.3.15}$$

(i) There exist $\delta_0>0, \varepsilon_0(\Delta)>0$ and $C_T(b)>0$ independent of $x, y, x_j, \tilde{x}_j, \xi_j$ such that if $\delta\le\delta_0$ and $\varepsilon\le\varepsilon_0(\Delta)$ then for any $b>0$,

$$\begin{aligned} &P\big(\max_{1\le j\le N}|\Xi_j^\varepsilon((x,y),\tilde{x}_{j-1})-\xi_j|<\delta\big) \\ &\le \exp\big\{-\tfrac{1}{\varepsilon}\big(\textstyle\sum_{j=1}^N (t_j-t_{j-1})L_b(\tilde{x}_{j-1},\xi_j)-\eta_{b,T}(\varepsilon,\Delta)-C_T(b)(d+\delta)\big)\big\} \end{aligned}\tag{2.3.16}$$

where $d=|x-x_0|+\max_{0\le j\le N}|x_j-\tilde{x}_j|$, $\eta_{b,T}(\varepsilon,\Delta)$ does not depend on $x, x_j, \tilde{x}_j, \xi_j$ and

$$\lim_{\Delta\to 0}\limsup_{\varepsilon\to 0}\eta_{b,T}(\varepsilon,\Delta)=0. \tag{2.3.17}$$

In particular, if for each $j=1,...,N$ there exists $\beta_j\in\mathbb{R}^d$ such that

$$L(\tilde{x}_j,\xi_j)=\langle\beta_j,\xi_j\rangle-H(\tilde{x}_j,\beta_j) \tag{2.3.18}$$

and

$$\max_{1\le j\le N}|\beta_j|\le b<\infty \tag{2.3.19}$$

then (2.3.16) holds true with $L(\tilde{x}_j,\xi_j)$ in place of $L_b(\tilde{x}_j,\xi_j)$, $j=1,...,N$.

(ii) For any $b,\lambda,\delta,q>0$ there exist $\Delta_0=\Delta_0(b,\lambda,\delta,q)>0$ and $\varepsilon_0=\varepsilon_0(b,\lambda,\delta,q,\Delta)$, the latter depending also on $\Delta>0$, such that if ξ_j and β_j satisfy (2.3.18) and (2.3.19), $\max_{1\le j\le N}|\xi_j|\le q$, $\Delta<\Delta_0$ and $\varepsilon<\varepsilon_0$ then

$$\begin{aligned} &P\big(\max_{1\le j\le N}|\Xi_j^\varepsilon((x,y),\tilde{x}_{j-1})-\xi_j|<\delta\big) \\ &\ge \exp\big\{-\tfrac{1}{\varepsilon}\big(\textstyle\sum_{j=1}^N (t_j-t_{j-1})L(\tilde{x}_{j-1},\xi_j)-\eta_{b,T}(\varepsilon,\Delta)+C_T(b)d+\lambda\big)\big\} \end{aligned}\tag{2.3.20}$$

with some $C_T(b)>0$ depending only on b and T.

PROOF. (i) Introduce the events

$$\Gamma^j(r)=\big\{\big|\Xi_j(v,\tilde{x}_{j-1})-\xi_j\big|<r\big\},\ j=1,...,N$$

so that we have

$$P\{\max_{1\le j\le n}|\Xi_j^\varepsilon(v,\tilde{x}_{j-1})-\xi_j|<r\}=P\big(\cap_{j=1}^n\Gamma^j(r)\big). \tag{2.3.21}$$

Now for $v=(x,y)$ by the Markov property

$$\begin{aligned} &P\big(\cap_{j=1}^n\Gamma^j(\delta)\big) \\ &=E\,\mathbb{I}_{\cap_{j=1}^{n-1}\Gamma^j(\delta)}P_{X_{x,y}^\varepsilon(t_{n-1}\varepsilon^{-1}),Y_{x,y}^\varepsilon(t_{n-1}\varepsilon^{-1})}\big\{\big|(t_n-t_{n-1})^{-1} \\ &\times\textstyle\int_0^{t_n-t_{n-1}}B\big(\tilde{x}_{n-1},Y^\varepsilon_{X_{x,y}^\varepsilon(t_{n-1}\varepsilon^{-1}),Y_{x,y}^\varepsilon(t_{n-1}\varepsilon^{-1})}(s/\varepsilon)\big)ds-\xi_n\big|<\delta\big\}. \end{aligned}\tag{2.3.22}$$

If $\omega\in\cap_{j=1}^{n-1}\Gamma^j(\delta)$ then $X_{x,y}^\varepsilon(t_{n-1}\varepsilon^{-1},\omega)=Z_{x,y}^\varepsilon(t_{n-1},\omega)$ in view of (2.3.13) and (2.3.14) satisfies

$$\begin{aligned} &\big|X_{x,y}^\varepsilon(t_{n-1}\varepsilon^{-1},\omega)-\tilde{x}_{n-1}\big|\le d_{n-1}=(e^{Kt_{n-1}}Kt_{n-1}+1) \\ &\times(|x-x_0|+\max_{0\le j\le n-1}|x_j-\tilde{x}_j|+K\Delta+(n-1)\Delta\delta). \end{aligned}\tag{2.3.23}$$

Since $H(x',x'',\beta)$ is Lipschitz continuous in x' and x'' it follows from (2.3.22) that

$$\big|H\big(X_{x,y}^\varepsilon(t_{n-1}\varepsilon^{-1},\omega),\tilde{x}_{n-1},\beta_n^{(a)}\big)-H(\tilde{x}_{n-1},\beta_n^{(a)})\big|\le C(a)d_{n-1} \tag{2.3.24}$$

provided $\omega \in \cap_{j=1}^{n-1}\Gamma^j(\delta)$ where $C(a) > 0$ depends only on a. In view of Assumption 2.2.1 we can estimate from above the probability in the right hand side of (2.3.22) by means of Lemma 2.3.1(i) which together with (2.3.24) yield that

$$\begin{aligned}(2.3.25)\qquad & P\big(\cap_{j=1}^{n}\Gamma^j(\delta)\big) \le P\big(\cap_{j=1}^{n-1}\Gamma^j(\delta)\big)\\ &\times \exp\big(-\tfrac{(t_n-t_{n-1})}{\varepsilon}(L_a(\tilde{x}_{n-1},\xi_n)-\tilde{\eta}_{a,T}(\varepsilon,\Delta)-C(a)d_{n-1}-ra)\big)\end{aligned}$$

where $\tilde{\eta}_{a,T}(\varepsilon,\Delta) \to 0$ as, first, $\varepsilon \to 0$ and then $\Delta \to 0$. Applying (2.3.25) for $n = N, N-1, ..., 2$ and estimating $P(\Gamma^1(\delta))$ by means of Lemma 2.3.1(i) we derive (2.3.16) in view of (2.3.21).

(ii) In order to obtain (2.3.20) we rely on Assumption 2.2.1 and Lemma 2.3.1(iii) estimating from below the probability in the right hand side of (2.3.22) which together with (2.3.23) yield

$$\begin{aligned}(2.3.26)\qquad & P\big(\cap_{j=1}^{n}\Gamma^j(\delta)\big) \ge P\big(\cap_{j=1}^{n-1}\Gamma^j(\delta)\big)\\ &\times \exp\big(-\tfrac{(t_n-t_{n-1})}{\varepsilon}L_a(\tilde{x}_{n-1},\xi_n)\big)g_{n,b}(\varepsilon,\delta,\varsigma,\sigma)\end{aligned}$$

where

$$g_{n,b}(\varepsilon,\Delta,\varsigma,\sigma) = \exp\Big(-\tfrac{(t_n-t_{n-1})}{\varepsilon}\big(\tilde{\eta}_{b,T}(\varepsilon,\Delta)+C_T(b)d_{n-1}+\varsigma b\big)\Big)$$
$$\times\Big(1-\exp\big(-\tfrac{(t_n-t_{n-1})}{\varepsilon}(d(b)-\tilde{\eta}_{b,T}(\varepsilon,\Delta)-\sigma b-\sigma)\big)\Big),$$

$$d(b) = \min_{1\le j\le N}\tilde{L}_b^{\beta_j}(\tilde{x}_{j-1},\mathcal{K}_{\varsigma,C}(\xi_j)),\ \tilde{L}_b^{\beta}(x,\mathcal{K}) = \inf_{\alpha\in\mathcal{K}}\tilde{L}_b^{\beta}(x,\alpha),$$

$$\mathcal{K}_{\varsigma,C}(\alpha) = \bar{U}_C(0)\setminus U_\varsigma(\alpha),\ \tilde{L}_b^{\beta}(x,\alpha) = L_b(x,\alpha)-\langle\beta,\alpha\rangle+H(x,\beta),\ C_T(b)>0,$$

and $\tilde{\eta}_{b,T}(\varepsilon,\Delta) \to 0$ as, first, $\varepsilon \to 0$ and then $\Delta \to 0$. Employing (2.3.26) for $n = N, N-1, ..., 2$ and estimating $P(\Gamma^1(\delta))$ by means of Lemma 2.3.1(iii) we obtain from (2.3.21) that

$$\begin{aligned}(2.3.27)\qquad & P\big\{\max_{1\le j\le n}\big|\Xi_j^\varepsilon(v,\tilde{x}_{j-1})-\xi_j\big|<\delta\big\}\\ &\ge \exp\Big(-\tfrac{1}{\varepsilon}\big(\textstyle\sum_{j=1}^{N}(t_j-t_{j-1})L(\tilde{x}_{j-1},\xi_j)+C(\rho,\delta)\varepsilon\Delta^{-1}\big)\Big)\\ &\qquad\times\textstyle\prod_{n=1}^{N}g_{n,b}(\varepsilon,\Delta,\varsigma,\sigma)\end{aligned}$$

for some $C(\rho,\delta) > 0$ provided, say, $NC_1\varepsilon\Delta^{-1} \le 2TC_1\varepsilon\Delta^{-2} \le \frac{\delta}{2}$ and $T\varepsilon\Delta^{-1} < C\rho/2$. Since $H(x,\beta)$ is differentiable in β then

$$\tilde{L}(\tilde{x}_j,\alpha) = L(\tilde{x}_j,\alpha)-\langle\beta_j,\xi_j\rangle+H(\tilde{x}_j,\beta_j) > 0$$

for any $\alpha \ne \xi_j$ (see Theorems 23.5 and 25.1 in [**70**]), and so by the lower semicontinuity of $L(x,\alpha)$ in α (and, in fact, also in x),

$$\tilde{L}^{\beta_j}(\tilde{x}_{j-1},\mathcal{K}_{\varsigma,C}(\xi_j)) = \inf_{\alpha\in\mathcal{K}_{\varsigma,C}(\xi_j)}\tilde{L}^{\beta_j}(\tilde{x}_{j-1},\alpha) > 0.$$

This together with Lemma 2.3.2 yield that $d(b)$ appearing in the definition of $g_{n,b}(\varepsilon,\Delta,\varsigma,\sigma)$ is positive provided b is sufficiently large. In fact, it follows from the lower semicontinuity of $L(x,\alpha)$ that $d(b)$ is bounded away from zero by a positive constant independent of $\tilde{x}_j$ and ξ_j, $j = 1, ..., N$ if these points vary over fixed compact sets and (2.3.18) together with (2.3.19) hold true. Now, given $\lambda > 0$ choose, first, sufficiently large b as needed and then subsequently choosing small σ

and ς, then small Δ, and, finally, small enough ε we end up with an estimate of the form

$$g_{n,b}(\varepsilon,\Delta,\varsigma,\sigma) \geq \exp\big(-\frac{(t_n-t_{n-1})}{\varepsilon}(\eta_{b,\rho,T}(\varepsilon,T)+C_T(b)d+\lambda)\big) \tag{2.3.28}$$

where $C_T(b)>0$ and $\eta_{b,\rho,T}(\varepsilon,T)$ satisfies (2.3.17). Finally, (2.3.20) follows from (2.3.27) and (2.3.28). □

The remaining part of the proof of Theorem 2.2.2 contains mostly some convex analysis arguments and it repeats almost verbatim the corresponding part of the proof of Theorem 1.2.3 in Part 1 but for readers' convenience we exhibit it also here. We remark that some of the details below are borrowed from [**79**] but we believe that our exposition and the way of proof are more precise, complete and easier to follow. We start with the lower bound. Assume that $S_{0T}(\gamma)<\infty$, and so that γ is absolutely continuous, since there is nothing to prove otherwise. Then by (2.2.9), $L(\gamma_s,\dot\gamma_s)<\infty$ for Lebesgue almost all $s\in[0,T]$. By (2.2.1) and Assumption 2.2.1,

$$H(x,\beta)\leq \tilde K|\beta| \tag{2.3.29}$$

for some $\tilde K>0$, and so if $L(\gamma_s,\dot\gamma_s)<\infty$ it follows from (2.2.4) that $|\dot\gamma|\leq\tilde K$. Suppose that $\mathcal{D}(L_s)=\{\alpha:\ L(\gamma_s,\alpha)<\infty\}\neq\emptyset$ and let $\mathrm{ri}\mathcal{D}(L_s)$ be the interior of $\mathcal{D}(L_s)$ in its affine hull (see [**70**]). Then either $\mathrm{ri}\mathcal{D}(L_s)\neq\emptyset$ or $\mathcal{D}(L_s)$ (by its convexity) consists of one point and recall that $\dot\gamma_s\in\mathcal{D}(L_s)$ for Lebesgue almost all $s\in[0,T]$. By (2.2.6) and (2.3.29),

$$0=H(\gamma_s,0)=\inf_{\alpha\in\mathbb{R}^d}L(\gamma_s,\alpha). \tag{2.3.30}$$

This together with the nonnegativity and lower semi-continuity of $L(\gamma_s,\cdot)$ yield that there exists $\hat\alpha_s$ such that $L(\gamma_s,\hat\alpha_s)=0$ and by a version of the measurable selection (of the implicit function) theorem (see [**15**], Theorem III.38), $\hat\alpha_s$ can be chosen to depend measurably in $s\in[0,T]$. Of course, if $\mathrm{ri}\mathcal{D}(L_s)=\emptyset$ then $\mathcal{D}(L_s)$ contains only $\hat\alpha_s$ and in this case $\hat\alpha_s=\dot\gamma_s$ for Lebesgue almost all $s\in[0,T]$. Taking $\alpha_s=\hat\alpha_s$ and $\beta_s=0$ we obtain

$$L(\gamma_s,\alpha_s)=\langle\beta_s,\alpha_s\rangle-H(\gamma_s,\beta_s). \tag{2.3.31}$$

Observe that $\ell(s,\alpha)=L(\gamma_s,\alpha)$ is measurable as a function of s and α since it is obtained via (2.2.4) as a supremum in one argument of a family of continuous functions, and so this supremum can be taken there over a countable dense set of β's. Hence, the set $A=\{(s,\alpha):\ s\in[0,T],\ \alpha\in\mathcal{D}(L_s)\}=\ell^{-1}[0,\infty)$ is measurable, and so the set $B=A\setminus\{(s,\dot\gamma_s),\ s\in[0,T]\}$ is measurable, as well. Its projection $V=\{s\in[0,T]:\ (s,\alpha)\in B\text{ for some }\alpha\in\mathbb{R}^d\}$ on the first component of the product space is also measurable and V is the set of $s\in[0,T]$ such that $\mathcal{D}(L_s)$ contains more than one point. Employing Theorem III.22 from [**15**] we select $\bar\alpha_s\in\mathbb{R}^d$ measurably in $s\in V$ and such that $(s,\bar\alpha_s)\in B$. By convexity and lower semicontinuity of $L(\gamma_s,\cdot)$ it follows from Corollary 7.5.1 in [**70**] that

$$L(\gamma_s,\dot\gamma_s)=\lim_{p\uparrow\infty}L(\gamma_s,\alpha_s^{(p)})\ \text{ where }\ \alpha_s^{(p)}=(1-p^{-1})\dot\gamma_s+p^{-1}\bar\alpha_s. \tag{2.3.32}$$

For each $\delta>0$ set

$$n_\delta(s)=\min\{n\in\mathbb{N}:\ |L(\gamma_s,\dot\gamma_s)-L(\gamma_s,\alpha_s^{(n)})|+|\dot\gamma_s-\alpha_s^{(n)}|<\delta\}.$$

Then, clearly, $n_\delta(s)$ is a measurable function of s, and so $\alpha_s=\alpha_s^{(\delta)}=\alpha_s^{(n_\delta(s))}$ and $L(\gamma_s,\alpha_s)$ are measurable in s, as well. By Theorems 23.4 and 23.5 from [**70**] for each $\alpha_s = \alpha_s^{(\delta)}$ there exists $\beta_s = \beta_s^{(\delta)} \in \mathbb{R}^d$ such that (2.3.31) holds true. Given $\delta', \lambda > 0$ take $\delta = \min(\delta', \lambda/3)$ and for $s \in [0,T] \setminus V$ set $\alpha_s = \hat{\alpha}_s$. Then

$$\int_0^T \big|L(\gamma_s, \dot{\gamma}_s) - L(\gamma_s, \alpha_s)\big| ds < \lambda/3 \text{ and } \int_0^T |\dot{\gamma}_s - \alpha_s| ds < \delta'. \tag{2.3.33}$$

For each $b > 0$ set $\alpha_s^b = \alpha_s$ if the corresponding β_s in (2.3.31) satisfies $|\beta_s| \le b$ and $\alpha_s^b = \hat{\alpha}_s$, otherwise. Note, that (2.3.31) remains true with α_s^b in place of α_s with $\beta_s = 0$ if $\alpha_s^b = \hat{\alpha}_s$. As observed above $|\alpha| \le K$ whenever $L(z,\alpha) < \infty$, and so $|\hat{\alpha}_s| \le K$ for Lebesgue almost all $s \in [0,T]$. We recall also that $|\dot{\gamma}_s - \alpha_s| < \delta$ and $\dot{\gamma}_s \le K$ for Lebesgue almost all $s \in [0,T]$. Since $S_{0T}(\gamma) < \infty$, $|L(\gamma_s, \dot{\gamma}_s) - L(\gamma_s, \alpha_s)| < \delta$, and $L(\gamma_s, \alpha_s^b) \uparrow L(\gamma_s, \alpha_s)$ as $b \uparrow \infty$ for Lebesgue almost all $s \in [0,T]$, we conclude from (2.3.33) and the above observations that for b large enough

$$\int_0^T \big|L(\gamma_s, \alpha_s) - L(\gamma_s, \alpha_s^b)\big| ds < \lambda/3 \text{ and } \int_0^T |\alpha_s - \alpha_s^b| ds < \delta'. \tag{2.3.34}$$

Next, we apply Lemma 2.3.3 to conclude that there exists a sequence $m_j \to \infty$ such that for each $\Delta_j = T/m_j$ and Lebesgue almost all $c \in [0,T)$,

$$\int_0^T \big|L(\gamma_s, \alpha_s^b) - L(\gamma_{q_j(s,c)}, \alpha^b_{q_j(s,c)})\big| ds < \lambda/3 \text{ and } \int_0^T |\alpha_s^b - \alpha^b_{q_j(s,c)}| ds < \delta'. \tag{2.3.35}$$

where $q_j(s,c) = [(s+c)\Delta_j^{-1}]\Delta_j - c$, $[\cdot]$ denotes the integral part and we assume $L(\gamma_s, \alpha_s^b) = 0$ and $\alpha_s^b = 0$ if $s < 0$.

Choose $c = c_j \in [\frac{1}{3}\Delta_j, \frac{2}{3}\delta_j]$ and set $\hat{\gamma}_s = x + \int_0^s \alpha^b_{q_j(u,c)} du$, $\psi_s = \gamma_{q_j(s,c)}$ where $\gamma_u = \gamma_0$ if $u < 0$, $x_0 = \tilde{x}_0 = x$, $x_N = \hat{\gamma}_T$, $\tilde{x}_N = \gamma_T$ and $x_k = \hat{\gamma}_{k\Delta_j - c}$, $\tilde{x}_k = \gamma_{k\Delta_j - c}$ for $k = 1, ..., N-1$ and $\xi_k = \alpha^b_{(k-1)\Delta_j - c}$ for $k = 1, 2, ..., N$ where $N = \min\{k : k\Delta_j - c > T\}$. Since $|\dot{\gamma}_s| \le \tilde{K}$ for Lebesgue almost all $s \in [0,T]$ then $\mathbf{r}_{0T}(\gamma, \psi) \le \tilde{K}\Delta_j$ and, in addition, $\mathbf{r}_{0T}(\gamma, \hat{\gamma}) \le 3\delta'$ by (2.3.33)–(2.3.35). This together with (2.3.13) and (2.3.14) yield that for $v = (x,y)$,

$$\begin{aligned} \mathbf{r}_{0T}(Z_v^\varepsilon, \gamma) &\le \mathbf{r}_{0T}(Z_v^\varepsilon, \psi) + \tilde{K}\Delta_j \le (KTe^{KT}+1)\mathbf{r}_{0T}(Z_v^{\varepsilon,\psi}, \psi) + \tilde{K}\Delta_j \\ &\le (KTe^{KT}+1)\big(3\delta' + \tilde{K}\Delta_j + (T+1)\max\nolimits_{1\le k\le N} \big|\Xi_k^\varepsilon(v, \tilde{x}_{k-1}) - \xi_k\big|\big) + \tilde{K}\Delta_j \end{aligned} \tag{2.3.36}$$

provided $\Delta_j \le 1$ where $Z_v^{\varepsilon,\psi}$ and $\Xi_k^\varepsilon(v,x)$ are the same as in Lemma 2.3.4, the latter is defined with $t_k = k\Delta_j - c$, $k = 1, ..., N-1$ and $t_N = T$. Choose δ' so small and m_j so large that

$$(KTe^{KT}+1)\big(3\delta' + \tilde{K}\Delta_j + (T+1)\delta'\big) + \tilde{K}\Delta_j < \delta$$

then by (2.3.36),

$$\big\{\mathbf{r}_{0T}(Z^\varepsilon_{x,y}, \gamma) < \delta\big\} \supset \big\{\max_{1\le k\le N} \big|\Xi_k^\varepsilon(v, \tilde{x}_{k-1}) - \xi_k\big| < \delta'\big\}. \tag{2.3.37}$$

By (2.3.33)–(2.3.35),

$$\sum_{k=1}^N (t_k - t_{k-1}) L(\tilde{x}_{k-1}, \xi_k) \le S_{0T}(\gamma) + \lambda \tag{2.3.38}$$

and by the construction above the conditions of the assertion (ii) of Proposition 2.3.5 hold true, so choosing m_j sufficiently large we derive (2.2.6) (with 2λ in place of λ) from (2.3.20), (2.3.37) and (2.3.38) provided ε is small enough.

Next, we pass to the proof of the upper bound (2.2.7). Assume that (2.2.7) is not true, i.e. there exist $a, \lambda, \delta > 0$ and $x \in \mathcal{X}_T$ such that for some sequence $\varepsilon_k \to 0$ as $k \to \infty$,

$$P\{\mathbf{r}_{0T}(Z^{\varepsilon_k}_{x,y}, \Psi^a_{0T}(x)) \geq 3\delta\} > \exp\big(-\frac{1}{\varepsilon_k}(a-\lambda)\big). \tag{2.3.39}$$

Since $|B(x,y)| \leq K$ by (2.2.1) all paths of $Z^\varepsilon_{x,y}(t)$, $t \in [0,T]$ and of $Z^{\varepsilon,\psi}_{v,x}(t)$, $t \in [0,T]$ given by (2.3.11) (the latter for any measurable ψ) belong to a compact set $\tilde{\mathcal{K}}^x \subset C_{0T}$ which consists of curves starting at x and satisfying the Lipschitz condition with the constant K. Let $\tilde{U}^x_\rho$ denotes the open ρ-neighborhood of the compact set $\Psi^a_{0T}(x)$ and $\mathcal{K}^x_\rho = \tilde{\mathcal{K}}^x \setminus \tilde{U}^x_\rho$. For any small $\delta' > 0$ choose a δ'-net $\gamma_1, ..., \gamma_n$ in $\mathcal{K}^x_{2\delta}$ where $n = n(\delta')$. Since

$$\{\mathbf{r}_{0T}(Z^{\varepsilon_k}_{x,y}, \Psi^a_{0T}(x)) \geq 3\delta\} \subset \bigcup_{n \geq j \geq 1} \{\mathbf{r}_{0T}(Z^{\varepsilon_k}_{x,y}, \gamma_j) \leq \delta'\}$$

then there exists j and a subsequence of $\{\varepsilon_k\}$, for which we use the same notation, such that

$$P\{\mathbf{r}_{0T}(Z^{\varepsilon_k}_{x,y}, \gamma_j) \leq \delta'\} > n^{-1}\exp\big(-\frac{1}{\varepsilon_k}(a-\lambda)\big). \tag{2.3.40}$$

Denote such γ_j by $\gamma^{\delta'}$, choose a sequence $\delta_l \to 0$ and set $\gamma^{(l)} = \gamma^{\delta_l}$. Since $\mathcal{K}^x_{2\delta}$ is compact there exists a subsequence $\gamma^{(l_j)}$ converging in C_{0T} to $\hat{\gamma} \in \mathcal{K}^x_{2\delta}$ which together with (2.3.40) yield

$$\limsup_{\varepsilon \to 0} \varepsilon \ln P\{\mathbf{r}_{0T}(Z^\varepsilon_{x,y}, \hat{\gamma}) \leq \delta'\} > -a + \lambda \tag{2.3.41}$$

for all $\delta' > 0$.

We claim that (2.3.41) contradicts (2.3.12) and the assertion (i) of Proposition 2.3.5. Indeed, set

$$S^\psi_{b,0T}(\gamma) = \int_0^T L_b(\psi(s), \dot{\gamma}(s))ds \text{ and } S_{b,0T}(\gamma) = S^\gamma_{b,0T}(\gamma).$$

By the monotone convergence theorem

$$S^\psi_{b,0T}(\gamma) \uparrow S^\psi_{0T}(\gamma) \text{ and } S_{b,0T}(\gamma) \uparrow S_{0T}(\gamma) \text{ as } b \uparrow \infty. \tag{2.3.42}$$

Similarly to our remark in Section 2.2 it follows from the results of Section 9.1 of [**41**] that the functionals $S^\psi_{b,0T}(\gamma)$, $S^\psi_{0T}(\gamma)$ and $S_{b,0T}(\gamma)$, $S_{0T}(\gamma)$ are lower semicontinuous in ψ and γ (see also Section 7.5 in [**31**]). This together with (2.3.42) enable us to apply Lemma 2.3.2 in order to conclude that

$$\lim_{b \to \infty} S_{b,0T}(\mathcal{K}^x_\delta) = S_{0T}(\mathcal{K}^x_\delta) = \inf_{\gamma \in \mathcal{K}^x_\delta} S_{0T}(\gamma) > a \tag{2.3.43}$$

where $S_{b,0T}(\mathcal{K}^x_\delta) = \inf_{\gamma \in \mathcal{K}^x_\delta} S_{b,0T}(\gamma)$. The last inequality in (2.3.43) follows from the lower semicontinuity of S_{0T}. Thus we can and do choose $b > 0$ such that

$$S_{b,0T}(\mathcal{K}^x_\delta) > a - \lambda/8. \tag{2.3.44}$$

By the lower semicontinuity of $S^\psi_{b,0T}(\gamma)$ in ψ there exists a function $\delta(\gamma) > 0$ on $\mathcal{K}^x_\delta$ such that for each $\gamma \in \mathcal{K}^x_\delta$,

$$S^\psi_{b,0T}(\gamma) > a - \lambda/4 \text{ provided } \mathbf{r}_{0T}(\gamma, \psi) < \delta_\lambda(\gamma). \tag{2.3.45}$$

Next, we restrict the set of functions ψ to make it compact. Namely, we allow from now on only functions ψ for which there exists $\gamma \in \mathcal{K}_\delta^x$ such that either $\psi \equiv \gamma$ or $\psi(t) = \gamma(kT/m)$ for $t \in [kT/m, (k+1)T/m)$, $k = 0, 1, ..., m-1$ and $\psi(T) = \gamma(T)$ where m is a positive integer. It is easy to see that the set of such functions ψ is compact with respect to the uniform convergence topology in C_{0T} and it follows that $\delta_\lambda(\gamma)$ in (2.3.45) constructed with such ψ in mind is lower semicontinuous in γ. Hence

$$\delta_\lambda = \inf_{\gamma \in \mathcal{K}_\delta^x} \delta_\lambda(\gamma) > 0. \tag{2.3.46}$$

Now take $\hat{\gamma}$ satisfying (2.3.41) and for any integer $m \geq 1$ set $\Delta = \Delta_m = T/m$, $x_k = x_k^{(m)} = \hat{\gamma}(k\Delta)$, $k = 0, 1, ..., m$ and $\xi_k = \xi_k^{(m)} = \Delta^{-1}\big(\hat{\gamma}(k\Delta) - \hat{\gamma}((k-1)\Delta)\big)$, $k = 1, ..., m$. Define a piecewise linear χ_m and a piecewise constant ψ_m by

$$\chi_m(t) = x_k + \xi_k \Delta \text{ and } \psi_k(t) = x_k \text{ for } t \in [k\Delta, (k+1)\Delta) \tag{2.3.47}$$

and $k = 0, 1, ..., m-1$ with $\chi_m(T) = \psi_m(T) = \hat{\gamma}(T)$. Since $\hat{\gamma}$ is Lipschitz continuous with the constant $\tilde{K}$ then

$$\mathbf{r}_{0T}(\chi_m, \psi_m) \leq \tilde{K}\Delta \text{ and } \mathbf{r}_{0T}(\hat{\gamma}, \psi_m) \leq \tilde{K}\Delta. \tag{2.3.48}$$

If m is large enough and $\varepsilon > 0$ is sufficiently small then

$$\Delta < \tilde{K}^{-1} \min(\delta/2, \delta_\lambda) \text{ and } \eta_{b,T}(\varepsilon, \Delta) < \lambda/8 \tag{2.3.49}$$

where $\eta_{b,T}(\varepsilon, \Delta)$ is the same as in (2.3.16). Since $\hat{\gamma} \in \mathcal{K}_{2\delta}^x$ it follows from (2.3.48) and (2.3.49) that $\chi_m \in \mathcal{K}_\delta^x$ and by (2.3.45) and the first inequality in (2.3.49) we obtain that

$$S_{b,0T}^{\psi_m}(\chi_m) = \Delta \sum_{k=0}^{m-1} L_b(x_k, \xi_k) > a - \frac{\lambda}{4}. \tag{2.3.50}$$

Hence, by (2.3.16) and the second inequality in (2.3.49) for all ε small enough,

$$P\{\max_{1 \leq k \leq m} |\Xi_k^\varepsilon((x,y), x_{k-1}) - \xi_k| < \rho\} \leq e^{-\frac{1}{\varepsilon}(a - \lambda/2)} \tag{2.3.51}$$

provided $C_T(b)\rho < \lambda/8$ (taking into account that $x_0 = x$). By (2.3.12) and the definition of vectors ξ_k for any $v \in \mathcal{W}$,

$$\begin{aligned} |\Xi_k^\varepsilon(v, x_{k-1}) - \xi_k| &\leq |\Xi_k^\varepsilon(v, x_{k-1}) - \Delta^{-1}\big(Z_v^\varepsilon(k\Delta) - Z_v^\varepsilon((k-1)\Delta)\big)| \\ &+ 2\Delta^{-1}\mathbf{r}_{0T}(Z_v^\varepsilon, \hat{\gamma}) \leq (K + 2\Delta^{-1})\mathbf{r}_{0T}(Z_v^\varepsilon, \hat{\gamma}) + \tfrac{1}{2}\tilde{K}^2\Delta. \end{aligned} \tag{2.3.52}$$

Therefore,

$$\begin{aligned} &\{\mathbf{r}_{0T}(Z_{x,y}^\varepsilon, \hat{\gamma}) \leq \delta'\} \\ &\subset \{\max\nolimits_{1 \leq k \leq m} |\Xi_k^\varepsilon((x,y), x_{k-1}) - \xi_k| \leq (\tilde{K} + 2\Delta^{-1})\delta' + \tfrac{1}{2}\tilde{K}^2\Delta\}. \end{aligned} \tag{2.3.53}$$

Choosing, first, m large enough so that Δ satisfies (2.3.49) with all sufficiently small ε and also that $8C_T(b)\tilde{K}^2\Delta < \lambda$, and then choosing δ' so small that $16C_T(b)(\tilde{K} + 2\Delta^{-1})\delta' < \delta$, we conclude that (2.3.51) together with (2.3.53) contradicts (2.3.41), and so the upper bound (2.2.7) holds true, completing the proof of Theorem 2.2.2. □

2.3.6. REMARK. *In view of examples from* **[11]** *in the fully coupled setup we should not expect convergence (2.2.22) in the averaging principle with probability one in spite of exponentially fast convergence in probability (2.2.23) provided by the upper large deviations bound (2.2.11). Still, when derivatives of* X^ε *and* Y^ε *in*

ε grow not too fast we can derive convergence with probability one from (2.2.23). Indeed, consider, for instance, the following example

$$X^\varepsilon(t) = x + \varepsilon \int_0^t B(X^\varepsilon(s), Y^\varepsilon(s))ds \text{ and} \tag{2.3.54}$$
$$Y^\varepsilon(t) = y + cw_t + \int_0^t b(X^\varepsilon(s))ds \ (mod 1)$$

where $c \neq 0$ is a constant, w_t is the standard one dimensional Brownian motion, $B(x,y)$ satisfies (2.2.1) and it is 1-periodic in y and b has a bounded derivative in x. Set

$$x^\varepsilon(t) = \frac{dX^\varepsilon(t)}{d\varepsilon} \text{ and } y^\varepsilon(t) = \frac{dY^\varepsilon(t)}{d\varepsilon}.$$

Then

$$\frac{d}{dt}\begin{pmatrix} x^\varepsilon(t) \\ y^\varepsilon(t) \end{pmatrix} = \begin{pmatrix} \varepsilon \frac{\partial B\left(X^\varepsilon(t),Y^\varepsilon(t)\right)}{\partial x} & \varepsilon \frac{\partial B\left(X^\varepsilon(t),Y^\varepsilon(t)\right)}{\partial y} \\ \frac{\partial b\left(X^\varepsilon(t)\right)}{\partial x} & 0 \end{pmatrix} \begin{pmatrix} x^\varepsilon(t) \\ y^\varepsilon(t) \end{pmatrix}$$
$$+ \begin{pmatrix} B\left(X^\varepsilon(t), Y^\varepsilon(t)\right) \\ 0 \end{pmatrix}.$$

The solution of this linear equation is easy to estimate which yields that for some constant $C > 0$,

$$\sup_{0 \le t \le T/\varepsilon} \left|\frac{dX^\varepsilon(t)}{d\varepsilon}\right| \le C \exp(C/\sqrt{\varepsilon}). \tag{2.3.55}$$

Let μ^x be the invariant measure of the diffusion $Y_x(t) = y + w_t + tb(x)$ (mod 1) (which is unique since the Doeblin condition is satisfied here) and assume that

$$\int B(x,y)d\mu^x(y) = 0 \quad \text{for all } x \tag{2.3.56}$$

which does not harm the generality since we always can consider $B(x,y) - \int B(x,y)d\mu^x(y)$ in place of $B(x,y)$. Set $\varepsilon_k = \alpha(\delta)/2\ln k$ where $\alpha(\delta)$ is the same as in (2.2.23) written for our specific situation. Then $e^{-\alpha(\delta)/\varepsilon_k} = k^{-2}$ and by the Borel–Cantelli lemma we obtain that there exists $k_\delta(\omega)$, finite with probability one, so that for all $k \ge k_\delta(\omega)$,

$$\max_{0 \le t \le T/\varepsilon_k} |X^{\varepsilon_k}_{x,y}(t) - x| < \delta.$$

By (2.3.55) for $\varepsilon_{k+1} < \varepsilon \le \varepsilon_k$ and $k \ge 2$,

$$\max\nolimits_{0 \le t \le T/\varepsilon_{k+1}} |X^{\varepsilon_k}_{x,y}(t) - X^{\varepsilon}_{x,y}(t)| \le Ce^{C/\sqrt{\varepsilon_k}}(\varepsilon_k - \varepsilon_{k+1})$$
$$\le C\exp\left(C\sqrt{2(\ln k)/\alpha(\delta)}\right)\ln(1 + \tfrac{1}{k})(\ln k)^{-2} \longrightarrow 0 \text{ as } k \to \infty.$$

It follows that with probability one,

$$\max_{0 \le n \le T/\varepsilon} |X^{\varepsilon}_{x,y}(t) - x| \to 0 \text{ as } k \to \infty$$

which is what we need since in our case $X^{\varepsilon}_{x,y}(t) \equiv x$ in view of (2.3.56).

2.4. Verifying assumptions for random evolutions

In this section we will prove Proposition 2.2.4. Observe that $H(x,x',\beta)$ obtained by (2.2.18) is the principal eigenvalue of the operator $\mathcal{L}^x + \langle\beta, B(x',\cdot)\rangle$ acting on C^2 vector functions $f=(f_1,...,f_N)$ on the manifold M by the formula (see [**48**]),

$$\big((\mathcal{L}^x + \langle\beta, B(x',\cdot)\rangle)f\big)_k = \mathcal{L}^x_k f_k + \langle\beta, B_k(x',\cdot)\rangle f_k$$

where x, x' and β are considered as parameters. According to [**68**] this operator satisfies the strong maximum principle. Thus, the first part of Proposition 2.2.4 follows from the well known results on operators satisfying the maximum principle (see [**23**], [**24**] and [**48**]) and the results on the principle eigenvalue of positive operators (see [**58**], [**62**] and [**39**]) and of its smooth dependence on parameters which can be derived from the general perturbation theory of linear operators (see [**43**]).

Now we obtain from (2.2.18) that for $k=1,...,N$ uniformly in $z, x' \in \bar{\mathcal{X}}$, $v \in M$ and $|\beta| \le b$,

$$\Big|\frac{1}{s}\log E\exp\langle\beta, \int_0^s B_{\nu_{z,v,k}(u)}(x', \hat{Y}_{z,v,k}(u))du\rangle - H(z,x',\beta)\Big| \le \rho_b(s) \tag{2.4.1}$$

where $\rho_b(s) \to 0$ as $s \to \infty$. Next, we want to compare

$$Q_{z,v,k}(s) = E\exp\langle\beta, \textstyle\int_0^s B_{\nu_{z,v,k}(u)}(x', \hat{Y}_{z,v,k}(u))du\rangle \text{ and}$$
$$Q^\varepsilon_{z,v,k}(s) = E\exp\langle\beta, \textstyle\int_0^s B_{\nu^\varepsilon_{z,v,k}(u)}(x', \hat{Y}^\varepsilon_{z,v,k}(u))du\rangle.$$

In order to do this we introduce auxiliary random evolutions $W_{z,v,k}(s) = (\hat{W}_{z,v,k}(s), \eta_k(s))$ and $W^\varepsilon_{z,v,k}(s) = (\hat{W}^\varepsilon_{z,v,k}(s), \eta_k(s))$ governed by the stochastic differential equations

$$d\hat{W}_{z,v,k}(s) = \sigma_{\eta_k(s)}\big(z, \hat{W}_{z,v,k}(s)\big)dw_s + b_{\eta_k(s)}\big(z, \hat{W}_{z,v,k}(s)\big)ds \tag{2.4.2}$$

and

$$d\hat{W}^\varepsilon_{z,v,k}(s) = \sigma_{\eta_k(s)}\big(\tilde{X}^\varepsilon_{z,v,k}(s), \hat{W}^\varepsilon_{z,v,k}(s)\big)dw_s + b_{\eta_k(s)}\big(\tilde{X}^\varepsilon_{z,v,k}(s), \hat{W}^\varepsilon_{z,v,k}(s)\big)ds, \tag{2.4.3}$$

respectively, where $\hat{W}_{z,v,k}(0) = \hat{W}^\varepsilon_{z,v,k}(0) = v$,

$$\frac{d\tilde{X}^\varepsilon(s)}{dt} = \varepsilon B\big(\tilde{X}^\varepsilon(s), W^\varepsilon(s)\big),\quad \tilde{X}^\varepsilon_{z,v,k}(0) = z \tag{2.4.4}$$

and for $i \ne j$,

$$P\big\{\eta_k(s+\Delta) = j \big| \eta_k(s) = i\big\} = \Delta + o(\Delta),\quad \eta_k(0) = k. \tag{2.4.5}$$

According to [**27**] (which relies on Theorem 2 in §6, Ch. VII of [**34**]) the distributions in the path space of the processes $Y_{z,v,k}$ and $Y^\varepsilon_{z,v,k}$ are absolutely continuous with respect to the distributions in the path space of the processes $W_{z,v,k}$ and $W^\varepsilon_{z,v,k}$, respectively, with the densities

$$p_s(\hat{W}_{z,v,k}(\cdot), \eta) = \textstyle\prod_{i=0}^{n(s)-1} q_{\eta_k(\zeta_i)\eta_k(\zeta_{i+1})}(z, \hat{W}_{z,v,k}(\zeta_i)) \tag{2.4.6}$$
$$\times \exp\big(-\textstyle\sum_{i=0}^{n(s)} \int_{\zeta_i}^{\zeta_{i+1}\wedge s} (q_{\eta_k(\zeta_i)}(z, \hat{W}_{z,v,k}(u)) - N + 1)du\big)$$

and

(2.4.7) $p^\varepsilon_s(\tilde X^\varepsilon_{z,v,k}(\cdot),\hat W^\varepsilon_{z,v,k}(\cdot),\eta)=\prod_{i=0}^{n(s)-1}q_{\eta_k(\zeta_i)\eta_k(\zeta_{i+1})}(\tilde X^\varepsilon_{z,v,k}(\zeta_i),\hat W^\varepsilon_{z,v,k}(\zeta_i))$
$$\exp\big(-\sum_{i=0}^{n(s)}\int_{\zeta_i}^{\zeta_{i+1}\wedge s}(q_{\eta_k(\zeta_i)}(\tilde X^\varepsilon_{z,v,k}(u),\hat W^\varepsilon_{z,v,k}(u))-N+1)du\big),$$
respectively, where
$$q_k(z,y)=\sum_{l=1,l\neq k}^{N}q_{kl}(z,y),$$
$\zeta_0=0$, $\zeta_{i+1}=\inf\{u>\zeta_i:\eta_k(u)\neq\eta_k(\zeta_i)\}$ and $n(s)=\max\{i:\eta_i\leq s\}$.

Thus, we have to compare

$$Q_{z,v,k}(s)=Ep_s(\hat W_{z,v,k}(\cdot),\eta)\exp\langle\beta,\int_0^s B_{\eta_k(u)}(x',\hat W_{z,v,k}(u))du\rangle \tag{2.4.8}$$

and

$$Q^\varepsilon_{z,v,k}(s)=Ep^\varepsilon_s(\tilde X^\varepsilon_{z,v,k}(\cdot),\hat W_{z,v,k}(\cdot),\eta)\exp\langle\beta,\int_0^s B_{\eta_k(u)}(x',\hat W^\varepsilon_{z,v,k}(u))du\rangle. \tag{2.4.9}$$

Observe that by (2.2.1),

$$\sup_{0\leq u\leq s}|\tilde X^\varepsilon_{z,v,k}(u)-z|\leq\varepsilon sK. \tag{2.4.10}$$

Let K_q be both an upper bound for $|q_{ij}(x,y)|$ and their Lipschitz constant then we see from (2.2.1) and (2.4.6)–(2.4.10) that

$$\begin{aligned}|Q_{z,v,k}(s)-Q^\varepsilon_{z,v,k}(s)|&\leq e^{(N+K_q+K|\beta|)s}E\Big(n(s)K_q^{n(s)}\Big(\varepsilon sKK_q\\&+K_q\sup_{0\leq u\leq s}\operatorname{dist}(\hat W_{z,v,k}(s),\hat W^\varepsilon_{z,v,k}(s))+2\big|\exp\big(\varepsilon sK(K_q+|\beta|)\\&+s(K_q+K|\beta|)\sup_{0\leq u\leq s}\operatorname{dist}(\hat W_{z,v,k}(s),\hat W^\varepsilon_{z,v,k}(s))\big)-1\big|\Big)\Big).\end{aligned} \tag{2.4.11}$$

Employing the Witney theorem embed smoothly M as a compact submanifold in an Euclidean space $\mathbb{R}^D$ of a sufficiently high dimension D and extend the operator $\mathcal{L}^x+\langle\beta,B(x',\cdot)\rangle$ from M to $\mathbb{R}^d$ so that its coefficients remain C^2 and they vanish outside a relatively compact set containing M (cf. [**38**]). Now we can view (2.4.2) and (2.4.3) as stochastic differential equations in $\mathbb{R}^d$ keeping the same notations for their coefficients and processes there. Then using standard martingale moment estimates for stochastic integrals (see, for instance, [**42**]) together with (2.4.10) and the Lipschitz continuity of coefficients in (2.4.2) and (2.4.3) we obtain

$$\begin{aligned}&E\sup_{0\leq u\leq s}|\hat W_{z,v,k}(u)-\hat W^\varepsilon_{z,v,k}(u)|^2\\&\leq C_1(1+s)\big(\varepsilon^2s^2K^2+\textstyle\int_0^s E\sup_{0\leq r\leq u}|\hat W_{z,v,k}(r)-\hat W^\varepsilon_{z,v,k}(r)|^2du\big)\end{aligned}$$

for some $C_1>0$ independent of t,x,v,k and ε. Hence, by the Gronwall inequality

$$\begin{aligned}&E\sup_{0\leq u\leq s}\big(\operatorname{dist}(\hat W_{z,v,k}(u),\hat W^\varepsilon_{z,v,k}(u))\big)^2\\&=E\sup_{0\leq u\leq s}|\hat W_{z,v,k}(u)-\hat W^\varepsilon_{z,v,k}(u)|^2\leq C_1(1+s)\varepsilon^2s^2K^2e^{C_1(1+s)s}.\end{aligned} \tag{2.4.12}$$

Observe also that the distribution of $n(s)$ can be written explicitly as (see §55 in [**33**]),

$$P\{n(s)=k\}=e^{-(N-1)s}\frac{((N-1)s)^k}{k!}. \tag{2.4.13}$$

In order to estimate the last expression in the right hand side of (2.4.11) we note that for any random variable ξ,

$$|e^{\xi}-1|\le 2|\xi|+(1+e^{\xi})\mathbb{I}_{|\xi|>1},$$

and so by the Cauchy–Schwarz and the Chebyshev's inequalities

$$E(e^{\xi}-1)^2\le 4E\xi^2+2\big(E(1+e^{\xi})^4\big)^{1/2}(E\xi^2)^{1/2}. \tag{2.4.14}$$

Now by (2.4.11)–(2.4.14) together with the Cauchy–Schwarz inequality we obtain that for $k=1,...,N$ uniformly in $x,x'\in\bar{\mathcal{X}}$, $v\in M$ and $|\beta|\le b$,

$$|Q_{z,v,k}(s)-Q^{\varepsilon}_{z,v,k}(s)|\le C_2(1+b+s)(\varepsilon s+\sqrt{\varepsilon s})e^{C_2(1+b+s)s} \tag{2.4.15}$$

for another constant $C_2>0$ independent of z,x',v,β,t and ε.

Choose $s=s(\varepsilon)=(\log(1/\varepsilon))^{1/3}$ and set $l(\varepsilon)=[t/(\log(1/\varepsilon))^{1/3}]$. By (2.2.21),

$$e^{-Kbs(\varepsilon)}Q^{\varepsilon}_{x,y,k}(l(\varepsilon)s(\varepsilon))\le Q^{\varepsilon}_{x,y,k}(t)\le e^{Kbs(\varepsilon)}Q^{\varepsilon}_{x,y,k}(l(\varepsilon)s(\varepsilon)). \tag{2.4.16}$$

If $|z-x|\le K\varepsilon t$ and $s\le t$ then by (2.4.1), (2.4.15) and the Lipschitz continuity of H we obtain that

$$\begin{aligned} -C_2(1+b+s)(\varepsilon s+\sqrt{\varepsilon s})e^{C_2(1+b+s)s}+\exp\big(s(H(x,x',\beta)\\ -C(b)K\varepsilon t-\rho_b(s))\big)\le Q^{\varepsilon}_{z,v,k}(s)\le C_2(1+b+s)(\varepsilon s+\sqrt{\varepsilon s})e^{C_2(1+b+s)s}\\ +\exp\big(s(H(x,x',\beta)-C(b)K\varepsilon t-\rho_b(s))\big)\end{aligned} \tag{2.4.17}$$

for some constant $C(b)>0$ independent of $z,x,x'\in\bar{\mathcal{X}}$, $v\in M$, $|\beta|\le b$, t and ε. Observe that by the Markov property,

$$\begin{aligned} Q^{\varepsilon}_{x,y,k}(ls(\varepsilon))=E\exp\langle\beta,\textstyle\int_0^{(l-1)s(\varepsilon)}(B_{\nu^{\varepsilon}_{x,y,k}(u)}(x',\hat{Y}^{\varepsilon}_{x,y,k}(u))du\rangle\\ \times Q^{\varepsilon}_{X^{\varepsilon}_{x,y,k}((l-1)s(\varepsilon)),Y^{\varepsilon}_{x,y,k}((l-1)s(\varepsilon)),\nu^{\varepsilon}_{x,y,k}((l-1)s(\varepsilon))}\big(s(\varepsilon)\big).\end{aligned} \tag{2.4.18}$$

Now by (2.4.10) and (2.4.17) applying (2.4.18) for $l=l(\varepsilon),l(\varepsilon)-1,...,2$ we obtain that

$$\Big|\frac{1}{t}\log Q^{\varepsilon}_{x,y,k}(t)-H(x,x',\beta)\Big|\le\tilde{C}(b)\big(\varepsilon^{1/3}+\varepsilon t+\rho_b(\min(t,s(\varepsilon))\big) \tag{2.4.19}$$

for some $\tilde{C}(b)>0$ independent of $x,x'\in\bar{\mathcal{X}}$, $y\in M$, $|\beta|\le b$, t and ε, which yields (2.2.3) completing the proof of Proposition 2.2.4. □

2.5. Further properties of S-functionals

In this section we study essential properties of the functionals S_{0T} which will be needed in the proofs of Theorems 2.2.5 and 2.2.7 in the next sections. The following result which follows from [**67**] is a basic step in our analysis of functionals $S_{0t}(\gamma)$ and our thanks go to R. Pinsky who quickly produced on our request [**67**] deriving some properties of functionals $I_x(\mu)$ needed here.

2.5.1. LEMMA. *For each $x\in\bar{\mathcal{X}}$ and any vector measure $\mu=(\mu_1,...,\mu_N)$ on M with $\sum_{k=1}^N\mu_k(M)=1$, $I_x(\mu)<\infty$ if and only if each μ_k, $k=1,...,N$ has density $g_k=d\mu_k/dm$ with respect to the Riemannian volume m on M such that*

$$\int_M\|\nabla\sqrt{g_k}\|^2dm<\infty,\ k=1,...,N \tag{2.5.1}$$

where ∇ is the Riemannian gradient and $\|\cdot\|$ is a corresponding norm. Furthermore, there exists $C > 0$ such that for any $x \in \bar{\mathcal{X}}$ and each μ as above for which (2.5.1) holds true,

(2.5.2)

$$C^{-1}\sum_{k=1}^{N} a_k \int_M \|\nabla\sqrt{g_k}\|^2 dm - C \le I_x(\mu) \le C \sum_{1\le k\le N} a_k \int_M \|\nabla\sqrt{g_k}\|^2 dm + C$$

where $a_k = \mu_k(M)$, and if $z \in \bar{\mathcal{X}}$ is another point then

$$|I_x(\mu) - I_z(\mu)| \le C|x-z| \sum_{k=1}^{N} a_k \int_M \|\nabla\sqrt{g_k}\|^2 dm. \tag{2.5.3}$$

Next, using Lemma 2.5.1 we are able to show that each point where B is complete can be connected with close points by curves with small S-functionals which, in particular, enables us to obtain important examples of S-compacts.

2.5.2. Lemma. *(i) There exists $C > 0$ and for each $x \in \bar{\mathcal{X}}$ where the vector field B is complete there exists $r = r(x) > 0$ such that if $|z_1 - x| < r$ and $|z_2 - x| < r$ then we can construct $\gamma \in C_{0t}$ with $t \le C|z_1 - z_2|$ satisfying*

$$\gamma_0 = z_1,\ \gamma_t = z_2 \text{ and } S_{0t}(\gamma) \le C|z_1 - z_2|.$$

It follows that $R(\tilde{z}, z)$ and $R(z, \tilde{z})$ are locally Lipschitz continuous in z belonging to the open r-neighborhood of x when $\tilde{z}$ is fixed.

(ii) Let $\mathcal{O} \subset \mathcal{X}$ be a compact Π^t-invariant set which either contains a dense in $\mathcal{O}$ orbit of Π^t or $R(x,z) = 0$ for any pair $x, z \in \mathcal{O}$. Suppose that B is complete at each point of $\mathcal{O}$. Then $\mathcal{O}$ is an S-compact.

(iii) Assume that for any $\eta > 0$ there exists $T(\eta) > 0$ such that for each $x \in \mathcal{O}$ its orbit $\{\Pi^t x,\ t \in [0, T(\eta)]\}$ of length $T(\eta)$ forms an η-net in $\mathcal{O}$ and suppose that B is complete at a point of $\mathcal{O}$. Then $\mathcal{O}$ is an S-compact.

Proof. (i) Fix some $x \in \bar{\mathcal{X}}$ and assume that B is complete at x. Then we can find a simplex Δ_x with vertices in $\Gamma_x = \{\bar{B}_\mu(x) : I_x(\mu) < \infty\}$ such that $\{\alpha\Delta_x,\ \alpha \in [0,1]\}$ contains an open neighborhood of 0 in $\mathbb{R}^d$ and

$$\Delta_x = \{\sum_{i=1}^{k} \lambda_i \bar{B}_{\mu^{(i)}}(x) : \sum_{i=1}^{k} \lambda_i = 1,\ \lambda_i \ge 0\,\forall i\}$$

for some $\mu^{(i)}$ with $I_x(\mu^{(i)}) < \infty$. By compactness of Δ_x it follows that

$$\text{dist}(\Delta_x, 0) = d_x > 0.$$

By (2.2.1) there exists a small $r(x) > 0$ such that if $|z - x| \le r(x)$ then each simplex

$$\Delta_z = \{\sum_{i=1}^{k} \lambda_i \bar{B}_{\mu^{(i)}}(z) : \sum_{i=1}^{k} \lambda_i = 1,\ \lambda_i \ge 0\,\forall i\}$$

intersects and not at 0 with any ray emanating from $0 \in \mathbb{R}^d$ or, in other words, $\{\alpha\Delta_z,\ \alpha \in [0,1]\}$ contains an open neighborhood of 0 in $\mathbb{R}^d$ and, moreover,

$$\text{dist}(0, \Delta_z) \ge \frac{1}{2} d_x.$$

It follows that for any z in the $r(x)$-neighborhood of x and any vector ξ there exist $\lambda_1, ..., \lambda_k \geq 0$ with $\lambda_1 + \cdots + \lambda_k = 1$ such that

$$\sum_{i=1}^{k} \lambda_i \bar{B}_{\mu^{(i)}}(z) = \bar{B}_{\sum_{1\leq i\leq k} \lambda_i \mu^{(i)}}(z) = \xi.$$

Observe that by (2.5.2) and convexity of I_z,

$$I_z\big(\sum_{1\leq i\leq k} \lambda_i \mu^{(i)}\big) \leq \max_{1\leq i\leq k} I_z(\mu^{(i)}) \leq \tilde{C}\big(\max_{1\leq i\leq k} I_z(\mu^{(i)}) + 1\big)$$

for some $\tilde{C} > 0$. Hence, any two points z_1 and z_2 from the open $r(x)$-neighborhood of x can be connected by a curve γ lying on the interval connecting z_1 and z_2 with $K \geq |\dot{\gamma}_s^{(1)}| \geq \frac{1}{2}d_x$, i.e. $\gamma_0 = z_1$, $\gamma_t = z_2$ with some $t \in [K^{-1}|z_1 - z_2|, 2d_x^{-1}|z_1 - z_2|]$ and by (2.2.9),

$$S_{0t}(\gamma) \leq 2\tilde{C}d_x^{-1}|z_1 - z_2|\big(\max_{1\leq i\leq k} I_x(\mu^{(i)}) + 1\big).$$

In view of the triangle inequality for R what we have proved yields the continuity of $R(\tilde{z}, z)$ and $R(z, \tilde{z})$ in z belonging to the open $r(x)$-neighborhood of x when $\tilde{z}$ is fixed. Covering $\bar{\mathcal{X}}$ by $r(x)-$neighborhoods of points $x \in \bar{\mathcal{X}}$ and choosing a finite subcover we obtain (i) with the same constant $C > 0$ for all points in $\bar{\mathcal{X}}$.

For the proof of sufficient conditions (ii) and (iii) of S-compacthess see Lemma 1.6.2(ii)–(iii) in Part 1. □

2.5.3. LEMMA. *For any $\eta > 0$ and $T > 0$ there exists $\zeta > 0$ such that if $\gamma \in C_{0T}$, $\gamma \subset \mathcal{X}$, $S_{0T}(\gamma) < \infty$, $\gamma_0 = x_0$, and $|z_0 - x_0| < \zeta$ then we can find $\tilde{\gamma} \in C_{0T}$, $\tilde{\gamma} \subset \mathcal{X}$ with $\tilde{\gamma}_0 = z_0$ satisfying*

$$\mathbf{r}_{0T}(\gamma, \tilde{\gamma}) < \eta \text{ and } |S_{0T}(\tilde{\gamma}) - S_{0T}(\gamma)| < \eta. \tag{2.5.4}$$

PROOF. By (2.2.9), (2.2.19) and the lower semicontinuity of the functionals $I_z(\nu)$ there exist measures $\nu_t \in \mathcal{M}_{\gamma_t}$, $t \in [0, T]$ such that $\dot{\gamma}_t = \bar{B}_{\nu_t}(\gamma_t)$ for Lebesgue almost all $t \in [0, T]$ and $I_{\gamma_t}(\nu_t) = L(\gamma_t, \dot{\gamma}_t)$ for Lebesgue almost all $t \in [0, T]$. Recall also that $\dot{\gamma}_t$ is measurable in t. Introduce the (measurable) map $q: [0, T] \times \mathcal{P}(\bar{\mathcal{W}}) \to \mathbb{R} \cup \{\infty\} \times \mathbb{R}^d$ defined by $q(t, \nu) = \big(I_{\gamma_t}(\nu), \bar{B}_\nu(\gamma_t)\big)$. Recall that $\dot{\gamma}_t$ is measurable in t, and so another map $r: [0, T] \to \mathbb{R} \cup \{\infty\} \times \mathbb{R}^d$ defined by $r(t) = \big(L(\gamma_t, \dot{\gamma}_t), \dot{\gamma}_t\big)$ is also measurable in $t \in [0, T]$. Then $q(t, \nu_t) = r(t)$ and it follows from the measurable selection in the implicit function theorem (see [**15**], Theorem III.38) that measures ν_t satisfying this condition can be chosen to depend measurably on $t \in [0, T]$. Since $S_{0T}(\gamma) < \infty$ and the I-functionals are nonnegative then $I_{\gamma_t}(\nu_t) < \infty$ for Lebesgue almost all $t \in [0, T]$ (and, actually, without loss of generality we can assume that $I_{\gamma_t}(\nu_t)$ is finite for all $t \in [0, T]$).

Now let

$$\tilde{\gamma}_t = z_0 + \int_0^t \bar{B}_{\nu_s}(\tilde{\gamma}_s)ds, \; t \in [0, T],$$

which in view of (2.2.1) determines $\tilde{\gamma} \in C_{0T}$. Then by (2.2.1),

$$\mathbf{r}_{0t}(\gamma, \tilde{\gamma}) \leq \zeta + K \int_0^t \mathbf{r}_{0s}(\gamma, \tilde{\gamma})ds$$

and by Gronwall's inequality

$$\mathbf{r}_{0T}(\gamma, \tilde{\gamma}) \leq \zeta e^{KT}.$$

This together with (2.5.2) and (2.5.3) yields that

$$\Big|\int_0^T I_{\gamma_t}(\nu_t)dt - \int_0^T I_{\tilde\gamma_t}(\nu_t)dt\Big| \le \tilde C\zeta e^{KT}S_{0T}(\gamma)$$

for some $\tilde C > 0$ independent of ζ and γ. Exchanging γ and $\tilde\gamma$, applying the same argument and using the inequality $S_{0T}(\tilde\gamma) \le \int_0^T I_{\tilde\gamma_t}(\nu_t)dt$ we conclude that

$$\big|S_{0T}(\gamma) - S_{0T}(\tilde\gamma)\big| \le \tilde C\zeta e^{KT}\max\big(S_{0T}(\gamma), S_{0T}(\tilde\gamma)\big) \le \tilde C\zeta e^{KT}(1+\tilde C\zeta e^{KT})S_{0T}(\gamma).$$

Choosing ζ small enough we arrive at (2.5.4). □

The following result will enable us to control the time which the slow motion can spend away from the ω-limit set of the averaged motion.

2.5.4. LEMMA. *Let $G \subset \mathcal{X}$ be a compact set not containing entirely any forward semi-orbit of the flow Π^t. Then there exist positive constants $a = a_G$ and $T = T_G$ such that for any $x \in G$ and $t \ge 0$,*

$$\inf\big\{S_{0t}(\gamma) : \gamma \in C_{0t} \text{ and } \gamma_s \in G \text{ for all } s \in [0,t]\big\} \ge a[t/T]$$

where $[c]$ denotes the integral part of c.

PROOF. The result is a simple consequence of lower semicontinuity of functionals S_{0t} and the fact that $S_{0T}(\gamma) = 0$ if and only if γ is a part of an orbit of the flow Π^t. Further details of the argument can be found in Lemma 1.6.4 in Part 1 and in Lemma 2.2(a), Chapter 4 of [**31**]. □

For the proof of the following result see Lemma 1.6.5 in Part 1.

2.5.5. LEMMA. *Let V be a connected open set with a piecewise smooth boundary and assume that (2.2.25) holds true. Then the function $R_\partial(x)$ is upper semicontinuous at any $x_0 \in V$ for which $R_\partial(x_0) < \infty$. Let $\mathcal{O} \subset V$ be an S-compact.*

(i) Then for each $z \in \bar V$ the function $R(x,z)$ takes on the same value $R^{\mathcal{O}}(z)$ for all $x \in \mathcal{O}$, and so $R_\partial(x)$ takes on the same value R_∂ for all $x \in \mathcal{O}$ and the set $\partial_{\min}(x) = \{z \in \partial V : R(x,z) = R_\partial\}$ coincides with the same (may be empty) set $\partial_{\min}$ for all $x \in \mathcal{O}$. Furthermore, for each $\delta > 0$ there exists $T(\delta) > 0$ such that for any $x \in \mathcal{O}$ we can construct $\gamma^x \in C_{0t_x}$ with $t_x \in (0, T(\delta)]$ satisfying

$$\gamma_0^x = x,\ \gamma_{t_x}^x \in \partial V \text{ and } S_{0t_x}(\gamma^z) \le R_\partial + \delta. \tag{2.5.5}$$

(ii) Suppose that $R_\partial < \infty$ and dist$(\Pi^t x, \mathcal{O}) \le d(t)$ for some $x \in V$ and $d(t) \to 0$ as $t \to \infty$. Then $R_\partial(x) \le R_\partial$ and for any $\delta > 0$ there exist $T_{\delta,d} > 0$ (depending only on δ and the function d but not on x) and $\hat\gamma^x \in C_{0s_x}$ with $s_x \in (0, T_{\delta,d}]$ satisfying

$$\hat\gamma_0^x = x,\ \hat\gamma_{s_x}^x \in \partial V \text{ and } S_{0s_x}(\hat\gamma^x) \le R_\partial + \delta. \tag{2.5.6}$$

In particular, if $R_\partial < \infty$ then $R_\partial(x) < \infty$ and if $\mathcal{O}$ is an S-attractor of the flow Π^t then $R_\partial(x) < \infty$ for all $x \in V$.

(iii) Suppose that for any open set $U \supset \mathcal{O}$ the compact set $\bar V \setminus U$ does not contain entirely any forward semi-orbit of the flow Π^t. Then the function $R^{\mathcal{O}}(z)$ is lower semicontinuous in $z \in \bar V$, $R^{\mathcal{O}}(z) \to 0$ as dist$(z, \mathcal{O}) \to 0$, and $\partial_{\min}$ is a nonempty compact set.

2.6. "Very long" time behavior: exits from a domain

We start with the following result which will not only yield Theorem 2.2.5 but also will play an important role in the proof of Theorem 2.2.7.

2.6.1. PROPOSITION. *Let V be a connected open set with a piecewise smooth boundary ∂V such that $\bar{V} = V \cup \partial V \subset \mathcal{X}$. Assume that for each $z \in \partial V$ there exist $\iota = \iota(z) > 0$ and a probability measure μ with $I_z(\mu) < \infty$ so that*

$$z + s\bar{B}_\mu(z) \in \mathbb{R}^d \setminus \bar{V} \text{ for all } s \in (0, \iota], \tag{2.6.1}$$

i.e. $\bar{B}_\mu(z) \neq 0$ and it points out into the exterior of $\bar{V}$.

(i) Suppose that for some $A_1, T > 0$ and any $z \in \bar{V}$ there exists $\varphi^z \in C_{0T}$ such that for some $t = t(z) \in (0, T]$,

$$\varphi_0^z = z,\ \varphi_t^z \notin V \text{ and } S_{0t}(\varphi^z) \le A_1. \tag{2.6.2}$$

Then for any $x \in V$ uniformly in $y \in \mathbf{M}$,

$$\limsup_{\varepsilon \to 0} \varepsilon \log E\tau_{x,y}^\varepsilon(V) \le A_1 \tag{2.6.3}$$

and for any $\alpha > 0$ there exists $\lambda(\alpha) = \lambda(x, \alpha) > 0$ such that uniformly in $y \in \mathbf{M}$ for all small $\varepsilon > 0$,

$$P\{\tau_{x,y}^\varepsilon(V) \ge e^{(A_1+\alpha)/\varepsilon}\} \le e^{-\lambda(\alpha)/\varepsilon}. \tag{2.6.4}$$

(ii) Assume that there exists an open set G such that V contains its closure $\bar{G}$ and the intersection of $\bar{V} \setminus G$ with the ω-limit set of the flow Π^t is empty. Let Γ be a compact subset of ∂V such that

$$\inf_{x \in G, z \in \Gamma} R(x, z) \ge A_2 \tag{2.6.5}$$

for some $A_2 > 0$. Then for some $T > 0$ and any $\beta > 0$ there exists $\lambda(\beta) > 0$ such that uniformly in $y \in \mathbf{M}$ for each $x \in V$ and any small $\varepsilon > 0$,

$$\begin{aligned} &P\{Z_{x,y}^\varepsilon(\tau_{x,y}^\varepsilon(V)) \in \Gamma,\ \tau_{x,y}^\varepsilon(V) \le e^{(A_2-\beta)/\varepsilon}\} \\ &\le P\{Z_{x,y}^\varepsilon(\tau_{x,y}^\varepsilon(V)) \in \Gamma,\ \tau_{x,y}^\varepsilon(V) < T\} + e^{-\lambda(\beta)/\varepsilon}. \end{aligned} \tag{2.6.6}$$

Suppose that for some $x \in V$,

$$a(x) = \inf_{t \ge 0} dist(\Pi^t x, \partial V) > 0. \tag{2.6.7}$$

Then $R_\partial(x) > 0$ and for each $T > 0$ there exists $\hat{\lambda}(T) = \hat{\lambda}(T, x) > 0$ such that uniformly in $y \in \mathbf{M}$ for all small $\varepsilon > 0$,

$$P\{\tau_{x,y}^\varepsilon(V) < T\} \le e^{-\hat{\lambda}(T)/\varepsilon} \tag{2.6.8}$$

and if the set Γ from (2.6.5) coincides with the whole ∂V then for all $x \in V$ uniformly in $y \in \mathbf{M}$,

$$\liminf_{\varepsilon \to 0} \varepsilon \log E\tau_{x,y}^\varepsilon(V) \ge A_2. \tag{2.6.9}$$

PROOF. In order to prove (i) we observe, first, that the assumption (2.6.1) above together with Lemma 2.5.2(i) and the compactness of ∂V considerations enable us to extend any φ^z, $z \in V$ slightly so that it will exit some fixed neighborhood of V with only slight increase in its S-functional. Hence, from the beginning we assume that for each $\beta > 0$ there exists $\delta = \delta(\beta) > 0$ such that for any $z \in V$ we can find $T > 0$, $\varphi^z \in C_{0T}$ and $t = t(z) \in (0, T]$ satisfying

$$\varphi_0^z = z,\ \varphi_t^z \notin V_\delta \text{ and } S_{0t}(\varphi^z) \leq A_1 + \beta$$

where $V_\delta = \{x : \text{dist}(x, V) \leq \delta\}$. Employing the Markov property we obtain that for any $x \in V$, $y \in \mathbf{M}$, $n \geq 1$,

$$\begin{aligned}(2.6.10)\qquad & P\{\tau^\varepsilon_{x,y}(V) > nT\} = P\{Z^\varepsilon_{x,y}(t) \in V,\ \forall t \in [0, nT]\}\\ &= P\{\tau^\varepsilon_{Z^\varepsilon_{x,y}(kT)}(V) > T,\ \forall k = 0, 1, ..., n-1\} \leq \big(\sup_{w \in V\times\mathbf{M}} P\{\tau^\varepsilon_w(V) > T\}\big)^n.\end{aligned}$$

From (2.2.10) and (2.6.2) it follows that

(2.6.11)

$$P\{\tau^\varepsilon_w(V) > T\} \leq P\{\mathbf{r}_{0T}(Z^\varepsilon_w, \varphi^z) \geq \delta \text{ for any } z \in V\} \leq 1 - \exp\big(-(A_1+\beta+\lambda)/\varepsilon\big).$$

By (2.6.10) and (2.6.11),

$$(2.6.12)\qquad P\{\tau^\varepsilon_w(V) > e^{(A_1+\beta)/\varepsilon}\} < e^{-c(\beta)/\varepsilon}$$

and

$$\begin{aligned}(2.6.13)\qquad & E\tau^\varepsilon_w(V) \leq \textstyle\sum_{n=0}^\infty (n+1)T\big(P\{\tau^\varepsilon_w(V) > nT\} - P\{\tau^\varepsilon_w(V)\\ & > (n+1)T\}\big) = T\textstyle\sum_{n=0}^\infty P\{\tau^\varepsilon_w(V) > nT\} \leq T\exp\big(\frac{1}{\varepsilon}(A_1+\beta+\lambda)\big)\end{aligned}$$

yielding (2.6.3) and (2.6.4) since β and λ in (2.6.13) can be chosen arbitrarily small as $\varepsilon \longrightarrow \infty$.

Next, we derive the assertion (ii). Let $t > 0$ and n be the integral part of t/T where $T > 0$ will be chosen later. Let, again, $w = (x, y)$ with $x \in V$ and $y \in \mathbf{M}$. Then

$$\begin{aligned}(2.6.14)\qquad & P\{Z^\varepsilon_w(\tau^\varepsilon_w) \in \Gamma,\ \tau^\varepsilon_w(V) < t\}\\ &\leq P\{Z^\varepsilon_w(\tau^\varepsilon_w(V)) \in \Gamma,\ \tau^\varepsilon_w(V) < (n+1)T\}\\ &= \textstyle\sum_{k=0}^n P\{Z^\varepsilon_w(\tau^\varepsilon_w(V)) \in \Gamma,\ kT \leq \tau^\varepsilon_w(V) < (k+1)T\}.\end{aligned}$$

Let K be the intersection of the ω-limit set of the flow Π^t with $\bar{V}$. Then K is a compact set and by our assumption $K \subset G$. Hence,

$$\delta = \frac{1}{3}\inf\{|x - z| : x \in K,\ z \in \bar{V} \setminus G\} > 0$$

and if we set $U_\eta = \{z \in V : \text{dist}(z, K) < \eta\}$ then $U_{3\delta} \subset G$. Now suppose that $kT \leq \tau^\varepsilon_{x,v}(V) < (k+1)T$ for some $k \geq 1$ and $Z^\varepsilon_{x,v}(\tau^\varepsilon_{x,v}(V)) \in \Gamma$ with $x \in V$ and $v \in \mathbf{M}$. Then either there is $t_1 \in [(k-1)T, kT]$ such that $Z^\varepsilon_{x,v}(t) \in \bar{V} \setminus U_{2\delta}$ for all $t \in [t_1, t_1 + T]$ or there exist $t_2, t_3 > 0$ such that $(k-1)T \leq t_2 < t_3 < (k+1)T$ and $Z^\varepsilon_{x,v}(t_2) \in U_{2\delta}$ while $Z^\varepsilon_{x,v}(t_3) \in \Gamma$. Set $\mathcal{T}_z = \{\gamma \in C_{0,2T} : \gamma_0 = z$ and either there is $t_1 \in [0, T]$ so that $\gamma_t \in \bar{V} \setminus U_{2\delta}$ for all $t \in [t_1, t_1 + T]$ or $\gamma_{t_2} \in U_{2\delta}$ and $\gamma_{t_3} \in \Gamma$ for some $0 \leq t_2 < t_3 < 2T\}$. Then for any $k \geq 1$,

$$\begin{aligned}(2.6.15)\qquad & \{Z^\varepsilon_w(\tau^\varepsilon_w(V)) \in \Gamma,\ kT \leq \tau^\varepsilon_w(V) < (k+1)T\}\\ &\subset \big\{Z^\varepsilon_w(\tau^\varepsilon_w(V)) \in \Gamma,\ Z^\varepsilon_{Z^\varepsilon_w((k-1)T)} \in \mathcal{T}_{Z^\varepsilon_w((k-1)T)}\big\}.\end{aligned}$$

For each $q > 0$ set $\mathcal{T}_z^q = \{\gamma \in C_{0,2T} : \gamma_0 = z$ and $\mathbf{r}_{0,2T}(\gamma, \mathcal{T}_z) \leq q\}$ and suppose that for some $\eta > 0$ there is $d_\eta \geq 0$ so that

$$(2.6.16)\qquad \inf_{z\in V}\inf_{\gamma\in\mathcal{T}_z^{2\eta}} S_{0,2T}(\gamma) > d_\eta.$$

Then $\mathcal{T}_z^{2\eta} \cap \Psi_{0,2T}^{d_\eta}(z) = \emptyset$, where $\Psi_{0,t}^a(z)$ is the same as in Theorem 2.2.2, and so

$$\mathcal{T}_z^\eta \subset \{\gamma \in C_{0,2T} : \gamma_0 = z \text{ and } \mathbf{r}_{0,2T}(\gamma, \Psi_{0,2T}^{d_\eta}(z)) \geq \eta\}. \tag{2.6.17}$$

From (2.2.10) and (2.6.15)–(2.6.17) we obtain that for any $\beta > 0$ and all sufficiently small ε,

$$P\{Z_w^\varepsilon(\tau_w^\varepsilon(V)) \in \Gamma,\ kT \leq \tau_w^\varepsilon(V) < (k+1)T\} \leq \hat{C}e^{-(d_\eta - \beta)/\varepsilon} \tag{2.6.18}$$

for some $\hat{C} > 0$.

Next, we will specify d_η in (2.6.16) choosing $\eta \leq \frac{1}{2}\delta$. For each $z \in V$ we can write

$$\mathcal{T}_z^{2\eta} \subset \tilde{\mathcal{T}}_z^\eta \cup \hat{\mathcal{T}}_z^\eta \tag{2.6.19}$$

where $\tilde{\mathcal{T}}_z^\eta = \{\gamma \in C_{0,2T} : \gamma_0 = z,\ \gamma_{t_2} \in U_{3\delta} \text{ and } \gamma_{t_3} \in \Gamma_{2\eta} \text{ for some } 0 \leq t_2 < t_3 < 2T\}$ with $\Gamma_r = \{z : \operatorname{dist}(z, \Gamma) \leq r\}$ and $\hat{\mathcal{T}}_z = \{\gamma \in C_{0,2T} : \gamma_0 = z$ and there is $t_1 \in [0, T]$ so that $\gamma_t \in V_{2\eta} \setminus U_\delta$ for all $t \in [t_1, t_1 + T]\}$. By (2.6.5) and the lower semicontinuity of the functional $S_{0,2T}$ it follows that for any $\zeta > 0$ we can choose $\eta > 0$ small enough so that

$$\inf_{z \in V} \inf_{\gamma \in \tilde{\mathcal{T}}_z^\eta} S_{0,2T}(\gamma) > A_2 - \zeta. \tag{2.6.20}$$

Since $\bar{V} \setminus U_\delta$ is disjoint with the ω-limit set of the flow Π^t and the latter is closed then if η is sufficiently small $V_{2\eta} \setminus U_\delta$ is also disjoint with this ω-limit set and, in particular, it does not contain any forward semi-orbit of Π^t. Hence we can apply Lemma 2.5.4 which in view of (2.2.9) implies that there exists $a > 0$ such that for all small $\eta > 0$,

$$\inf_{z \in V} \inf_{\gamma \in \hat{\mathcal{T}}_z} S_{0,2T}(\gamma) > aT \tag{2.6.21}$$

which is not less than A_2 if we take $T = A_2/a$. Now, (2.6.20) and (2.6.21) produce (2.6.16) with $d = A_2 - \zeta$, and so (2.6.18) follows with such d_η. This together with (2.6.14) yield that for any $\beta > 0$ we can choose sufficiently small $\zeta, \lambda > 0$ and then $\eta > 0$ so that for all ε small enough

$$\begin{aligned} &P\{Z_v^\varepsilon(\tau_v^\varepsilon) \in \Gamma,\ \tau_v^\varepsilon(V) \leq e^{(A_2 - \beta)/\varepsilon}\} \\ &\leq P\{Z_v^\varepsilon(\tau_v^\varepsilon) \in \Gamma,\ \tau_v^\varepsilon(V) < T\} + e^{-\lambda/2\varepsilon} \end{aligned} \tag{2.6.22}$$

and (2.6.6) follows.

Now assume that (2.6.7) holds true for some $x \in V$. Recall, that $S_{0T}(\gamma) = 0$ implies that γ is a piece of an orbit of the flow Π^t. Since no $\gamma \in C_{0T}$ satisfying

$$\gamma_0 = x \text{ and } \inf_{t \in [0,T]} \operatorname{dist}(\gamma_t, \partial V) \leq a(x)/2 \tag{2.6.23}$$

can be such piece of an orbit we conclude by the lower semicontinuity of S_{0T} that $S_{0T}(\gamma) > c(x)$ whenever (2.6.23) holds true for some $c(x) > 0$ independent of γ (but depending on x). Hence, by (2.2.11),

$$\begin{aligned} P\{\tau_v^\varepsilon < T\} \leq P\{&\mathbf{r}_{0T}(Z_v^\varepsilon, \Psi_{0T}^{c(x)}(x)) \\ &\geq a(x)/2\} \leq \exp(-c(x)/2\varepsilon) \end{aligned} \tag{2.6.24}$$

provided ε is small enough and (2.6.8) follows. Observe also that any $\gamma \in C_{0t}$ with $\gamma_0 = x \in V$ and $\gamma_t \in \partial V$ should contain a piece which either belongs to some $\tilde{\mathcal{T}}_z^\eta$ or to $\hat{\mathcal{T}}_z^\eta$, as above, or to satify (2.6.23). By (2.6.20), (2.6.21), and the above remarks it follows that $S_{0t}(\gamma) \geq q(x)$ for such γ where $q(x) > 0$ depends only on x, and so $R_\partial(x) \geq q(x)$. If $\Gamma = \partial V$ then by (2.6.6) and (2.6.8),

$$(2.6.25) \qquad E\tau^\varepsilon_{x,y}(V) \geq e^{(A_2-\beta)/\varepsilon} P\{\tau^\varepsilon_{x,y}(V) \geq e^{(A_2-\beta)/\varepsilon}\} \geq e^{(A_2-\beta)/\varepsilon}(1 - e^{-\lambda(\beta)/\varepsilon} - e^{-\hat{\lambda}(T)/\varepsilon})$$

and, since $\beta > 0$ is arbitrary, (2.6.9) follows completing the proof of Proposition 2.6.1. □

Now we will derive Theorem 2.2.5 from Proposition 2.6.1. Assume, first, that $R_\partial < \infty$. Then by Lemma 2.5.5, $R_\partial(x)$ is finite in the whole V. Moreover, since $\mathcal{O}$ is an S-attractor the conditions of Lemma 2.5.5 are satisfied with some $d(t) \to 0$ as $t \to \infty$ the same for all points of V which yields the conditions of Proposition 2.6.1(i) with $A_1 = R_\partial + \delta$ for any $\delta > 0$. Hence, (2.6.3) and (2.6.4) hold true with $A_1 = R_\partial$. Since $\mathcal{O}$ is an S-attractor of the flow Π^t and its basin contains $\bar{V}$ then the intersection of $\bar{V} \setminus \mathcal{O}$ with the ω-limit set of Π^t is empty. By the definition of an S-attractor for any $\zeta > 0$ there exists an open set $U_\zeta \supset \mathcal{O}$ such that $R(x,z) \leq \zeta$ whenever $x \in \mathcal{O}$ and $z \in U_\zeta$. Hence, by the triangle inequality for the function R and Lemma 2.5.5 for any set $\Gamma \subset \partial V$,

$$(2.6.26) \qquad \inf_{z \in U_\zeta, \tilde{z} \in \Gamma} R(z, \tilde{z}) \geq \inf_{\tilde{z} \in \Gamma} R^{\mathcal{O}}(\tilde{z}) - \zeta.$$

If $\Gamma = \partial V$ then by Lemma 2.5.5 the right hand side of (2.6.26) equals $A_2 = R_\partial - \zeta$. Assuming that $R_\partial < \infty$ we can apply Proposition 2.6.1(ii) with such A_2 yielding (2.6.6), (2.6.8) and since $\zeta > 0$ is arbitrary (2.2.26) and (2.2.27) follow in this case. If $R_\partial = \infty$ then (2.2.27) is trivial and by (2.6.26), $R(z, \tilde{z}) = \infty$ for any $z \in U_\zeta$ and $\tilde{z} \in \partial V$, and so we can apply Proposition 2.6.1(ii) with any A_2 which sais that the left hand side in (2.6.9) equals ∞, and so (2.2.26) holds true in this case, as well.

Next, we establish (2.2.28). For small $\delta, \beta > 0$ and large $T > 0$ which will be specified later on set $t_\varepsilon = T + e^{\beta/\varepsilon}$ and define the event

$$\Xi^\varepsilon_T(n) = \{\tau^\varepsilon_{Z^\varepsilon_v(t_\varepsilon n+T), Y^\varepsilon_v((t_\varepsilon n+T)/\varepsilon)}(U_\delta(\mathcal{O})) \leq e^{\beta/\varepsilon}\}.$$

Then

$$(2.6.27) \qquad \Theta^\varepsilon_v((n+1)t_\varepsilon \wedge \tau^\varepsilon_v(V)) - \Theta^\varepsilon_v(nt_\varepsilon \wedge \tau^\varepsilon_v(V)) \leq T + t_\varepsilon \mathbb{I}_V(Z^\varepsilon_v(t_\varepsilon n))\big(\mathbb{I}_{V \setminus U_{\delta/2}(\mathcal{O})}(Z^\varepsilon_v(t_\varepsilon n + T)) + \mathbb{I}_{U_{\delta/2}(\mathcal{O})}(Z^\varepsilon_v(t_\varepsilon n + T))\mathbb{I}_{\Xi^\varepsilon_T(n)}\big).$$

If δ is sufficiently small then V_δ is still contained in the basin of $\mathcal{O}$ with respect to the flow Π^t, and so we can choose T (depending only on δ) so that

$$\Pi^T V_\delta \subset U_{\delta/4}(\mathcal{O}).$$

Then for some $a > 0$,

$$\inf\{S_{0T}(\gamma) : \gamma \in C_{0T}, \gamma_0 \in V_\delta, \gamma_T \notin U_{\delta/3}(\mathcal{O})\} > a,$$

and so if $\gamma_0 \in V_\delta$ and $\gamma_T \notin U_{\delta/2}(\mathcal{O})$ then $\mathrm{dist}(\gamma, \Psi^a_{0T}(z)) \geq \delta/6$ for any $z \in V_\delta$. Relying on (2.2.11) and the Markov property we obtain that for any $v = (z,y)$ with $z \in V$,

$$(2.6.28) \qquad P\{Z^\varepsilon_v(t_\varepsilon n) \in V \text{ and } Z^\varepsilon_v(t_\varepsilon n + T) \in V \setminus U_{\delta/2}(\mathcal{O})\} \leq e^{-a/2\varepsilon}$$

provided ε is small enough. Next, the same arguments which yield (2.6.22) and (2.6.24) together with the Markov property enable us to conclude that if $\beta > 0$ is small enough then for any $v = (z, y)$ with $z \in V$,

$$P\{Z_v^\varepsilon(t_\varepsilon n + T) \in U_{\delta/2}(\mathcal{O}) \text{ and } \Xi_T^\varepsilon(n)\} \le e^{-\beta/\varepsilon}. \tag{2.6.29}$$

Applying (2.6.27)–(2.6.29) we conclude that for sufficiently small β and any much smaller ε,

$$E\big(\Theta_v^\varepsilon((n+1)t_\varepsilon \wedge \tau_v^\varepsilon(V)) - \Theta_v^\varepsilon(nt_\varepsilon \wedge \tau_v^\varepsilon(V))\big) \le t_\varepsilon e^{-\beta/\varepsilon}(T+1). \tag{2.6.30}$$

Finally, (2.2.27) and (2.6.30) together with the Chebyshev inequality yield that for $n(\varepsilon) = [e^{(R_\partial+\beta/4)/\varepsilon} t_\varepsilon^{-1}]$, each $x \in V$, a small $\beta > 0$ and any much smaller $\varepsilon > 0$,

$$\begin{aligned} &P\{\Theta_{x,w}^\varepsilon(\tau_{x,w}^\varepsilon(V)) \ge e^{-\beta/4\varepsilon}\tau_{x,w}^\varepsilon(V)\} \\ &\le P\{\Theta_{x,w}^\varepsilon((n(\varepsilon)+1)t_\varepsilon) \ge e^{-\beta/4\varepsilon}e^{(R_\partial-\beta/4)/\varepsilon}\} \\ &+P\{\tau_{x,w}^\varepsilon(V) < e^{(R_\partial-\beta/4)/\varepsilon} \text{ or } \tau_{x,w}^\varepsilon(V) > e^{(R_\partial+\beta/4)/\varepsilon}\} \\ &\le \tilde{C}e^{-\beta/4\varepsilon}\big(1 + e^{-(R_\partial+\beta/4)/\varepsilon}(T + e^{\beta/\varepsilon})\big) + e^{-\lambda(\beta/4)/\varepsilon}. \end{aligned} \tag{2.6.31}$$

Since $R_\partial > 0$ and we can choose β to be arbitrarily small, (2.6.31) yields (2.2.28).

In order to complete the proof of Theorem 2.2.5 we have to derive (2.2.29). If $\partial_{\min} = \partial V$ then there is nothing to prove, so we assume that $\partial_{\min}$ is a proper subset of ∂V and in this case, clearly, $R_\partial < \infty$. Since $\Gamma = \{z \in \partial V : \text{dist}(z, \partial_{\min}) \ge \delta\}$ is compact and disjoint with $\partial_{\min}$ which is also compact then by the lower semicontinuity of $R^{\mathcal{O}}(z)$ established in Lemma 2.5.5(iii) it follows that $R^{\mathcal{O}}(z) \ge R_\partial + \beta$ for some $\beta > 0$ and all $z \in \Gamma$. Then by (2.6.26), $R(z, \tilde{z}) \ge R_\partial + \beta/2$ for any $z \in U_{\beta/2}$ and $\tilde{z} \in \Gamma$. Hence, applying Proposition 2.6.1 we obtain that

$$P\{\tau_{x,y}^\varepsilon(V) \ge e^{(R_\partial+\frac{1}{3}\beta)/\varepsilon}\} \le e^{-\lambda/\varepsilon}$$

and

$$P\{Z_{x,y}^\varepsilon(\tau_{x,y}^\varepsilon(V)) \in \Gamma,\ \tau_{x,y}^\varepsilon(V) \le e^{(R_\partial+\frac{1}{3}\beta)/\varepsilon}\} < e^{-\lambda/\varepsilon}$$

for some $\lambda > 0$ and all ε small enough yielding (2.2.29) and completing the proof of Theorem 2.2.5. □

2.7. Adiabatic transitions between basins of attractors

In this section we will prove Theorem 2.2.7 relying, again, on Proposition 2.6.1 together with Markov and strong Markov property of the Markov process $(X^\varepsilon(t), Y^\varepsilon(t))$. In view of (2.2.32) and Lemma 2.5.2(i) any curve $\gamma \in C_{0t}$ starting at $\gamma_0 = x \in V_{j_1}$ and ending at $\gamma_t = z \in \cap_{1\le i\le k}\partial V_{j_i}$, $k \le \ell$ can be extended into each V_{j_i}, $i = 1, ..., k$ with arbitrarily small increase in its S-functional. Hence,

$$R_\partial^{(i)} = \min_{j\ne i} R_{ij} \tag{2.7.1}$$

where $R_\partial^{(i)} = \inf\{R(x, z) : x \in \mathcal{O}_i,\ z \in \partial V_i\}$. Let Q be an open ball of radius at least r_0 centered at the origin of $\mathbb{R}^d$. By Assumption 2.2.6 the slow motion $Z_{x,y}^\varepsilon$ cannot exit Q provided $x \in Q$ and $y \in \mathcal{W}$. Furthermore, it is clear that Q contains the ω-limit set of the averaged flow Π^t. Assumption 2.2.6 enables us to deal only with restricted basins $V_i^Q = V_i \cap Q$ and though the boundaries ∂V^Q of V_i^Q may include now parts of the boundary ∂Q of Q it makes no difference since Z^ε cannot reach ∂Q if it starts in Q. Set $V^{(i)} = Q \setminus \cup_{j\ne i} U_\delta(\mathcal{O}_j)$ where $\delta > 0$ is small enough.

We claim that in view of (2.2.32) each V_i satisfies conditions of Proposition 2.6.1(i) for any $\beta > 0$ with $A_1 = R_\partial^{(i)} + \beta$ and some $T = T_\beta$ depending on β. Indeed, set

$$\partial(\eta) = \{v \in Q : \text{dist}(v, \cup_{1\le j\le \ell}\partial V_j) \le \eta\},\ \eta > 0.$$

In view of (2.2.32) and Lemma 2.5.1 there exists $L > 0$ such that if η is small enough and $z \in \partial(\eta)$ we can construct a curve $\varphi^z \in C_{0,L\eta}$ with $S_{0,L\eta}(\varphi^z) \le L\eta$, $\varphi_0^z = z$, $\varphi_t^z \in V_j \setminus \partial(\eta)$ for some $t \in [0, L\eta]$ and $j = 1, ..., \ell$. Since V_j is the basin of $\mathcal{O}_j$ there exists $T = T_{\eta,\delta}$ such that $\Pi^T \varphi_t^z \in U_\delta(\mathcal{O}_j)$ and extending φ^z by the piece of the orbit of Π^t we obtain a curve $\tilde\varphi^z \in C_{0,L\eta+T}$ starting at z, entering $U_\delta(\mathcal{O}_j)$ and satisfying $S_{0,L\eta+T}(\tilde\varphi^z) \le L\eta$. Hence, for $z \in \partial(\eta)$ the condition (2.6.2) holds true with $V = V^{(i)}$ and $A_1 = L\eta$. Since the ω-limit set of the flow Π^t is contained in $Q \cap \big(\cup_{1\le j\le \ell}(\partial V_j \cup \mathcal{O}_j)\big)$ it follows from Assumption 2.2.6 and compactness considerations that there exists $\tilde T = \tilde T_{\eta,\delta}$ such that for any $z \in Q \setminus V_i$ we can find $t_z \in [0, \tilde T]$ with $\Pi^{t_z} z \in \partial(\eta) \cup \big(\cup_{j\ne i} U_\delta(\mathcal{O}_j)\big)$. If $\Pi^{t_z} z \in \cup_{j\ne i} U_\delta(\mathcal{O}_j)$ then we take $\varphi_t^z = \Pi^t z$, $t \in [0, \tilde T]$ to satisfy (2.6.2) for $V = V^{(i)}$ and $A_1 = 0$. If $\Pi^{t_z} z \in \partial(\eta)$ then we extend the curve $\varphi_t^z = \Pi^t z$, $t \in [0, t_z]$ as in the above argument which yields a curve $\tilde\varphi^z$ starting at z, ending in some $U_\delta(\mathcal{O}_j)$, $j \ne i$ and having its S-functional not exceeding $L\eta$. Finally, in the same way as in the proof of Theorem 2.2.5 for any $\beta > 0$ there exists $\hat T = \hat T_{\eta,\delta,\beta}$ such that whenever $z \in V_i(\eta) = V_i \cap Q \setminus \partial(\eta)$ we can construct $\varphi^z \in C_{0\hat T}$ such that (2.6.2) holds true with $V = V_i(\eta)$ and $A_1 = R_\partial^{(i)} + \beta/2$ and, moreover, $\text{dist}(\varphi_t^z, V_j) \le \eta$ for some $t \le \hat T$ and $j \ne i$ with $R_{ij} = R_\partial^{(i)}$. Then in the same way as above we can extend φ^z to some $\tilde\varphi^z \subset C_{\hat T + \tilde T}$ so that $\tilde\varphi_t^z \in U_\delta(V_j)$ for some j as above, $t \le \hat T + \tilde T$ and $S_{0,\hat T+\tilde T}(\tilde\varphi^z) \le R_\partial^{(i)} + \beta/2 + L\eta$ which gives (2.6.2) for all $z \in V = V^{(i)}$ with $A_1 = R_\partial^{(i)} + \beta$ provided η is small enough. Hence, Proposition 2.6.1(i) yields the estimates (2.6.3) and (2.6.4) for $\tau_{x,y}^\varepsilon(i)$ in place of $\tau_{x,y}^\varepsilon(V)$ with $A_1 = R_\partial^{(i)}$. In order to obtain the corresponding bounds in the other direction observe that in view of (2.2.32),

$$R_\partial^{(i)}(\delta) = \inf\{R(x, z) : x \in \mathcal{O}_i,\ z \notin V_i(\eta)\} \to R_\partial^{(i)} \text{ as } \delta \to 0. \tag{2.7.2}$$

Since $\overline{V_i(\eta)}$ is contained in the basin of $\mathcal{O}_i$ we can apply to $V_i(\eta)$ the same estimates as in Theorem 2.2.5 which together with (2.7.2) and the fact that the exit time of Z^ε from $V_i(\eta)$ is smaller than its exit time from V_i provide the remaining bounds yielding (2.2.33) and (2.2.34).

Next, we derive (2.2.35) similarly to (2.2.28) but taking into account that $\cup_{1\le j\le \ell}\partial V_j$ may contain parts of the ω-limit set of the flow Π^t which allows the slow motion Z^ε to stay long time near these boundaries. Still, set

$$\theta_v^\varepsilon = \inf\{t \ge 0 : Z_v^\varepsilon(t) \in \cup_{1\le j\le \ell} U_{\delta/3}(\mathcal{O}_j)\}.$$

Using the same arguments as above we conclude that for any $\eta > 0$ there exists $T = T_{\eta,\delta}$ such that whenever $z \in Q$ we can construct $\varphi^z \in C_{0T}$ with $\varphi_0^z = z$, $\varphi_T^z \in \cup_{1\le j\le \ell} U_\delta(\mathcal{O}_j)$ and $S_{0T}(\varphi^z) \le \eta$. This together with (2.6.12) and Assumption 2.2.6 yield that

$$P\{\theta_v^\varepsilon \ge e^{2\eta/\varepsilon}\} \le e^{-\lambda(\eta)/\varepsilon}$$

for some $\lambda(\eta) = \lambda(x, \eta) > 0$ and all small ε. Set

$$\Gamma_1(v) = \{Z_v^\varepsilon(e^{2\eta/\varepsilon}) \in Q \setminus \cup_{1\le j\le \ell} U_{\delta/2}(\mathcal{O}_j)\},$$

$$\Gamma_2(v) = \{\tau_v^\varepsilon\big(\cup_{1\le j\le \ell} U_\delta(\mathcal{O}_j)\big) \le e^{\beta/\varepsilon}\}$$

and $t_\varepsilon = e^{2\eta/\varepsilon} + e^{\beta/\varepsilon}$ where η is much smaller than β. Then proceeding similarly to the proof of (2.2.28) as in (2.6.28)–(2.6.31) above we arrive at (2.2.35).

Next, we obtain (2.2.36) relying on additional assumptions specified in the statement of Theorem 2.2.7. Let V_i^Q be the same as above and $\partial_0^{(i)}(x) = \{z \in \partial V_i^Q : R(x,z) = R_\partial^{(i)}\}$. Since $\mathcal{O}_i$ is an S-attractor it follows from Lemma 2.5.5(i) that $R(x,z)$ and $\partial_0^{(i)}(x)$ coincide with the same function $R^{\mathcal{O}_i}(z)$ and the same (in general, may be empty) set $\partial_0^{(i)}$, respectively, for all $x \in \mathcal{O}_i$. By Lemma 2.5.2(i), our assumption that B is complete on ∂V_i implies that $R^{\mathcal{O}_i}(z)$ is continuous in a neighborhood of ∂V_i, and so $\partial_0^{(i)}$ is a nonempty compact set. Since we assume that $\iota(i) \neq i$ is the unique index j for which $R_{ij} = R_{i\iota(i)} = R_\partial^{(i)}$ then by (2.2.32),

$$\min_{j \neq i, \iota(i)} \inf_{z \in \partial_0^{(i)}} \operatorname{dist}(z, \partial V_j) > 0.$$

Observe that if $\tilde{\mathcal{O}} \subset \partial V_i$ is an S-compact then either $\tilde{\mathcal{O}} \subset \partial_0^{(i)}$ or $\tilde{\mathcal{O}} \cap \partial_0^{(i)} = \emptyset$. Denote by L_Π the ω-limit set of the averaged flow Π^t. Since $L_\Pi \cap \partial V_i$ consists of a finite number of S-compacts it follows that

$$\inf\{|z - \tilde{z}| : z \in L_\Pi \cap \partial_0^{(i)},\ \tilde{z} \in L_\Pi \setminus \partial_0^{(i)}\} > 0.$$

By the continuity of $R^{\mathcal{O}_i}(z)$ in $z \in \partial V_i$ there exists $a > 0$ such that

$$\inf\left\{R^{\mathcal{O}_i}(z) : z \in \left(\cup_{j \neq i, \iota(i)} (\partial V_i \cap \partial V_j)\right) \cup \left((L_\Pi \setminus \partial_0^{(i)}) \cap \partial V_i\right)\right\} \geq R_\partial^{(i)} + 9a.$$

These considerations enable us to construct a connected open set G with a piecewise smooth boundary ∂G such that

$$\bar{G} \subset V_i \cup (V_{\iota(i)} \setminus \mathcal{O}_{\iota(i)}) \cup \left((\partial V_i \cap \partial V_{\iota(i)}) \setminus (L_\Pi \setminus \partial_0^{(i)})\right)$$

and for $\Gamma = \partial G \setminus U_\delta(\mathcal{O}_{\iota(i)})$ and some $a(\delta) > 0$,

$$\inf_{z \in \Gamma} R^{\mathcal{O}_i}(z) \geq R_\partial^{(i)} + 8a \tag{2.7.3}$$

provided $a \leq a(\delta)$. The idea of this construction is that if $Z_{x,y}^\varepsilon(\tau_{x,y}^\varepsilon(i)) \notin V_{\iota(i)}$ then the slow motion should exit G through the part Γ of its boundary. Somewhat similarly to the proof of Proposition 2.6.1(ii) we will show that "most likely" this can only occur after the time $\exp\left((R_\partial^{(i)} + 2a)/\varepsilon\right)$ and, on the other hand, we conclude from (2.2.34) that except for small probability the exit time $\tau_{x,y}^\varepsilon(i)$ does not exceed $\exp\left((R_\partial^{(i)} + a)/\varepsilon\right)$.

Let U_0 be a sufficiently small open neighborhood of $\partial_0^{(i)}$ so that, in particular,

$$\sup_{z \in U_0} R^{\mathcal{O}_i}(z) \leq R_\partial^{(i)} + a$$

and set

$$\tau_{x,y}^\varepsilon(G) = \inf\{t \geq 0 : Z_{x,y}^\varepsilon(\tau_{x,y}^\varepsilon(G)) \notin G\}.$$

Then

$$\begin{aligned} &\{\tau_v^\varepsilon(G) \leq e^{(R_\partial^{(i)}+a)/\varepsilon},\ Z_v^\varepsilon(\tau_v^\varepsilon(G)) \in \Gamma\} \subset \textstyle\bigcup_{0 \leq n \leq n(\varepsilon)+1} \big(A^{(1)}(n) \cup \\ &\quad \textstyle\bigcup_{(n-1)t_\varepsilon \leq k \leq (n+1)t_\varepsilon} \big(A^{(2)}(k) + A^{(3)}(k) + \bigcup_{k-2t_\varepsilon \leq m \leq k-2T} A^{(4)}(m) \cap A^{(5)}(k)\big)\big) \end{aligned} \tag{2.7.4}$$

where $t_\varepsilon = e^{\beta/\varepsilon}$ for some small $\beta > 0$, $n(\varepsilon) = \left[e^{(R_\partial^{(i)}+a-\beta)/\varepsilon}\right]$, $A^{(1)}(n) = \{Z_v^\varepsilon(t) \in G \setminus \left(U_\eta(\mathcal{O}_i) \cup U_\delta(\mathcal{O}_{\iota(i)})\right)$ for all $t \in [(n-1)t_\varepsilon, nt_\varepsilon]\}$ for a sufficiently small $\eta > 0$, $A^{(2)}(k) = \{\exists t_1, t_2 \text{ with } k \leq t_1 < t_2 < k + 3T,\ Z_v^\varepsilon(t_1) \in U_\eta(\mathcal{O}_i),\ Z_v^\varepsilon(t_2) \in \Gamma\}$,

$A^{(3)}(k) = \{Z_v^\varepsilon(t) \in G \setminus (U_0 \cup U_\eta(\mathcal{O}_i) \cup U_\delta(\mathcal{O}_{\iota(i)}))$ for all $t \in [k, k+T]\}$, $A^{(4)}(m) = \{\exists t_1, t_2 \text{ with } m \le t_1 < t_2 < m+T,\ Z_v^\varepsilon(t_1) \in U_\eta(\mathcal{O}_i),\ Z_v^\varepsilon(t_2) \in U_0\}$, and $A^{(5)}(k) = \{\exists t_3, t_4 \text{ with } k \le t_3 < t_4 < k+T,\ Z_v^\varepsilon(t_3) \in U_0,\ Z_v^\varepsilon(t_4) \in \Gamma\}$. Observe that $G \setminus (U_\eta(\mathcal{O}_i) \cup U_\delta(\mathcal{O}_{\iota(i)}))$ satisfies conditions of Proposition 2.6.1(i) with arbitrarily small A_1, so similarly to (2.6.12) we can estimate

$$P(A^{(1)}(n)) \le \exp(-\frac{1}{2}e^{\beta/\varepsilon}). \tag{2.7.5}$$

Similarly to the proof of Proposition 2.6.1(ii) we obtain also that

$$\max\big(P(A^{(2)}(k)), P(A^{(3)}(k))\big) \le e^{-(R_\partial^{(i)}+3a)/\varepsilon} \tag{2.7.6}$$

where we, first, choose η small and then T large enough.

Next, relying on the Markov property and the arguments similar to the proof of Proposition 2.6.1(ii) we estimate

$$P\big(A^{(4)}(m) \cap A^{(5)}(k)\big) \le e^{-(R_\partial^{(i)}+3a)/\varepsilon} \tag{2.7.7}$$

provided $m \le k - 2T$ and ε is small enough. Summing in m, k and n we obtain from (2.7.4)–(2.7.7) that for a small β and all sufficiently small ε,

$$P\{\tau_v^\varepsilon(G) \le e^{(R_\partial^{(i)}+a)/\varepsilon},\ Z_v^\varepsilon(\tau_v^\varepsilon(G)) \in \Gamma\} \le e^{-a/\varepsilon}. \tag{2.7.8}$$

Employing Proposition 2.6.1(i) we derive that

$$P\{\tau_v^\varepsilon(G) > e^{(R_\partial^{(i)}+a)/\varepsilon}\} \le e^{-\lambda/\varepsilon}$$

for some $\lambda > 0$ and all ε small enough which together with (2.7.8) yield (2.2.36).

In order to complete the proof of Theorem 2.2.7 it remains to derive (2.2.37) and (2.2.38). Both statements hold true for $n = 1$ in view of (2.2.34) and (2.2.36) and we proceed by induction. Set

$$H(n, \alpha) = \{\Sigma_i^\varepsilon(k, -\alpha) \le \tau_v(i, k) \le \Sigma_i^\varepsilon(k, \alpha)\ \ \forall k \le n\}$$

and

$$G(n) = \{Z_v^\varepsilon(\tau_v(i, k)) \in V_{\iota_k(i)}\ \ \forall k \le n\}.$$

As the induction hypotesis we assume that for any $\alpha > 0$ there exist $\lambda(\alpha) > 0$ and $\lambda > 0$ such that for all small ε,

$$P\big(H(n, \alpha)\big) \ge 1 - ne^{-\lambda(\alpha)/\varepsilon} \text{ and } m\big(G(n)\big) \ge 1 - ne^{-\lambda/\varepsilon}. \tag{2.7.9}$$

By (2.2.36) and the strong Markov property

$$P\big(G(n) \setminus G(n+1)\big) = P\big(\{Z_v^\varepsilon(\tau_v(i, n+1)) \notin V_{\iota_{n+1}(i)}\} \cap G(n)\big) \tag{2.7.10}$$
$$E\mathbb{I}_{Z_v^\varepsilon(\tau_v(i,n)) \in \partial U_\delta(\mathcal{O}_{\iota_n(i)})} P\{Z^\varepsilon_{Z_v^\varepsilon(\tau_v(i,n))}(\tau_{Z_v^\varepsilon(\tau_v(i,n))}(\iota_n(i)) \notin V_{\iota_{n+1}(i)}\} \le e^{-\lambda/\varepsilon}$$

which implies (2.2.38). Similarly, by (2.2.34) and the strong Markov property

$$P\big((H(n, \alpha) \setminus H(n+1, \alpha)) \cap G(n)\big) \le E\mathbb{I}_{Z_v^\varepsilon(\tau_v(i,n)) \in \partial U_\delta(\mathcal{O}_{\iota_n(i)})} \tag{2.7.11}$$
$$\times P\{\tau_{Z_v^\varepsilon(\tau_v(i,n))}(\iota_n(i)) > \Sigma_i^\varepsilon(n, \alpha) \text{ or } \tau_{Z_v^\varepsilon(\tau_v(i,n))}(\iota_n(i)) < \Sigma_i^\varepsilon(n, -\alpha)\} < e^{-\lambda(\alpha)/\varepsilon}$$

proving (2.2.37) and completing the proof of Theorem 2.2.7. □

Finally, we prove Theorem 2.2.8 employing the arguments similar to §2 and §3 in Ch. 6 of [**31**]. Namely, in order to obtain the upper bound in (2.2.39) observe that for any $h > 0$ there are $\rho_0, \delta_0 > 0$ such that if $\rho < \rho_0$, $\delta < \delta_0$ and a curve $\gamma \in C_{0t}$ satisfies $\gamma_0 \in \partial U_\delta(\mathcal{O}_i)$ and $\text{dist}(\gamma_t, \partial U_\delta(\mathcal{O}_j)) < \rho$ then $S_{0t}(\gamma) \ge R_{ij} - h$.

Using Lemma 2.5.4 and the upper bound of large deviations (2.2.11) we can choose $t = T_1$ such that for all small ε and any $v \in \cup_{1\le j\le \ell}\partial U_{2\delta}(\mathcal{O}_j)$,

$$P\{\sigma_v^{\varepsilon,\delta}(1) > T_1\} \le e^{-\frac{R_{ij}}{\varepsilon}}. \tag{2.7.12}$$

Any path of Z^ε starting at a point of $\partial U_{2\delta}(\mathcal{O}_i)$ and reaching $\partial U_\delta(\mathcal{O}_j)$ at time $\sigma_v^{\varepsilon,\delta}(1)$ either spends the time T_1 without touching the set $\cup_{1\le k\le \ell}\partial U_\delta(\mathcal{O}_k)$ or arrives at Γ_j during the time T_1. In the latter case $\mathbf{r}_{0T_1}(Z^\varepsilon, \Psi_{0T_1}^{R_{ij}-h}(x)) \ge \rho$ and by (2.2.11) and (2.7.12) for any $v = (x,y)$ with $x \in \partial U_{2\delta}(\mathcal{O}_i)$, all ε small enough and $j \ne i$,

$$\begin{aligned}P\{Z_v^\varepsilon(\sigma_v^{\varepsilon,\delta}(1)) \in \partial U_\delta(\mathcal{O}_j)\} &\le P\{\sigma_v^{\varepsilon,\delta}(1) > T_1\}\\ +P\{\mathbf{r}_{0T_1}(Z^\varepsilon, \Psi_{0T_1}^{R_{ij}-h}(x)) \ge \rho\} &\le \exp\big(-(R_{ij}-h-\tilde\beta)/\varepsilon\big)\end{aligned} \tag{2.7.13}$$

for some $\tilde\beta > 0$ independent of ε. Any path of Z_v^ε starting at $x \in \partial U_\delta(\mathcal{O}_i)$ and reaching $\partial U_\delta(\mathcal{O}_j)$ at the time $\sigma_v^{\varepsilon,\delta}(1)$ must first hit at time $\hat\sigma_v^{\varepsilon,\delta}(1)$ the set $\partial U_{2\delta}(\mathcal{O}_i)$, and so (2.7.13) together with the Markov property yields the upper bound in (2.2.39).

In order to derive the lower bound in (2.2.39) observe that using the definition of S-attractors and Lemma 2.5.3 (similarly to the proof of Lemma 1.6.5(ii) in Part 1 and see also §2 in Ch. 6 of [**31**]) we conclude that for any $h > 0$ there exists δ_0 such that if $\delta < \delta_0$ then for any $v = (x,y) \in \Gamma_i$ there exists a curve $\gamma \in C_{0t}$ such that $\gamma_0 = x, \gamma_s \in U_{2\delta}(\mathcal{O}_i)$ for $s \in [0, s_1]$, $\gamma_s \notin \cup_{1\le k\le \ell, k\ne j}\partial U_\delta(\mathcal{O}_k)$ for $s > s_1$, $\gamma_t \in U_{\delta/2}(\mathcal{O}_j)$ and, finally, $S_{0t}(\gamma) \le R_{ij} + h$. Then by (2.2.10) for all small $\varepsilon > 0$,

$$P\{Z_v^\varepsilon(\sigma_v^{\varepsilon,\delta}(1)) \in \partial U_\delta(\mathcal{O}_j)\} \ge P\{\mathbf{r}_{0t}(Z^\varepsilon,\gamma) < \delta/2\} \ge \exp\big(-(R_{ij}+h+\tilde\beta)/\varepsilon\big) \tag{2.7.14}$$

for some $\tilde\beta > 0$ independent of ε which together with (2.7.13) yields (2.2.39).

Now, (2.2.40) follows from (2.2.39) and the estimates for invariant measures of Markov chains from §3, Ch. 6 in [**31**]. □

2.8. Averaging in difference equations

Theorem 2.2.10 follows by a slight modification (essentially, by simplification) of the proof of Theorems 2.2.2, in particular, the standard Gronwall inequality required in the proof of Lemma 2.3.4 should be replaced by its discrete time version from [**26**]. We have also to check that (2.2.42) holds true here which is easier to do than in the continuous time case. Indeed,

$$\begin{aligned}Q_k^\varepsilon(x',x,y) = E\exp\langle\beta, \textstyle\sum_{j=1}^k B(x', Y_{x,y}^\varepsilon(j))\rangle &= \textstyle\int_{\mathbf{M}}\cdots\int_{\mathbf{M}} dm(y_1)p^x(y,y_1)\\ \times \exp\langle\beta, B(x',y_1)\rangle dm(y_2)p^{x_1}(y_1,y_2)&\exp\langle\beta, B(x',y_2)\rangle\\ \times\cdots\times dm(y_k)p^{x_{k-1}}(y_{k-1},y_k)&\exp\langle\beta, B(x',y_k)\rangle\end{aligned}$$

where $x_{k+1} = x_k + \varepsilon B(x_k, y_k)$, $k = 0,1,...,k-1$, $x_0 = x, y_0 = y$. By (2.2.1),

$$|x_j - x| \le Kk\varepsilon \text{ for } j = 1,...,k.$$

Since all $y_j \in \mathbf{M}$ which is compact and $|x_j - x| \le KT$, i.e. all x_j stay also in a compact set, we obtain from our assumptions on transition densities that for all $y, \tilde y \in \mathbf{M}$ and $j = 0,1,...,k-1$,

$$1 - CKk\varepsilon \le \frac{p^{x_j}(y,\tilde y)}{p^x(y,\tilde y)} \le 1 + CKk\varepsilon$$

for some $C > 0$ independent of $\varepsilon, k, y, \tilde{y}$ and x staying in a compact set. Hence,

$$(1 - CKk\varepsilon)^k \leq \frac{Q_k^\varepsilon(x', x, y)}{Q_k(x', x, y)} \leq (1 + CKk\varepsilon)^k$$

where $Q_k(x', x, y) = Q_k^0(x', x, y)$ is obtained from $Q_k^\varepsilon(x', x, y)$ by replacing $Y_{x,y}^\varepsilon$ in the latter by $Y_{x,y}^0 = Y_{x,y}$. It follows from standard facts on principal eigenvalues of positive operators (see, for instance, [**62**] and [**39**]) that uniformly in $y \in \mathbf{M}$ and $x, x' \in \bar{\mathcal{X}}$ the limit

$$\lim_{k\to\infty} \frac{1}{k} \log Q_k((x', x, y) = H(x, x', \beta),$$

exists and it satisfies the conditions of Assumption 2.2.9, and so taking the logarithm in the ineguality above and dividing by k we arrive at (2.2.42).

Theorem 2.2.12 also follows by a slight modification of proofs of Theorems 2.2.5 and 2.2.7, only we have to derive a result which replaces Lemma 2.5.1 providing required properties of I-functionals given by (2.2.45). Since, without loss of generality, we can assume that $C^{-1} \leq p^x(y, v) \leq C$ for some $C > 0$ and by (2.2.45),

$$I_x(\mu) = \sup_{u>0} \int_{\mathbf{M}} \log \frac{u(y)}{\int_{\mathbf{M}} p^x(y, v)u(v)dm(v)} d\mu(y),$$

where the supremum is taken over positive continuous functions u. Then

$$\sup_{u>0} \int_{\mathbf{M}} \log \frac{C^{-1}u(y)}{\int_{\mathbf{M}} u(v)dm(v)} d\mu(y) \leq I_x(\mu) \leq \sup_{u>0} \int_{\mathbf{M}} \log \frac{Cu(y)}{\int_{\mathbf{M}} u(v)dm(v)} d\mu(y).$$

It is easy to see from here that $I_x(\mu) < \infty$ if and only if $d\mu(y) = g(y)dm(y)$ and the density g is bounded. Hence, in this case,

$$I_x(\mu) \leq \sup_{u>0} \int_{\mathbf{M}} \sup g \log \tfrac{Cu(y)}{\int_{\mathbf{M}} u(v)dm(v)} dm(y) \leq \sup g \big(\log C$$
$$+ \sup_{u>0} \big(\int_{\mathbf{M}} \log u(y)dm(y) - \log \int_{\mathbf{M}} u(y)dm(y)\big)\big) \leq \sup g \log C.$$

Since

$$|p^z(y, v) - p^x(y, v)| \leq D|x - z| \leq CD|x - z| \min(p^z(y, v), p^x(y, v))$$

for some $D > 0$, we obtain

$$|I_x(\mu) - I_z(\mu)| \leq \sup_{u>0} \int_{\mathbf{M}} \big| \log \big(\tfrac{\int_{\mathbf{M}} p^z(y,v)u(v)dm(v)}{\int_{\mathbf{M}} p^x(y,v)u(v)dm(v)} \big) \big| d\mu(y)$$
$$\leq \log(1 + CD|x - z|).$$

Two last inequalities provide all properties of I-fuctionals which are needed in order to replace Lemma 2.5.1 and to proceed with arguments of Sections 2.5–2.7 in the discrete time case.

Theorem 2.2.10 provides, in particular, an approximation of the slow motion by the averaged one in probability but, in general, we do not have convergence in (2.1.4) also with probability one (see [**11**]). Sometimes, we can derive this almost sure convergence from the upper large deviations bound estimating the derivative in ε of the slow motion as in the following example. Let $B(x, y)$ be a bounded 1-periodic in y function on $\mathbb{R}^1 \times \mathbb{R}^1$ with bounded derivatives and let $\xi_1, \xi_2, ...$ be a sequence of independent identically distributed (i.i.d.) random variables. Define recursively

$$X_v^\varepsilon(n + 1) = X_v^\varepsilon(n) + \varepsilon B\big(X_v^\varepsilon(n), Y_v^\varepsilon(n)\big), \tag{2.8.1}$$
$$Y_v^\varepsilon(n + 1) = Y_v^\varepsilon(n) + X_v^\varepsilon(n) + \xi_{n+1}$$

where $v = (z, w)$ and $X_v^\varepsilon(0) = z$, $Y_v^\varepsilon(0) = w$. Then

$$\frac{dX_v^\varepsilon(n+1)}{d\varepsilon} = \frac{dX_v^\varepsilon(n)}{d\varepsilon} + B\big(X_v^\varepsilon(n), Y_v^\varepsilon(n)\big) \tag{2.8.2}$$
$$+\varepsilon\frac{\partial B\big(X_v^\varepsilon(n),Y_v^\varepsilon(n)\big)}{\partial x}\frac{dX_v^\varepsilon(n)}{d\varepsilon} + \varepsilon\frac{\partial B\big(X_v^\varepsilon(n),Y_v^\varepsilon(n)\big)}{\partial y}\frac{dY_v^\varepsilon(n)}{d\varepsilon},$$
$$\frac{dY_v^\varepsilon(n+1)}{d\varepsilon} = \frac{dY_v^\varepsilon(n)}{d\varepsilon} + \frac{dX_v^\varepsilon(n)}{d\varepsilon}.$$

Set

$$A_v^\varepsilon(n) = \begin{pmatrix} \frac{\partial B\big(X_v^\varepsilon(n),Y_v^\varepsilon(n)\big)}{\partial x} & \frac{\partial B\big(X_v^\varepsilon(n),Y_v^\varepsilon(n)\big)}{\partial y} \\ 0 & 0 \end{pmatrix} \text{ and } q_v^\varepsilon(n) = \begin{pmatrix} B\big(X_v^\varepsilon(n), Y_v^\varepsilon(n)\big) \\ 0 \end{pmatrix}$$

which are sequences of bounded matrices and vectors. Taking into account the equalities

$$\frac{dX_v^\varepsilon(0)}{d\varepsilon} = \frac{dY_v^\varepsilon(0)}{d\varepsilon} = 0 \text{ and } \begin{pmatrix} 1 & 0 \\ 1 & 1 \end{pmatrix}^k = \begin{pmatrix} 1 & 0 \\ k & 1 \end{pmatrix}$$

we obtain from (2.8.2) by induction (with the agreement $\prod_{j=n}^{n-1} = 1$) that

$$\begin{pmatrix} \frac{dX_v^\varepsilon(n)}{d\varepsilon} \\ \frac{dY_v^\varepsilon(n)}{d\varepsilon} \end{pmatrix} = \sum_{k=0}^{n-1} \prod_{j=k+1}^{n-1} \Big(\begin{pmatrix} 1 & 0 \\ 1 & 1 \end{pmatrix} + \varepsilon A_v^\varepsilon(j)\Big) q_v^\varepsilon(k)$$
$$= \sum_{k=0}^{n-1} \Big(\sum_{l=0}^{n-k-2} \begin{pmatrix} 1 & 0 \\ l & 1 \end{pmatrix}$$
$$\times \varepsilon^{n-k-l-2} \sum_{k+1\le j_1<\ldots<j_{n-k-l-2}\le n-1} \prod_{i=1}^{n-k-l-2} A_v^\varepsilon(j_i)\Big) q_v^\varepsilon(k).$$

Since $A_v^\varepsilon(j)$ and $q_v^\varepsilon(k)$ are bounded we obtain that

$$\Big|\frac{dX_v^\varepsilon(n)}{d\varepsilon}\Big| \le Cn \sum_{k=0}^{n-1} (1 + C\varepsilon)^{n-k-2} \le n\varepsilon^{-1}(1 + C\varepsilon)^n \tag{2.8.3}$$

for some $C > 0$ independent of n and ε.

Since $B(x, y)$ is 1-periodic in y we can replace the second equality in (2.8.1) by

$$Y_v^\varepsilon(n + 1) = Y_v^\varepsilon(n) + X_v^\varepsilon(n) + \xi_{n+1} \text{ (mod 1)}, \tag{2.8.4}$$

i.e. we consider now $Y_v^\varepsilon(n)$ evolving on the interval $[0, 1]$ with 0 and 1 identified which makes it the circle of radius $1/2\pi$. Suppose that the distribution of ξ_1 has a C^1 density $p(y)$ with respect to the Lebesgue measure which is positive on $[0, 1]$. Now we have the family of Markov chains $Y_{x,y}(n)$, $n \ge 0$ with transition probabilities

$$P_x(y, \Gamma) = P\{Y_{x,y}(1) \in \Gamma\} = P\{x+y+\xi_1 \text{ (mod 1)} \in \Gamma\} = \int_\Gamma p(z-x-y)dz. \tag{2.8.5}$$

Thus we are in the framework of our main model satisfying Assumption 2.2.9, and so the assertion of Theorem 2.2.10 holds true. Let μ^x be the invariant measure of the Markov chain Y_x (which is unique since the Doeblin condition is satisfied here) and assume that

$$\int B(x, y)d\mu^x(y) = 0 \quad \text{for all } x \tag{2.8.6}$$

which is, essentially, not a restriction since we always can consider $B(x, y) - \int B(x, y)d\mu^x(y)$ in place of $B(x, y)$. This means that $\bar X_x^\varepsilon(n) \equiv x$ and we derive from Theorem 2.2.10 that for any $\delta > 0$ there exists $\alpha(\delta) > 0$ such that for all small ε,

$$P\{\max_{0\le n\le T/\varepsilon} |X^\varepsilon_{x,y}(n) - x| \ge \delta\} \le e^{-\alpha(\delta)/\varepsilon}. \tag{2.8.7}$$

Set $\varepsilon_k = \alpha(\delta)/2\ln k$ then $e^{-\alpha(\delta)/\varepsilon_k} = k^{-2}$ and by the Borel–Cantelli lemma we obtain that there exists $k_\delta(\omega)$ finite with probability one so that for all $k \ge k_\delta(\omega)$,

$$\max_{0\le n\le T/\varepsilon_k} |X^{\varepsilon_k}_{x,y}(n) - x| < \delta. \tag{2.8.8}$$

By (2.8.3) for $\varepsilon_{k+1} < \varepsilon \le \varepsilon_k$ and $k \ge 2$,

$$\begin{aligned}(2.8.9)\ \max\nolimits_{0\le n\le T/\varepsilon_{k+1}} |X^{\varepsilon_k}_{x,y}(n) - X^\varepsilon_{x,y}(n)| &\le T\varepsilon^{-2}_{k+1}(1+C\varepsilon_k)^{T/\varepsilon_{k+1}}(\varepsilon_k - \varepsilon_{k+1})\\ &\le 2Te^{2CT}(\alpha(\delta))^{-1}\ln(1+\tfrac{1}{k}) \longrightarrow 0 \text{ as } k\to\infty.\end{aligned}$$

It follows that with probability one,

$$\max_{0\le n\le T/\varepsilon} |X^\varepsilon_{x,y}(n) - x| \to 0 \text{ as } k\to\infty. \tag{2.8.10}$$

The conditions above can be relaxed a bit but this method will not already work if, for instance, the second equality in (2.8.1) is replaced by

$$Y^\varepsilon_v(n+1) = 2Y^\varepsilon_v(n) + X^\varepsilon_v(n) + \delta\xi_{n+1}$$

since in this case the derivative $\frac{dX^\varepsilon_v(n)}{d\varepsilon}$ may grow exponentially in n and, indeed, we show in [**11**] that for the latter example there is no convergence with probability one in (2.8.10) provided $\delta > 0$ is small enough.

Next, we exhibit two examples of computations which demonstrate adiabatic transitions between attractors of the averaged system via the statistics of proportions of time the slow motion spends in basins of different attractors. The fast motions Y^ε_v in both examples are given by the second equation in (2.8.1) where $\xi_1, \xi_2, \ldots$ are i.i.d. random variables with the uniform distribution on $[0,1]$. The slow motion X^ε_v is given by the first equation in (2.8.1) where in the first example

$$B(x,y) = B_1(x,y) = x(x^2-4)(1-x^2) + 50\sin 2\pi y$$

and in the second example

$$B(x,y) = B_2(x,y) = x(x^2-4)(1-x)(1.5+x) + 50\sin 2\pi y.$$

The Markov chains Y_x preserve here the Lebesgue measure on $[0,1]$ which is the unique invariant measure for them, and so the averaged equation (2.1.6) for $\bar Z(t) = \bar X^\varepsilon(t/\varepsilon)$ has the right hand side $\bar B(x) = \bar B_1(x) = x(x^2-4)(1-x^2)$ in the first case and, $\bar B(x) = \bar B_2(x) = x(x^2-4)(1-x)(1.5+x)$ in the second case. The one dimensional vector field $\bar B(x)$ has three attracting fixed points $\mathcal{O}_1 = 2, \mathcal{O}_2 = 0, \mathcal{O}_3 = -2$ and two repelling fixed points 1 and -1, while $\bar B_2(x)$ has the same attracting fixed points but one repelling fixed point moves now from -1 to $-3/2$ making the basin of -2 smaller which makes it easier for the slow motion to escape from there. It is easy to see that B_1 and B_2 are complete at the fixed points of the averaged system, and so Theorem 2.2.12 is applicable in this situation. According to the corresponding part of Theorem 2.2.12 the transitions between $\mathcal{O}_1, \mathcal{O}_2$, and $\mathcal{O}_3$ are determined by R_{ij}, $i,j = 1,2,3$ which are obtained via the functionals $S_{0t}(\gamma)$ given by (2.2.9) but even here these functionals are not easy to compute. The functionals $S_{0t}(\gamma)$ yield non classical variational problems and the effective ways of their computation remain for further research .

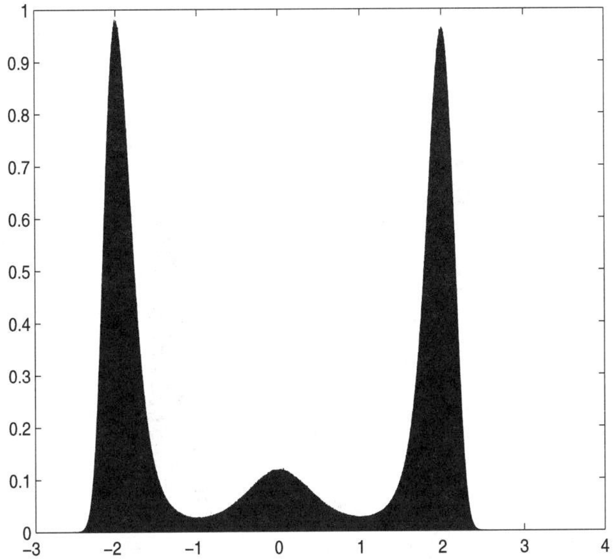

FIGURE 2.8.1. Symmetrical basins case

In the first example we plot above the histogram with 10^4 intervals of a single orbit of the slow motion $X^{\varepsilon}_{x,y}(n)$, $n = 0, 1, 2, ..., 10^8$ with $\varepsilon = 10^{-3}$ and the initial values $x = 0$, $y = 0$. The histogram shows that most of the points of the orbit stay near the attractors $\mathcal{O}_1$, $\mathcal{O}_2$ and $\mathcal{O}_3$ and $X^{\varepsilon}_{x,y}(n)$ hops between basins of attraction of these points. The form of the histogram indicates the equality $R_{21} = R_{23}$, which follows also by the symmetry considerations, but in this case Theorem 2.2.12 cannot specify whether the slow motion mostly exits from the basin of $\mathcal{O}_2$ to the basin of $\mathcal{O}_1$ or to the basin of $\mathcal{O}_3$.

In the second example the basin of attraction of -2 becomes smaller while the left interval of the basin of attraction of 0 becomes larger. The latter leads to the inequality $R_{23} > R_{21}$ which according to Theorem 2.2.12 makes it more difficult for the slow motion to exit to the left from the basin of $\mathcal{O}_2$ than to the right. In the histogram below (which has again 10^4 intervals) we plot $X^{\varepsilon}_{x,y}(n)$, $n = 0, 1, 2, ..., 10^8$ with $\varepsilon = 10^{-3}$ and the initial values $x = -2$, $y = 0$. In compliance with Theorem 2.2.12 the histogram demonstrates that the slow motion leaves the basin of $\mathcal{O}_3$ and after arriving at the basin of $\mathcal{O}_2$ it exits mostly to the basin of $\mathcal{O}_1$, and so the slow motion hops mostly between basins of $\mathcal{O}_1$ and $\mathcal{O}_2$ staying most of the time in small neighborhoods of these points. Still, a complete rigorous explanation of these histograms even for our simple examples requires nontrivial additional arguments. It is interesting to observe that these histograms have the same form as in Section 1.9 of Part 1 where randomness is generated by the expanding (chaotic) map $y \to 3y$ instead of adding uniformly distributed random variables as we do it here.

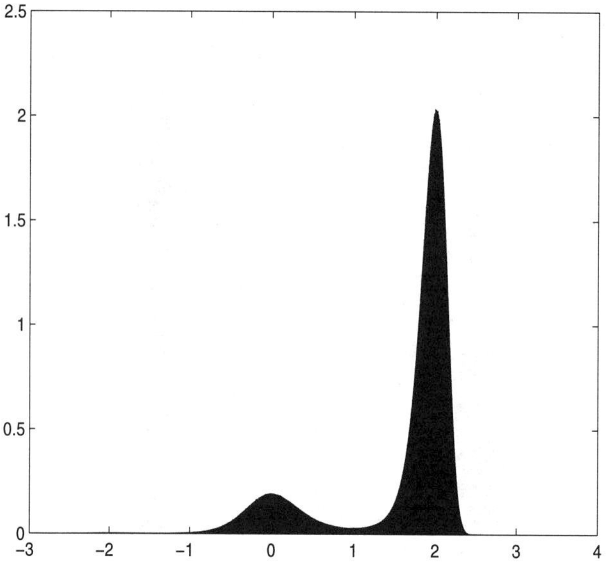

FIGURE 2.8.2. Asymmetrical basins case

2.9. Extensions: stochastic resonance

The scheme for the stochastic resonance type phenomenon described below is a slight modification of the model suggested by M.Freidlin (cf. [**30**]) and it can be demonstrated in the setup of three scale systems

$$
\begin{aligned}
\frac{dV^{\varepsilon,\delta}(t)}{dt} &= \delta\varepsilon A(V^{\varepsilon,\delta}(t), X^{\varepsilon,\delta}(t), Y^{\varepsilon,\delta}(t)) \\
\frac{dX^{\varepsilon,\delta}(t)}{dt} &= \varepsilon B(V^{\varepsilon,\delta}(t), X^{\varepsilon,\delta}(t), Y^{\varepsilon,\delta}(t)) \qquad (2.9.1)\\
dY^{\varepsilon,\delta}(t) &= \sigma(V^{\varepsilon,\delta}(t), X^{\varepsilon,\delta}(t), Y^{\varepsilon,\delta}(t))dw_t + b(V^{\varepsilon,\delta}(t), X^{\varepsilon,\delta}(t), Y^{\varepsilon,\delta}(t))dt,
\end{aligned}
$$

$V^{\varepsilon,\delta} = V^{\varepsilon,\delta}_{v,x,y}$, $X^{\varepsilon,\delta} = X^{\varepsilon,\delta}_{v,x,y}$, $Y^{\varepsilon,\delta} = Y^{\varepsilon,\delta}_{v,x,y}$ with initial conditions $V^{\varepsilon,\delta}(0) = v$, $X^{\varepsilon,\delta}(0) = x$ and $Y^{\varepsilon,\delta}(0) = y$ and the last equation in (2.9.1) is a stochastic differential equation coupled with first two ordinary differential equations though together they should be considered as a system of stochastic differential equations (with a degeneration in the first two). We assume that $V^{\varepsilon,\delta} \in \mathbb{R}^l$, $X^{\varepsilon,\delta} \in \mathbb{R}^d$ while $Y^{\varepsilon,\delta}$ evolves on a compact $n_{\mathbf{M}}$-dimensional C^2 Riemannian manifold $\mathbf{M}$ and the coefficients A, B, b are bounded smooth vector fields on $\mathbb{R}^l$, $\mathbb{R}^d$ and $\mathbf{M}$, respectively, depending on other variables as parameters. We suppose also that $a = \sigma\sigma^*$ is a uniformly positive definite smooth matrix field on $\mathbf{M}$. In the same way as in Section 2.2 we can generalize the setup taking $Y^{\varepsilon,\delta}$ to be random evolutions but in order to simplify the notations we restrict ourselves to fast motions $Y^{\varepsilon,\delta}$ being diffusions. The solution of (2.9.1) determines a Markov diffusion process which the triple $(V^{\varepsilon,\delta}, X^{\varepsilon,\delta}, Y^{\varepsilon,\delta})$. Taking $\varepsilon = \delta = 0$ we arrive at the (unperturbed) process $(v, x, Y^{0,0}_{v,x,y})$ where $Y = Y_{v,x,y} = Y^{0,0}_{v,x,y}$ solves the unperturbed stochastic differential equation

$$dY(t) = \sigma(v, x, Y(t))dw_t + b(v, x, Y(t))dt. \tag{2.9.2}$$

It is natural to view the diffusion $Y(t)$ as describing an idealized physical system where parameters $v = (v_1, ..., v_l)$ and $x = (x_1, ..., x_d)$ are assumed to be constants of motion while the perturbed process $(V^{\varepsilon,\delta}, X^{\varepsilon,\delta}, Y^{\varepsilon,\delta})$ is regarded as describing a real system where evolution of these parameters is also taken into consideration but unlike the averaging setup (2.1.1) we have now two sets of parameters moving with very different speeds.

Let $\mu_{v,x}$ be the unique invariant measure of the diffusion $Y(t) = Y_{v,x,y}(t)$. Set

$$\bar{B}_v(x) = \bar{B}(v, x) = \int B(v, x, y)d\mu_{v,x}(y) \tag{2.9.3}$$

and let $\bar{X}^{(v)}$ be the solution of the averaged equation

$$\frac{d\bar{X}^{(v)}(t)}{dt} = \bar{B}_v(\bar{X}^{(v)}(t)). \tag{2.9.4}$$

First, we apply averaging and large deviations estimates in averaging from the previous sections to two last equations in (2.9.1) freezing the slowest variable v (i.e. taking for a moment $\delta = 0$). Namely, set $\hat{X}^\varepsilon(t) = X^{\varepsilon,0}_{v,x,y}(t/\varepsilon)$ and $\hat{Y}^\varepsilon(t) = Y^{\varepsilon,0}_{v,x,y}(t/\varepsilon)$ so that

$$\begin{aligned} \frac{d\hat{X}^\varepsilon(t)}{dt} &= \varepsilon B(v, \hat{X}^\varepsilon(t), \hat{Y}^\varepsilon(t)) \\ \frac{d\hat{Y}^\varepsilon(t)}{dt} &= \sigma(v, \hat{X}^\varepsilon(t), Y^\varepsilon(t))dw_t + b(v, \hat{X}^\varepsilon(t), \hat{Y}^\varepsilon(t))dt. \end{aligned} \tag{2.9.5}$$

Suppose for simplicity that $l = d = 1$ (i.e. both $V^{\varepsilon,\delta}$ and $X^{\varepsilon,\delta}$ are one dimensional) and that the solution $\bar{X}^{(v)}(t)$ of (2.9.4) has the limit set consisting of two attracting points $\mathcal{O}_1$ and $\mathcal{O}_2$, which for simplicity we assume to be independent of v, and a repelling fixed point $\mathcal{O}_0^v$ depending on v and separating their basins. As an example of $\bar{B}$ we may have in mind $\bar{B}_v(x) = (x - v)(1 - x^2), -1 < v < 1$. Let $S^v_{0T}(\gamma)$ be the large deviations rate functional for the system of last two equations in (2.9.1) defined in (2.2.9) and set for $i, j = 1, 2$,

$$R_{ij}(v) = \inf\{S^v_{0T}(\gamma) : \gamma \in C_{0T},\ \gamma_0 = \mathcal{O}_i,\ \gamma_T = \mathcal{O}_j,\ T \geq 0\} \tag{2.9.6}$$

(cf. with R_{ij} in Theorem 2.2.7).

Set

$$\bar{A}_i(v) = \int A(v, \mathcal{O}_i, y)d\mu_{v,\mathcal{O}_i}(y) \tag{2.9.7}$$

and assume that for all v,

$$\bar{A}_1(v) < 0 \quad \text{and} \quad \bar{A}_2(v) > 0 \tag{2.9.8}$$

which means in view of the averaging principle (see Theorem 2.2.2 and the following it discussion) that $V^{\varepsilon,\delta}_{v,x,y}(t)$ decreases (increases) with high probability while $X^{\varepsilon,\delta}_{v,x,y}(t)$ stays close to $\mathcal{O}_1$ (to $\mathcal{O}_2$).

The following statement suggests a "nearly" periodic behavior of the slowest motion.

2.9.1. CONJECTURE. *Suppose that there exist strictly increasing and decreasing functions $v_-(r)$ and $v_+(r)$, respectively, so that*

$$R_{12}(v_-(r)) = R_{21}(v_+(r)) = r$$

and $v_-(\lambda) = v_+(\lambda) = v^*$ *for some* $\lambda > 0$ *while* $v_-(r) < v^* < v_+(r)$ *for* $r < \lambda$. *Assume that* $\delta \to 0$ *and* $\varepsilon \to 0$ *in such a way that*

$$\lim_{\varepsilon,\delta\to 0} \varepsilon \ln(\delta\varepsilon) = -\rho > -\lambda. \tag{2.9.9}$$

Then for any v, x *there exists* $t_0 > 0$ *so that the slowest motion* $\tilde{V}^{\varepsilon,\delta}_{v,x,y}(t+t_0) = V^{\varepsilon,\delta}_{v,x,y}\big((t+t_0)/\delta\varepsilon\big)$, $t \geq 0$ *converges in distribution (as* $\varepsilon, \delta \to 0$ *so that (2.9.9) holds true) to a periodic function* $\psi(t)$, $\psi(t+T) = \psi(t)$ *with*

$$T = T(\rho) = \int_{v_-(\rho)}^{v_+(\rho)} \frac{dv}{|\bar{A}_1(v)|} + \int_{v_-(\rho)}^{v_+(\rho)} \frac{dv}{|\bar{A}_2(v)|}.$$

The argument supporting this conjecture goes as follows. Since $V^{\varepsilon,\delta}$ moves much slower than $X^{\varepsilon,\delta}$ we can freeze the former and in place of (2.9.1) we can study, first, (2.9.5). Applying the arguments of Theorem 2.2.7 to the pair $\hat{X}, \hat{Y}$ from (2.9.5) we conclude from (2.2.35) that the intermediate motion $X^{\varepsilon,\delta}$ most of the time stays very close to either $\mathcal{O}_1$ or $\mathcal{O}_2$ before it exits from the corresponding basin, and so in view of an appropriate averaging principle (which follows, for instance, from Theorem 2.2.2) on bounded time intervals the slowest motion $V^{\varepsilon,\delta}$ mostly stays close to the corresponding averaged motion determined by the vector fields $\bar{A}_1$ and $\bar{A}_2$ given by (2.9.7). When $X^{\varepsilon,\delta}$ is close to $\mathcal{O}_1$ the slowest motion $V^{\varepsilon,\delta}$ decreases until $v = v_-(\rho)$ where $R_{12}(v) = \rho$. In view of (2.2.34) and the scaling (2.9.9) between ε and δ, a moment later $R_{12}(v)$ becomes less than ρ and $X^{\varepsilon,\rho}$ jumps immediately close to $\mathcal{O}_2$. There $\bar{A}_2(v) > 0$, and so $V^{\varepsilon,\delta}$ starts to grow until it reaches $v = v_+(\rho)$ where $R_{21}(v) = \rho$. A moment later $R_{21}(v)$ becomes smaller than ρ and in view of (2.2.34) the intermediate motion $X^{\varepsilon,\delta}$ jumps immediately close to $\mathcal{O}_1$. This leads to a nearly periodic behavior of $V^{\varepsilon,\delta}$. In order to make these arguments precise we have to deal here with an additional difficulty in comparison with the two scale setup considered in previous sections since now the large deviations S-functionals from Theorem 2.2.2 and the R-functions describing adiabatic fluctuations and transitions of Theorems 2.2.5 and 2.2.7 depend on another very slowly changing parameter. Still, we can use the technique of Sections 2.6 and 2.7 above applied on time intervals where changes in the v-variable can be neglected should work here but the details of this approach have not been worked out yet.

On the other hand, when the fast motion $Y^{\varepsilon,\delta}$ does not depend on the slow motions, i.e. when the coefficients σ and b in (2.9.1) depend only on the coordinate y (but not on v and x), then the above arguments can be made precise without much effort. Indeed, we can obtain estimates for transition times $\tau^\varepsilon(1)$ and $\tau^\varepsilon(2)$ of $X^{\varepsilon,\delta}(t/\varepsilon)$ between neighborhoods of $\mathcal{O}_1$ and $\mathcal{O}_2$ as in Theorem 2.2.7 applying the latter to $\hat{X}^\varepsilon$ and $\hat{Y}^\varepsilon$ from (2.9.5) with freezed v-variable. This is possible since the method of Proposition 2.6.1 requires us to make large deviations estimates, essentially, only for probabilities $P\{kT \leq \tau(i) < (k+1)T\}$, i.e. on bounded time intervals, and then combine them with the Markov property arguments. During such times the slowest motion $V^{\varepsilon,\delta}$ can move only a distance of order δT. Thus freezing v and using the Gronwall inequality for the equation of X^ε in order to estimate the resulting error we see that the latter is small enough for our purposes. Observe, that it would be much more difficult to justify freezing v in the coefficients σ and b of Y^ε, if we allow the latter to depend on v, since a strightforward application of the Gronwall inequality there would yield an error estimate of an exponential in $1/\varepsilon$ order which is comparable with $1/\delta$. Still, it may be possible to take care about the general case using methods of Section 2.3 since we produce large deviations estimates there by gluing large deviations estimates on smaller time intervals where the x-variable (and so, of course, v-variable) can be freezed. Next, set

$$V_{v,y}^{\varepsilon,\delta,i}(t) = v + \delta\varepsilon \int_0^t A(V_v^{\delta,\varepsilon,i}(s), \mathcal{O}_i, Y(s))ds$$

where now Y does not depend on ε and δ. Then by (2.9.1) together with the Gronwall inequality we obtain that

$$|V_{v,x,y}^{\varepsilon,\delta}(t) - V_{v,y}^{\varepsilon,\delta,i}(t)| \le K\delta\varepsilon e^{\delta\varepsilon Kt} \int_0^t |X_{v,x,y}^{\varepsilon,\delta}(s) - \mathcal{O}_i|ds$$

where K is the Lipschitz constant of A. If x belongs to the basin $\mathcal{O}_i$ then according to Theorem 2.2.7 $\hat{X}^\varepsilon$, and so also $X^{\varepsilon,\delta}$, stays most of the time near $\mathcal{O}_i$ up to its exit from the basin of the latter which yields according to the above inequality that $V^{\varepsilon,\delta}$ stays close to $V^{\varepsilon,\delta,i}$ during this time. But now we can employ the averaging principle for the pair $V^{\varepsilon,\delta,i}(t), Y(t)$ which sais that $V^{\varepsilon,\delta,i}(t)$ stays close on the time intervals of order $1/\delta\varepsilon$ to the averaged motion $\bar{V}_v^{\varepsilon,\delta,i}(t)$ defined by

$$\bar{V}_v^{\varepsilon,\delta,i}(t) = v + \int_0^t \bar{A}_i(\bar{V}_v^{\varepsilon,\delta,i}(s))ds$$

and in view of (2.9.9), $\bar{V}_v^{\varepsilon,\delta,1}(t)$ decreases while $\bar{V}_v^{\varepsilon,\delta,2}(t)$ increases which leads to the behavior described in Conjecture 2.9.1.

A similar conjecture can be made under the corresponding conditions for the discrete time case determined by a three scale difference system of equations of the form

$$V^{\varepsilon,\delta}(n+1) - V^{\varepsilon,\delta}(n) = \varepsilon\delta A(V^{\varepsilon,\delta}(n), X^{\varepsilon,\delta}(n), Y^{\varepsilon,\delta}(n)),\ \ V^{\varepsilon,\delta}(0) = v,$$
$$X^{\varepsilon,\delta}(n+1) - X^{\varepsilon,\delta}(n) = \varepsilon B(V^{\varepsilon,\delta}(n), X^{\varepsilon,\delta}(n), Y^{\varepsilon,\delta}(n)),\ \ X^{\varepsilon,\delta}(0) = x$$

where A and B are smooth vector functions and $Y^{\varepsilon,\delta}(n)) = Y_{v,x,y}^{\varepsilon,\delta}(n)$, $Y_{v,x,y}^{\varepsilon,\delta}(0) = y$ are coupled with $V^{\varepsilon,\delta}(n)$ and $X^{\varepsilon,\delta}(n)$ perturbations of a parametrized by v and x appropriate family of Markov chains having smooth transition densities similar to those considered in Theorem 2.2.12.

2.10. Young measures approach to averaging

In this section we derive the averaging principle and discuss corresponding large deviations in the sense of convergence of Young measures adapted to our probabilistic setup. For more detailed information about Young measures we refer the reader to [**3**] and references there.

Let μ belongs to the space $\mathcal{P}(\mathbb{R}^d \times \mathbf{M})$ of probability measures on $\mathbb{R}^d \times \mathbf{M}$. We consider a random Young measure ζ^ε from $([0,T] \times \mathbb{R}^d \times \mathbf{M}, \ell_T \times \mu)$ to $\mathcal{P}(\mathbb{R}^d \times \mathbf{M})$ which we define by

$$\zeta^\varepsilon(t,x,y) = \delta_{X_{x,y}^\varepsilon(t/\varepsilon), Y_{x,y}^\varepsilon(t/\varepsilon)}$$

where ℓ_T is the Lebesgue measure on $[0,T]$, δ_w is the unit mass at w, and X^ε, Y^ε are the same as in (2.1.1).

Suppose that $\mu \in \mathcal{P}(\mathbb{R}^d \times \mathbf{M})$ has a disintegration

$$d\mu(x,y) = d\mu_x(y)d\lambda(x),\ \ \lambda \in \mathcal{P}(\mathbb{R}^d) \tag{2.10.1}$$

such that for each Lipschitz continuous function g on $\mathbf{M}$ and any $x, z \in \mathbb{R}^d$,

$$|\int g d\mu_x - \int g d\mu_z| \le K_{L(g)}|x - z| \tag{2.10.2}$$

for some $K_L > 0$ depending only on L where $L(g)$ is both a Lipschitz constant of g and it also bounds $|g|$. Set

$$\bar{B}(x) = \int B(x, y) d\mu_x(y) \tag{2.10.3}$$

and assume that (2.2.1) holds true which together with (2.10.2) yields that $\bar{B}$ is bounded and Lipschitz continuous, and so there exists a unique solution $\bar{X}^\varepsilon(t) = \bar{X}_x^\varepsilon(t)$ of (2.1.3). For any bounded continuous function g on $\mathbb{R}^d \times \mathbf{M}$ define

$$\mathcal{E}_\varepsilon^g(t, \delta) = \{(x, y) \in \mathbb{R}^d \times \mathbf{M} : E\big|\frac{1}{t}\int_0^t g(x, Y_{x,y}^\varepsilon(u))du - \bar{g}(x)\big| > \delta\}$$

where $\bar{g}(x) = \int g(x, y) d\mu_x(y)$.

In the spirit of [**3**]) we say that the Young measures ζ^ε converge as $\varepsilon \to 0$ to a Young measure ζ^0 defined by

$$\zeta^0(t, x, y) = \delta_{\bar{Z}_x(t)} \times \mu_{\bar{Z}_x(t)} \in \mathcal{P}(\mathbb{R}^d \times \mathbf{M}),$$

$\bar{Z}_x(t) = \bar{X}_x^\varepsilon(t/\varepsilon)$, if for any bounded continuous function $f(t, x, y)$ on $([0, T] \times \mathbb{R}^d \times \mathbf{M}$,

$$E\big|\int d\mu(x, y)\int_0^T \big(f(s, X_{x,y}^\varepsilon(s/\varepsilon), Y_{x,y}^\varepsilon(s/\varepsilon)) - \bar{f}(s, \bar{Z}_x(s))\big)ds\big| \to 0 \text{ as } \varepsilon \to 0.$$

The following result provides a verifiable (in some interesting cases) criterion for even stronger convergence.

2.10.1. THEOREM. *Let $\mu \in \mathcal{P}(\mathbb{R}^d \times \mathbf{M})$ has the disintegration (2.10.1) satisfying (2.10.2). Then*

$$\lim_{\varepsilon \to 0} \int_{\mathbb{R}^d}\int_{\mathbf{M}} E \sup_{0 \le t \le T}\big|\int_0^t \big(f(s, X_{x,y}^\varepsilon(s/\varepsilon), Y_{x,y}^\varepsilon(s/\varepsilon)) - \bar{f}(s, \bar{Z}_x(s))\big)ds\big|d\mu(x, y) = 0 \tag{2.10.4}$$

for any bounded continuous function $f = f(t, x, y)$ on $[0, T] \times \mathbb{R}^d \times \mathbf{M}$ where $\bar{f}(t, x) = \int f(t, x, y) d\mu_x(y)$ if and only if for each $N \in \mathbb{N}$ and any finite collection $g_1, ..., g_N$ of bounded Lipschitz continuous functions on $\mathbb{R}^d \times \mathbf{M}$ there exists an integer valued function $n = n(\varepsilon) \to \infty$ as $\varepsilon \to 0$ such that for any $\delta > 0$ and $l = 1, ..., N$,

$$\lim_{\varepsilon \to 0}\max_{0 \le j < n(\varepsilon)} \int_{\mathbb{R}^d}\int_{\mathbf{M}} P\{\big(X_{x,y}^\varepsilon(jt(\varepsilon)), Y_{x,y}^\varepsilon(jt(\varepsilon))\big) \in \mathcal{E}_\varepsilon^{g_l}(t(\varepsilon), \delta)\}d\mu(x, y) = 0, \tag{2.10.5}$$

where $t(\varepsilon) = \frac{T}{\varepsilon n(\varepsilon)}$.

PROOF. First, we prove that (2.10.4) implies (2.10.5). Let $g_1, ..., g_N$ be bounded Lipschitz continuous functions on $\mathbb{R}^d \times \mathbf{M}$ and set

$$\rho_{x,y}^{\varepsilon,l}(t) = \varepsilon \int_0^t \big(g_l(X_{x,y}^\varepsilon(s), Y_{x,y}^\varepsilon(s)) - \bar{g}_l(\bar{X}_x^\varepsilon(s))\big)ds. \tag{2.10.6}$$

If

$$\rho_{x,y}^{\varepsilon,l} = \sup_{0 \le t \le T/\varepsilon} |\rho_{x,y}^{\varepsilon,l}(t)|$$

then by (2.10.4) for each $l = 1, ..., N$,

$$\rho_l^\varepsilon = \int_{\mathbb{R}^d} \int_{\mathbf{M}} E\rho_{x,y}^{\varepsilon,l} d\mu(x,y) \to 0 \text{ as } \varepsilon \to 0. \tag{2.10.7}$$

Choose an integer valued function $n(\varepsilon) \to \infty$ as $\varepsilon \to 0$ so that

$$n(\varepsilon) \max_{1\le l\le N} \rho_l^\varepsilon \to 0 \text{ as } \varepsilon \to 0 \tag{2.10.8}$$

and let $t(\varepsilon) = T/\varepsilon n(\varepsilon)$. Set $x_k^\varepsilon = X_{x,y}^\varepsilon(kt(\varepsilon))$, $y_k^\varepsilon = Y_{x,y}^\varepsilon(kt(\varepsilon))$ and $\bar{x}_k^\varepsilon = \bar{X}_x^\varepsilon(kt(\varepsilon))$, $k = 0, 1, ...$ Then by (2.10.6),

(2.10.9)

$$\rho_{x,y}^{\varepsilon,l}((j+1)t(\varepsilon)) - \rho_{x,y}^{\varepsilon,l}(jt(\varepsilon)) = \varepsilon \int_0^{t(\varepsilon)} \big(g_l(X_{x_j^\varepsilon,y_j^\varepsilon}^\varepsilon(u), Y_{x_j^\varepsilon,y_j^\varepsilon}^\varepsilon(u)) - \bar{g}_l(\bar{X}_{\bar{x}_j^\varepsilon}^\varepsilon(u))\big)du$$

where $X_{x_j^\varepsilon,y_j^\varepsilon}^\varepsilon(u) = X_{x,y}^\varepsilon(jt(\varepsilon) + u)$ and $Y_{x_j^\varepsilon,y_j^\varepsilon}^\varepsilon(u) = Y_{x,y}^\varepsilon(jt(\varepsilon) + u)$. By (2.2.1),

$$\begin{aligned} &\varepsilon\Big| \int_0^{t(\varepsilon)} \big(g_l(X_{x_j^\varepsilon,y_j^\varepsilon}^\varepsilon(u), Y_{x_j^\varepsilon,y_j^\varepsilon}^\varepsilon(u)) - g_l(x_j^\varepsilon, Y_{x_j^\varepsilon,y_j^\varepsilon}^\varepsilon(u))\big)du\Big| \\ &\le \varepsilon L_l \int_0^{t(\varepsilon)} |X_{x_j^\varepsilon,y_j^\varepsilon}^\varepsilon(u) - x_j^\varepsilon|du \le L_l K(\varepsilon n(\varepsilon))^2 \end{aligned} \tag{2.10.10}$$

where L_l is the Lipschitz constant of g_l. Similarly, by (2.2.1) and (2.10.2),

$$\varepsilon\Big| \int_0^{t(\varepsilon)} \big(\bar{g}_l(\bar{X}_{\bar{x}_j^\varepsilon}^\varepsilon(u)) - \bar{g}_l(\bar{x}_j^\varepsilon)\big)du\Big| \le (L_l + K_{L_l})K(\varepsilon t(\varepsilon))^2 \tag{2.10.11}$$

and

$$|\bar{g}_l(\bar{x}_j^\varepsilon) - \bar{g}_l(x_j^\varepsilon)| \le (L_l + K_{L_l})|\bar{x}_j^\varepsilon - x_j^\varepsilon| \le (L_l + K_{L_l})\rho_{x,y}^\varepsilon. \tag{2.10.12}$$

It follows from (2.10.9)–(2.10.12) that

$$\begin{aligned} &\Big|\tfrac{1}{t(\varepsilon)} \int_0^{t(\varepsilon)} g_l(x_j^\varepsilon, Y_{x_j^\varepsilon,y_j^\varepsilon}^\varepsilon(u))du - \bar{g}_l(x_j^\varepsilon)\Big| \\ &\le TK(2L_l + K_{L_l})/n(\varepsilon) + (L_l + K_{L_l} + 2T^{-1}n(\varepsilon))\rho_{x,y}^\varepsilon. \end{aligned} \tag{2.10.13}$$

Given $\delta > 0$ choose $\varepsilon_\delta > 0$ such that for all $\varepsilon \le \varepsilon_\delta$ and $l = 1, ..., N$,

$$TK(2L_l + K_{L_l})/n(\varepsilon) \le \delta/2.$$

Then by (9.13) and the Markov property,

$$\begin{aligned} &\big\{(x,y) \in \mathbb{R}^d \times \mathbf{M} : \big(X_{x,y}^\varepsilon(jt(\varepsilon)), Y_{x,y}^\varepsilon(jt(\varepsilon))\big) \in \mathcal{E}_\varepsilon^{g_l}(t(\varepsilon), \delta) \subset A_\varepsilon(\delta) \\ &= \big\{(x,y) \in \mathbb{R}^d \times \mathbf{M} : (L_l + K_{L_l} + 2T^{-1}n(\varepsilon))E_{x_j^\varepsilon,y_j^\varepsilon}\rho_{x,y}^{\varepsilon,l} > \delta/2\big\} \end{aligned}$$

where

$$E_{x_j^\varepsilon,y_j^\varepsilon} = E\big(\cdot\,|X_{x,y}^\varepsilon(jt(\varepsilon)), Y_{x,y}^\varepsilon(jt(\varepsilon))\big)$$

is the conditional expectation. By Chebyshev's inequality

$$E\mu(A_\varepsilon(\delta)) \le \frac{2}{\delta}(L_l + K_{L_l} + 2T^{-1}n(\varepsilon))E\rho_l^\varepsilon. \tag{2.10.14}$$

By (2.10.8) the right hand side of (2.10.14) tends to 0 as $\varepsilon \to 0$ yielding (2.10.5).

Next, we derive (2.10.4) from (2.10.5). Since f in (2.10.4) is a bounded function and λ is a probability measure it is easy to see that it suffices to prove (2.10.4) when the integration in x there is restricted to compact subsets of $\mathbb{R}^d$. But if x belongs to a compact set $G \subset \mathbb{R}^d$ in view of (2.1.1) and (2.2.1) the slow motion $X_{x,y}^\varepsilon(s)$, as well as the averaged one $\bar{X}_x^\varepsilon(s)$, stays during the time T/ε in a $KT-$neighborhood G_{KT} of G. But on $[0,T]\times G_{KT}\times\mathbf{M}$ we can approximate f uniformly by Lipschitz continuous functions. Thus, in place of (2.10.4) it suffices to show that for any compact set

$G \subset \mathbb{R}^d$ and a bounded Lipschitz continuous function f on $[0,T] \times G_{KT} \times \mathbf{M}$ with a Lipschitz constant $L = L(f)$ in all variables,

$$\text{(2.10.15)} \qquad \lim_{\varepsilon\to 0} \varepsilon \int_G \int_{\mathbf{M}} E \sup_{0\le t\le T/\varepsilon} \big| \int_0^t \big(f(\varepsilon s, X^\varepsilon_{x,y}(s), Y^\varepsilon_{x,y}(s)) \\ -\bar f(\varepsilon s, \bar X^\varepsilon_x(s))\big) ds \big| d\mu_x(y) d\lambda(x) = 0.$$

By (2.2.1), (2.10.2) and (2.10.3),

$$\text{(2.10.16)} \qquad \varepsilon \big| \int_0^t \big(f(\varepsilon s, X^\varepsilon_{x,y}(s), Y^\varepsilon_{x,y}(s)) - \bar f(\varepsilon s, \bar X^\varepsilon_x(s))\big) ds \big| \\ \le \varepsilon \big| \int_0^t \big(f(\varepsilon s, X^\varepsilon_{x,y}(s), Y^\varepsilon_{x,y}(s)) - \bar f(\varepsilon s, X^\varepsilon_x(s))\big) ds \big| \\ +(L + K_L) T \sup_{0\le s\le T/\varepsilon} |X^\varepsilon_x(s) - \bar X^\varepsilon_x(s)|.$$

By (2.1.1), (2.2.1), (2.10.2) and (2.10.3),

$$|X^\varepsilon_{x,y}(t) - \bar X^\varepsilon_x(t)| = \varepsilon \big| \int_0^t \big(B(X^\varepsilon_{x,y}(s), Y^\varepsilon_{x,y}(s)) - \bar B(\bar X^\varepsilon_x(s))\big) ds \big| \\ \le \varepsilon \big| \int_0^t \big(B(X^\varepsilon_{x,y}(s), Y^\varepsilon_{x,y}(s)) - \bar B(X^\varepsilon_{x,y}(s))\big) ds \big| \\ +(K + K_K) \varepsilon \int_0^t |X^\varepsilon_{x,y}(s) - \bar X^\varepsilon_x(s)| ds.$$

This together with the Gronwall inequality gives

$$\text{(2.10.17)} \qquad \sup_{0\le t\le T/\varepsilon} |X^\varepsilon_{x,y}(t) - \bar X^\varepsilon_x(t)| \\ \le e^{(K+K_K)T} \varepsilon \sup_{0\le t\le T/\varepsilon} \big| \int_0^t \big(B(X^\varepsilon_{x,y}(s), Y^\varepsilon_{x,y}(s)) - \bar B(X^\varepsilon_{x,y}(s))\big) ds \big|.$$

Now we see that the integral term in the right hand side of (2.10.17) is a particular case of the integral term in the right hand side of (2.10.16) with $f = B$, and so it suffices to estimate only the latter.

Set, again, $x^\varepsilon_k = X^\varepsilon_{x,y}(kt(\varepsilon))$, $y^\varepsilon_k = Y^\varepsilon_{x,y}(kt(\varepsilon))$, $\bar x^\varepsilon_k = \bar X^\varepsilon_x(kt(\varepsilon))$, $k = 0, 1, ...$ and fix a large $N \in \mathbb{N}$. Let $l = [\varepsilon j t(\varepsilon) N/T] = [jN/n(\varepsilon)]$ then by (2.2.1), (2.10.1) and (2.10.2),

$$\text{(2.10.18)} \qquad \varepsilon \big| \int_0^{t(\varepsilon)} \big(f(\varepsilon j t(\varepsilon) + \varepsilon u, X^\varepsilon_{x^\varepsilon_j, y^\varepsilon_j}(u), Y^\varepsilon_{x^\varepsilon_j, y^\varepsilon_j}(u)) \\ -f(lT/N, x^\varepsilon_j, Y^\varepsilon_{x^\varepsilon_j, y^\varepsilon_j}(u))\big) ds \big| \le LT^2((Nn(\varepsilon))^{-1} + (n(\varepsilon))^{-2}) \\ +L\varepsilon \int_0^{t(\varepsilon)} |X^\varepsilon_{x^\varepsilon_j, y^\varepsilon_j}(u) - x^\varepsilon_j| du \le LT^2/Nn(\varepsilon) + LT^2(1+K))(n(\varepsilon))^{-2}$$

and

$$\text{(2.10.19)} \qquad \varepsilon \big| \int_0^{t(\varepsilon)} \big(\bar f(\varepsilon j t(\varepsilon) + \varepsilon u, X^\varepsilon_{x^\varepsilon_j, y^\varepsilon_j}(u)) - \bar f(lT/N, x^\varepsilon_j)\big) du \big| \\ \le LT^2/Nn(\varepsilon) + T^2(L + LK + KK_L)(n(\varepsilon))^{-2}.$$

Now using (2.10.18), (2.10.19) together with the Markov property and assuming that $|f| \le \hat{L}_f$ for some constant $\hat{L}_f > 0$ we obtain

$$\text{(2.10.20)} \quad \varepsilon E \sup_{0\le t\le T/\varepsilon} \big| \int_0^t \big(f(\varepsilon s, X^\varepsilon_{x,y}(s), Y^\varepsilon_{x,y}(s)) - \bar{f}(\varepsilon s, X^\varepsilon_{x,y}(s))\big)ds\big| \le$$
$$2\hat{L}_f\varepsilon t(\varepsilon) + \varepsilon E \sum_{j=0}^{n(\varepsilon)-1} \big| \int_{jt(\varepsilon)}^{(j+1)t(\varepsilon)} \big(f(\varepsilon s, X^\varepsilon_{x,y}(s), Y^\varepsilon_{x,y}(s)) - \bar{f}(\varepsilon s, \bar{X}^\varepsilon_x(s))\big)ds\big|$$
$$\le 2\hat{L}_f\varepsilon t(\varepsilon) + \varepsilon E \sum_{j=0}^{n(\varepsilon)-1} \big| \int_0^{t(\varepsilon)} \big(f(\varepsilon jt(\varepsilon) + \varepsilon s, X^\varepsilon_{x^\varepsilon_j, y^\varepsilon_j}(s), Y^\varepsilon_{x^\varepsilon_j, y^\varepsilon_j}(s))$$
$$-\bar{f}(\varepsilon jt(\varepsilon) + \varepsilon s, X^\varepsilon_{x^\varepsilon_j, y^\varepsilon_j}(s))\big)ds\big|$$
$$\le 2LT^2/N + 2\big(\hat{L}_f T + T^2(L + LK + KK_L)\big)/n(\varepsilon) + \varepsilon t(\varepsilon)$$
$$\times \sum_{l=0}^{N-1} \sum_{ln(\varepsilon)/N \le j < (l+1)n(\varepsilon)/N, j\le n(\varepsilon)} E\big| \tfrac{1}{t(\varepsilon)} \int_0^{t(\varepsilon)} f(lT/N, x^\varepsilon_j, Y^\varepsilon_{x^\varepsilon_j, y^\varepsilon_j}(s))ds$$
$$-\bar{f}(lT/N, x^\varepsilon_j)\big| \le 2LT^2/N + 2\big(\hat{L}_f T + T^2(L + LK + KK_L)\big)/n(\varepsilon)$$
$$+\varepsilon t(\varepsilon)n(\varepsilon)\delta + 2\hat{L}_f\varepsilon t(\varepsilon) \sum_{l=0}^{N-1} \sum_{ln(\varepsilon)/N \le j < (l+1)n(\varepsilon)/N, j\le n(\varepsilon)}$$
$$P\big\{\big(X^\varepsilon_{x,y}(jt(\varepsilon)), Y^\varepsilon_{x,y}(jt(\varepsilon))\big) \in \mathcal{E}^{f_l}_\varepsilon(t(\varepsilon), \delta)\big\}$$

where $f_l(z, v) = f(lT/N, z, v)$. Integrating against μ both parts of (2.10.20) over $G \times \mathbf{M}$ we obtain

$$\text{(2.10.21)} \quad \varepsilon \int_G \int_{\mathbf{M}} E \sup_{0\le t\le T/\varepsilon} \big| \int_0^t \big(f(\varepsilon s, X^\varepsilon_{x,y}(s), Y^\varepsilon_{x,y}(s))$$
$$-\bar{f}(\varepsilon s, X^\varepsilon_{x,y}(s))\big)ds\big| d\mu(x,y) \le 2\big(\hat{L}_f T + T^2(L + LK + KK_L)\big)/n(\varepsilon)$$
$$+2LT^2/N + T\delta + 2\hat{L}_f \max_{0\le l\le N-1} \eta_l(\varepsilon, \delta)$$

where

$$\eta_l(\varepsilon, \delta) = \max_{0\le j\le n(\varepsilon)-1} \int_G \int_{\mathbf{M}} P\big\{\big(X^\varepsilon_{x,y}(jt(\varepsilon)), Y^\varepsilon_{x,y}(jt(\varepsilon))\big) \in \mathcal{E}^{f_l}_\varepsilon(t(\varepsilon), \delta)\big\} d\mu(x,y).$$

By the assumption there exists an integer valued function $n(\varepsilon) \to \infty$ as $\varepsilon \to 0$ such that (2.10.5) holds true for all $g = f_0, f_1, ..., f_{N-1}$ and then $\max_{0\le l\le N-1} \eta_l(\varepsilon, \delta) \to 0$ as $\varepsilon \to 0$. Hence, letting first $\varepsilon \to 0$, then $\delta \to 0$ and, finally, $N \to 0$ we obtain (2.10.15) in view of (2.10.16) and (2.10.17), completing the proof of Theorem 2.10.1. □

Observe that (2.10.4) holding true for all bounded continuous functions is, in principle, stronger than the averaging principle in the form

$$\text{(2.10.22)} \quad \lim_{\varepsilon\to 0} \varepsilon \int_G \int_{\mathbf{M}} E \sup_{0\le t\le T/\varepsilon} \big|X^\varepsilon_{x,y}(t) - \bar{X}^\varepsilon_x(t)\big| d\mu(x,y) \to 0 \text{ as } \varepsilon \to 0$$

since (2.10.22) is equivalent to (2.10.15) with $f = B$. In fact, if we require (2.10.5) only for one function $g = B$ then it will be equivalent to (2.10.22) which follows in the same way as the proof of Theorem 2.10.1 above. Still, the main interesting classes of systems, we are aware of, for which (2.10.4) holds true are the same for which (2.10.22) is satisfied though it is easy to construct examples of (somewhat degenerate) right hand sides B in (2.1.1) for which (2.10.22) holds true but (2.10.4) fails (since in the latter we require convergence for all functions f and in the former only for $f = B$).

It follows from [**51**] that the assumptions of Theorem 2.10.1 hold true when the unperturbed fast motions $Y_{x,y}(t)$ are diffusion processes on $\mathbf{M}$ so that μ_x is an invariant measure of Y_x on $\mathbf{M}$ ergodic for λ-almost all x, where λ is the normalized Lebesgue measure on a large compact in $\mathbb{R}^d$, and $\mu_x(U) = \int_U q(x,y)dm(y)$ with

$q(x,y) > 0$ differentiable in x and y. This can be extended to random evolutions considered in previous sections.

Observe that under assumptions of Theorem 2.2.2 we can obtain also large deviations bounds in the form (2.2.10) and (2.2.11) for

$$W^\varepsilon_{x,y}(t) = \int_0^t f(s, X^\varepsilon_{x,y}(s/\varepsilon), Y^\varepsilon_{x,y}(s/\varepsilon))ds$$

with the functional

$$\tilde{S}_{0T}(\tilde{\gamma}) = \inf\big\{S_{0T}(\gamma) : S_{0T}(\gamma) = \int_0^T I_{\gamma_t}(\nu_t)dt,$$

$$\dot{\gamma}_t = \bar{B}_{\nu_t}(\gamma_t),\ \tilde{\gamma}_t = \int_0^t \bar{f}_{\nu_s}(s,\gamma_s)ds\ \forall t \in [0,T]\big\},\ \bar{f}_\nu(s,x) = \int f(s,x,y)d\nu(y),$$

where f is a bounded Lipschitz continuous vector function. This assertion, actually, follows directly from Theorem 2.2.2 (together with the contraction principle, see [**25**]) applying it to the slow motion $(Z^\varepsilon,\ \tau^\varepsilon,\ W^\varepsilon)$ given by the equations

$$\frac{dZ^\varepsilon(t)}{dt} = B(Z^\varepsilon(t), Y^\varepsilon(t/\varepsilon)),\ \frac{\tau^\varepsilon(t)}{dt} = 1,\ \frac{dW^\varepsilon(t)}{dt} = f(\tau^\varepsilon(t), Z^\varepsilon(t), Y^\varepsilon(t/\varepsilon))$$

combined with the time changed equations (2.2.16) and (2.2.17) for the fast motion Y^ε. Analogous results can be obtained in the discrete time setup of difference equations (2.1.7).

Bibliography

[1] D.B. Anosov, *Averaging in systems of ordinary differential equations with fast oscillating solutions,* Izv. Acad. Nauk SSSR Ser. Mat. **24** (1960), 731–742 (in Russian).

[2] J.P. Aubin and I. Ekeland, *Applied Nonlinear Analysis,* (1984) Wiley, New York.

[3] Z. Artstein and M. Grinfeld, *Ergodicity and mixing via Young measures*, Ergod. Th.& Dynam. Sys. **22** (2002), 1001–1015.

[4] V.I. Bakhtin, *Asymptotics of superregular perturbations of fiber ergodic semigroups*, Stoch. and Stoch. Rep. **75** (2003), 295–318.

[5] V.I. Bakhtin, *Cramèr's asymptotics in systems with fast and slow motions*, Stoch. and Stoch. Rep. **75** (2003), 319–341.

[6] V.I. Bakhtin, *Foliated functions and an averaged weighted shift operator for perturbations of hyperbolic mappings*, Proc. Steklov Inst. Math. **244** (2004), 29–57.

[7] V.I. Bakhtin, *Cramér asymptotics in the averaging method for systems with fast hyperbolic motions*, Proc. Steklov Inst. Math. **244** (2004), 65–86.

[8] R. Bowen, *Periodic orbits for hyperbolic flows,* Amer. J. Math. **94** (1972), 1–30.

[9] R. Bowen, *Symbolic dynamics for hyperbolic flows,* Amer. J. Math. **95** (1973), 429–459.

[10] V.I. Bakhtin and Yu. Kifer, *Diffusion approximation for slow motion in fully coupled averaging,* Prob. Th. Rel. Fields **129** (2004), 157–181.

[11] V.I. Bakhtin and Yu. Kifer, *Nonconvergence examples in averaging*, Contemporary Math. **469** (2008), 1–17.

[12] N.N. Bogolyubov and Yu.A. Mitropol'skii, *Asymptotic Methods in the Theory of Nonlinear Oscillations,* (1961), Hindustan, Delhi.

[13] R. Bowen and D. Ruelle, *The ergodic theory of Axiom A flows,* Invent. Math. **29** (1975), 181–202.

[14] N. Chernov and D. Dolgopyat, *Brownian Brownian Motion*, Memoirs Amer. Math. Soc., to appear.

[15] C. Castaing and M. Valadier, *Convex Analysis and Measurable Multifunctions,* Lect. Notes in Math., **580** (1977), Springer, Berlin.

[16] G. Contreras, *Regularity of topological and metric entropy of hyperbolic flows,* Math. Z. **210** (1992), 97–111.

[17] W.J. Cowieson, *Stochastic stability for piecewise expanding maps in* $\mathbb{R}^d$, Nonlinearity **13** (2000), 1745–1760.

[18] S. Corti, F. Molteni, T.N. Palmer, *Signature of recent climate change in frequencies of natural atmospheric circulation regimes,* Nature **398** (1999), 799–802.

[19] J.L. Doob, *Stochastic Processes*, J. Wiley, New York, 1953.

[20] D. Dolgopyat, *Limit theorems for partially hyperbolic systems*, Trans. Amer. Math. Soc. **356** (2003), 1637–1689.

[21] D. Dolgopyat, *Averaging and invariant measures*, Moscow Math. J. **5** (2005), 537–576.

[22] J.-D. Deuschel and D.W. Stroock, *Large Deviations*, (1989), Academic Press, Boston.

[23] M.D. Donsker and S.R.S. Varadhan, *On a variational formula for the principal eigenvalue for operators with maximum principle*, Proc. Nat. Acad. Sci. U.S.A. **72** (1975), 780–783.

[24] M.D. Donsker and S.R.S. Varadhan, *Asymptotic evaluation of certain Markov process expectations for large time, I,* Comm. Pure Appl. Math. **28** (1975), 1–47.

[25] A. Dembo and O. Zeitouni, *Large Deviations Techniques and Applications*, 2nd ed., (1998), Springer, New York.

[26] S.N.Elaydi, *An Introduction to Difference Equations*, (1996), Springer, New York.

[27] A. Eizenberg and M. Freidlin, *On the Dirichlet problem for a class of second order PDE systems with small parameter*, Stoch. and Stoch. Rep. **33** (1990), 111–148.

[28] E. Franko, *Flows with unique equilibrium state,* Am. J. Math. **99** (1977), 486–514.

[29] M.I. Freidlin, *The averaging principle and theorems on large deviations,* Russ. Math. Surv., **33,** No.5 (1978), 107–160.

[30] M. Freidlin, *Quasi-deterministic approximation, metastability and stochastic resonance,* Physica D **137** (2000), 333–352.

[31] M.I. Freidlin and A.D. Wentzell, *Random Perturbations of Dynamical Systems,* 2nd ed., (1998), Springer, New York.

[32] G. Gallavotti, *Chaotic hypothesis and universal large deviations properties,* Doc. Math. J. DMV, Extra Volume ICM 1998, **I** (1998), 205–233.

[33] B.V. Gnedenko, *The Theory of Probability,* (1966), Chelsea, New York.

[34] I. Gikhman and A. Skorokhod, *The Theory of Stochastic Processes I,* (1980), Springer, Berlin.

[35] O.V. Gulinsky and A.Yu. Veretennikov, *Large deviations for discrete-time processes with averaging,* (1993), VSP, Utrecht.

[36] K. Hasselmann, *Stochastic climate models, Part I. Theory,* Tellus **28** (1976), 473–485.

[37] K. Hasselmann, *Linear and nonlinear signatures*, Nature **398** (1999), 755–756.

[38] E.P. Hsu, *Stochastic Analysis on Manifolds*, (2002), Amer. Math. Soc., Providence.

[39] H. Hennion and L. Herve, *Limit Theorems for Markov Chains and Stochastic Properties of Dynamical Systems by Quasi-Compactness*, Lecture Notes in Math. **1766** (2001) Springer–Verlag, Berlin.

[40] M. Hirsch, J. Palis, G. Pugh and M. Shub, *Neighborhoods of hyperbolic sets,* Invent. Math. **9** (1970), 121–134.

[41] A.D. Ioffe and V.M. Tikhomirov, *Theory of Extremal Problems,* (1979) North-Holland, Amsterdam.

[42] N. Ikeda and S. Watanabe, *Stochastic Differential Equations and Diffusion Processes,* (1981), North-Holland, Amsterdam.

[43] T. Kato, *Perturbation Theory for Linear Operators,* (1976) 2nd ed., Springer, New York.

[44] R.Z. Khasminskii, *On stochastic processes defined by differential equations with a small parameter,* Th. Probab. Appl., **11** (1966), 211–228.

[45] R.Z. Khasminskii, *On the averaging principle for Itô stochastic differential equations,* Kibernetika (Prague), **4** (1968), 260–279 (in Russian).

[46] Yu. Kifer, *Principal eigenvalues, topological pressure, and stochastic stability of equilibrium states,* Israel J. Math. **70** (1990), 1–47.

[47] Yu. Kifer, *Large deviations in dynamical systems and stochastic processes,* Trans. Amer. Math. Soc., **321** (1990), 505–524.

[48] Yu. Kifer, *Principal eigenvalues and equilibrium states corresponding to weakly coupled parabolic systems of PDE,* J. D'Analyse Math. **59** (1992), 89–102.

[49] Yu. Kifer, *Averaging in dynamical systems and large deviations,* Invent. Math., **110** (1992), 337–370.

[50] Yu. Kifer, *Limit theorems in averaging for dynamical systems,* Ergod. Th.& Dynam. Sys., **15** (1995), 1143–1172.

[51] Yu. Kifer, *Stochastic versions of Anosov and Neistadt's theorems on averaging,* Stoch. and Dynam. **1** (2001), 1–21.

[52] Yu. Kifer, *Averaging and climate models,* in: Stochastic Climate Models, Progress in Probability **49** (2001), 171–188, Birkhäuser, Basel.

[53] Yu. Kifer, *Averaging in difference equations driven by dynamical systems,* in: Geometric Methods in Dynamics (II), Astérisque **287** (2003), 103–123.

[54] Yu. Kifer, L^2 *diffusion approximation for slow motion in averaging,* Stoch. and Dynam. **3** (2003), 213–246.

[55] Yu. Kifer, *Averaging principle for fully coupled dynamical systems and large deviations,* Ergod. Th.& Dynam. Syst. **24** (2004), 847–871.

[56] Yu. Kifer, *Some recent advances in averaging,* in: Modern Dynamical Systems and Applications (2004) 385–403, Cambridge Univ. Press, Cambridge.

[57] Yu. Kifer, *Another proof of the averaging principle for fully coupled dynamical systems with hyperbolic fast motions,* Discrete Contin. Dyn. Syst. **13** (2005), 1187–1201.

[58] M.A. Krasnoselskii, *Positive Solutions of Operator Equations,* (1964), Noordhoff, Groningen.

[59] N.V. Krylov, *Introduction to the theory of random processes,* (2002), Amer. Math. Soc., Providence, RI.

[60] A. Katok and B. Hasselblatt, *Introduction to the Modern Theory of Dynamical Systems,* (1995), Cambridge Univ. Press, Cambridge.

[61] A. Katok, G. Knieper, M. Pollicott, H. Weiss, *Differentiability and analyticity of topological entropy for Anosov and geodesic flows,* Invent. Math. **98** (1989), 581–597.

[62] M.A. Krasnoselskii, E.A. Lifshitz and A.V. Sobolev, *Positive Linear Systems,* (1989) Heldermann Verlag, Berlin.

[63] P. Lochak and C. Meunier, *Multiple Averaging for Classical Systems,* (1988), Springer, New York.

[64] R. de la Llave, J. Marco and R. Moriyon, *Canonical perturbation theory for Anosov systems and regularity results for Livsic cohomology equations,* Ann. Math. **123** (1986), 537–611.

[65] R. Mañé, *Ergodic Theory of Differentiable Dynamics,* (1987) Springer, Berlin.

[66] E. Pardoux and A.Yu. Veretennikov, *On Poisson equation and diffusion approximation* II, Ann. Probab. **31** (2003), 1166–1192.

[67] R. Pinsky, *Regularity properties of the Donsker–Varadhan rate functional for non-reversible diffusions and random evolutions* , Stoch. Dynam. **7** (2007), 123–140.

[68] M.H. Protter and H.F. Weinberger, *Maximum Principles in Differential Equations,* (1984), Springer, New York.

[69] C. Robinson, *Structural stability of vector fields,* Ann. Math. **99** (1974), 154–175.

[70] R.T. Rockafeller, *Convex Analysis,* (1970) Princeton Univ. Press, Priceton, NJ.

[71] D. Ruelle, *Differentiation of SRB states,* Comm. Math. Phys., **187** (1997), 227–241.

[72] A.V. Skorokhod, *Asymptotic Methods in the Theory of Stochastic Differential Equations,* (1989), Amer. Math. Soc., Providence.

[73] J.A. Sanders, F.Verhulst and J. Murdock, *Averaging Methods in Nonlinear Dynamical Systems,* 2nd. ed., (2007), Springer, Berlin.

[74] I.A. Taimanov, *An example of jump from chaos to integrability in magnetic geodesic flows,* Math. Notes, **76** (2004), 587–589.

[75] I. Tamura, *Topology of Foliations: An Introduction,* (1992), Amer. Math. Soc., Providence.

[76] H. Totoki, *Time changes of flows,* Mem. Fac. Sci. Kyushu Univ. (Ser. A), **20** (1966), 27–55.

[77] A.Yu. Veretennikov, *On the averaging principle for systems of stochastic differential equations,* Math. USSR Sbornik, **69** (1991), 271–284.

[78] A.Yu. Veretennikov, *On large deviations in the averaging principle for SDEs with "full dependence",* Ann. Probab., **27** (1999), 284–296.

[79] A.Yu. Veretennikov, *On large deviations in the averaging principle for SDEs with "full dependence", correction,* Preprint (2005), ArXiv math.PR/0502098.

[80] P. Walters, *An introduction to ergodic theory,* Springer, New York, 1982.

Index

Editorial Information

To be published in the *Memoirs*, a paper must be correct, new, nontrivial, and significant. Further, it must be well written and of interest to a substantial number of mathematicians. Piecemeal results, such as an inconclusive step toward an unproved major theorem or a minor variation on a known result, are in general not acceptable for publication.

Papers appearing in *Memoirs* are generally at least 80 and not more than 200 published pages in length. Papers less than 80 or more than 200 published pages require the approval of the Managing Editor of the Transactions/Memoirs Editorial Board.

As of May 31, 2009, the backlog for this journal was approximately 11 volumes. This estimate is the result of dividing the number of manuscripts for this journal in the Providence office that have not yet gone to the printer on the above date by the average number of monographs per volume over the previous twelve months, reduced by the number of volumes published in four months (the time necessary for preparing a volume for the printer). (There are 6 volumes per year, each usually containing at least 4 numbers.)

A Consent to Publish and Copyright Agreement is required before a paper will be published in the *Memoirs*. After a paper is accepted for publication, the Providence office will send a Consent to Publish and Copyright Agreement to all authors of the paper. By submitting a paper to the *Memoirs*, authors certify that the results have not been submitted to nor are they under consideration for publication by another journal, conference proceedings, or similar publication.

Information for Authors

Memoirs are printed from camera copy fully prepared by the author. This means that the finished book will look exactly like the copy submitted.

Initial submission. The AMS uses Centralized Manuscript Processing for initial submissions. Authors should submit a PDF file using the Initial Manuscript Submission form found at `www.ams.org/peer-review-submission`, or send one copy of the manuscript to the following address: Centralized Manuscript Processing, MEMOIRS OF THE AMS, 201 Charles Street, Providence, RI 02904-2294 USA. If a paper copy is being forwarded to the AMS, indicate that it is for it Memoirs and include the name of the corresponding author, contact information such as email address or mailing address, and the name of an appropriate Editor to review the paper (see the list of Editors below).

The paper must contain a *descriptive title* and an *abstract* that summarizes the article in language suitable for workers in the general field (algebra, analysis, etc.). The *descriptive title* should be short, but informative; useless or vague phrases such as "some remarks about" or "concerning" should be avoided. The *abstract* should be at least one complete sentence, and at most 300 words. Included with the footnotes to the paper should be the 2000 *Mathematics Subject Classification* representing the primary and secondary subjects of the article. The classifications are accessible from `www.ams.org/msc/`. The list of classifications is also available in print starting with the 1999 annual index of *Mathematical Reviews*. The Mathematics Subject Classification footnote may be followed by a list of *key words and phrases* describing the subject matter of the article and taken from it. Journal abbreviations used in bibliographies are listed in the latest *Mathematical Reviews* annual index. The series abbreviations are also accessible from `www.ams.org/msnhtml/serials.pdf`. To help in preparing and verifying references, the AMS offers MR Lookup, a Reference Tool for Linking, at `www.ams.org/mrlookup/`.

Electronically prepared manuscripts. The AMS encourages electronically prepared manuscripts, with a strong preference for $\mathcal{A}_\mathcal{M}\mathcal{S}$-LaTeX. To this end, the Society has prepared $\mathcal{A}_\mathcal{M}\mathcal{S}$-LaTeX author packages for each AMS publication. Author packages include instructions for preparing electronic manuscripts, samples, and a style file that generates the particular design specifications of that publication series. Though $\mathcal{A}\mathcal{M}\mathcal{S}$-LaTeX is the highly preferred format of TeX, author packages are also available in $\mathcal{A}\mathcal{M}\mathcal{S}$-TeX.

Authors may retrieve an author package for *Memoirs of the AMS* from `www.ams.org/journals/memo/memoauthorpac.html` or via FTP to `ftp.ams.org` (login as `anonymous`, enter username as password, and type `cd pub/author-info`). The *AMS Author Handbook* and the *Instruction Manual* are available in PDF format from the author package link. The author package can also be obtained free of charge by sending email to `tech-support@ams.org` (Internet) or from the Publication Division, American Mathematical Society, 201 Charles St., Providence, RI 02904-2294, USA. When requesting an author package, please specify $\mathcal{A}\mathcal{M}\mathcal{S}$-LaTeX or $\mathcal{A}\mathcal{M}\mathcal{S}$-TeX and the publication in which your paper will appear. Please be sure to include your complete mailing address.

After acceptance. The final version of the electronic file should be sent to the Providence office (this includes any TeX source file, any graphics files, and the DVI or PostScript file) immediately after the paper has been accepted for publication.

Before sending the source file, be sure you have proofread your paper carefully. The files you send must be the EXACT files used to generate the proof copy that was accepted for publication. For all publications, authors are required to send a printed copy of their paper, which exactly matches the copy approved for publication, along with any graphics that will appear in the paper.

Accepted electronically prepared files can be submitted via the web at `www.ams.org/submit-book-journal/`, sent via FTP, or sent on CD-Rom or diskette to the Electronic Prepress Department, American Mathematical Society, 201 Charles Street, Providence, RI 02904-2294 USA. TeX source files, DVI files, and PostScript files can be transferred over the Internet by FTP to the Internet node `ftp.ams.org` (130.44.1.100). When sending a manuscript electronically via CD-Rom or diskette, please be sure to include a message identifying the paper as a Memoir.

Electronically prepared manuscripts can also be sent via email to `pub-submit@ams.org` (Internet). In order to send files via email, they must be encoded properly. (DVI files are binary and PostScript files tend to be very large.)

Electronic graphics. Comprehensive instructions on preparing graphics are available at `www.ams.org/authors/journals.html`. A few of the major requirements are given here.

Submit files for graphics as EPS (Encapsulated PostScript) files. This includes graphics originated via a graphics application as well as scanned photographs or other computer-generated images. If this is not possible, TIFF files are acceptable as long as they can be opened in Adobe Photoshop or Illustrator. No matter what method was used to produce the graphic, it is necessary to provide a paper copy to the AMS.

Authors using graphics packages for the creation of electronic art should also avoid the use of any lines thinner than 0.5 points in width. Many graphics packages allow the user to specify a "hairline" for a very thin line. Hairlines often look acceptable when proofed on a typical laser printer. However, when produced on a high-resolution laser imagesetter, hairlines become nearly invisible and will be lost entirely in the final printing process.

Screens should be set to values between 15% and 85%. Screens which fall outside of this range are too light or too dark to print correctly. Variations of screens within a graphic should be no less than 10%.

Inquiries. Any inquiries concerning a paper that has been accepted for publication should be sent to `memo-query@ams.org` or directly to the Electronic Prepress Department, American Mathematical Society, 201 Charles St., Providence, RI 02904-2294 USA.

Editors

This journal is designed particularly for long research papers, normally at least 80 pages in length, and groups of cognate papers in pure and applied mathematics. Papers intended for publication in the *Memoirs* should be addressed to one of the following editors. The AMS uses Centralized Manuscript Processing for initial submissions to AMS journals. Authors should follow instructions listed on the Initial Submission page found at www.ams.org/memo/memosubmit.html.

Algebra to ALEXANDER KLESHCHEV, Department of Mathematics, University of Oregon, Eugene, OR 97403-1222; email: ams@noether.uoregon.edu

Algebraic geometry to DAN ABRAMOVICH, Department of Mathematics, Brown University, Box 1917, Providence, RI 02912; email: amsedit@math.brown.edu

Algebraic geometry and its applications to MINA TEICHER, Emmy Noether Research Institute for Mathematics, Bar-Ilan University, Ramat-Gan 52900, Israel; email: teicher@macs.biu.ac.il

Algebraic topology to ALEJANDRO ADEM, Department of Mathematics, University of British Columbia, Room 121, 1984 Mathematics Road, Vancouver, British Columbia, Canada V6T 1Z2; email: adem@math.ubc.ca

Combinatorics to JOHN R. STEMBRIDGE, Department of Mathematics, University of Michigan, Ann Arbor, Michigan 48109-1109; email: JRS@umich.edu

Commutative and homological algebra to LUCHEZAR L. AVRAMOV, Department of Mathematics, University of Nebraska, Lincoln, NE 68588-0130; email: avramov@math.unl.edu

Complex analysis and harmonic analysis to ALEXANDER NAGEL, Department of Mathematics, University of Wisconsin, 480 Lincoln Drive, Madison, WI 53706-1313; email: nagel@math.wisc.edu

Differential geometry and global analysis to CHRIS WOODWARD, Department of Mathematics, Rutgers University, 110 Frelinghuysen Road, Piscataway, NJ 08854; email: ctw@math.rutgers.edu

Dynamical systems and ergodic theory and complex analysis to YUNPING JIANG, Department of Mathematics, CUNY Queens College and Graduate Center, 65-30 Kissena Blvd., Flushing, NY 11367; email: Yunping.Jiang@qc.cuny.edu

Functional analysis and operator algebras to DIMITRI SHLYAKHTENKO, Department of Mathematics, University of California, Los Angeles, CA 90095; email: shlyakht@math.ucla.edu

Geometric analysis to WILLIAM P. MINICOZZI II, Department of Mathematics, Johns Hopkins University, 3400 N. Charles St., Baltimore, MD 21218; email: trans@math.jhu.edu

Geometric topology to MARK FEIGHN, Math Department, Rutgers University, Newark, NJ 07102; email: feighn@andromeda.rutgers.edu

Harmonic analysis, representation theory, and Lie theory to ROBERT J. STANTON, Department of Mathematics, The Ohio State University, 231 West 18th Avenue, Columbus, OH 43210-1174; email: stanton@math.ohio-state.edu

Logic to STEFFEN LEMPP, Department of Mathematics, University of Wisconsin, 480 Lincoln Drive, Madison, Wisconsin 53706-1388; email: lempp@math.wisc.edu

Number theory to JONATHAN ROGAWSKI, Department of Mathematics, University of California, Los Angeles, CA 90095; email: jonr@math.ucla.edu

Number theory to SHANKAR SEN, Department of Mathematics, 505 Malott Hall, Cornell University, Ithaca, NY 14853; email: ss70@cornell.edu

Partial differential equations to GUSTAVO PONCE, Department of Mathematics, South Hall, Room 6607, University of California, Santa Barbara, CA 93106; email: ponce@math.ucsb.edu

Partial differential equations and dynamical systems to PETER POLACIK, School of Mathematics, University of Minnesota, Minneapolis, MN 55455; email: polacik@math.umn.edu

Probability and statistics to RICHARD BASS, Department of Mathematics, University of Connecticut, Storrs, CT 06269-3009; email: bass@math.uconn.edu

Real analysis and partial differential equations to DANIEL TATARU, Department of Mathematics, University of California, Berkeley, Berkeley, CA 94720; email: tataru@math.berkeley.edu

All other communications to the editors should be addressed to the Managing Editor, ROBERT GURALNICK, Department of Mathematics, University of Southern California, Los Angeles, CA 90089-1113; email: guralnic@math.usc.edu.

Titles in This Series

946 **Jay Jorgenson and Serge Lang,** Heat Eisenstein series on $SL_n(C)$, 2009

945 **Tobias H. Jäger,** The creation of strange non-chaotic attractors in non-smooth saddle-node bifurcations, 2009

944 **Yuri Kifer,** Large deviations and adiabatic transitions for dynamical systems and Markov processes in fully coupled averaging, 2009

943 **István Berkes and Michel Weber,** On the convergence of $\sum c_k f(n_k x)$, 2009

942 **Dirk Kussin,** Noncommutative curves of genus zero: Related to finite dimensional algebras, 2009

941 **Gelu Popescu,** Unitary invariants in multivariable operator theory, 2009

940 **Gérard Iooss and Pavel I. Plotnikov,** Small divisor problem in the theory of three-dimensional water gravity waves, 2009

939 **I. D. Suprunenko,** The minimal polynomials of unipotent elements in irreducible representations of the classical groups in odd characteristic, 2009

938 **Antonino Morassi and Edi Rosset,** Uniqueness and stability in determining a rigid inclusion in an elastic body, 2009

937 **Skip Garibaldi,** Cohomological invariants: Exceptional groups and spin groups, 2009

936 **André Martinez and Vania Sordoni,** Twisted pseudodifferential calculus and application to the quantum evolution of molecules, 2009

935 **Mihai Ciucu,** The scaling limit of the correlation of holes on the triangular lattice with periodic boundary conditions, 2009

934 **Arjen Doelman, Björn Sandstede, Arnd Scheel, and Guido Schneider,** The dynamics of modulated wave trains, 2009

933 **Luchezar Stoyanov,** Scattering resonances for several small convex bodies and the Lax-Phillips conjuecture, 2009

932 **Jun Kigami,** Volume doubling measures and heat kernel estimates of self-similar sets, 2009

931 **Robert C. Dalang and Marta Sanz-Solé,** Hölder-Sobolv regularity of the solution to the stochastic wave equation in dimension three, 2009

930 **Volkmar Liebscher,** Random sets and invariants for (type II) continuous tensor product systems of Hilbert spaces, 2009

929 **Richard F. Bass, Xia Chen, and Jay Rosen,** Moderate deviations for the range of planar random walks, 2009

928 **Ulrich Bunke,** Index theory, eta forms, and Deligne cohomology, 2009

927 **N. Chernov and D. Dolgopyat,** Brownian Brownian motion-I, 2009

926 **Riccardo Benedetti and Francesco Bonsante,** Canonical wick rotations in 3-dimensional gravity, 2009

925 **Sergey Zelik and Alexander Mielke,** Multi-pulse evolution and space-time chaos in dissipative systems, 2009

924 **Pierre-Emmanuel Caprace,** "Abstract" homomorphisms of split Kac-Moody groups, 2009

923 **Michael Jöllenbeck and Volkmar Welker,** Minimal resolutions via algebraic discrete Morse theory, 2009

922 **Ph. Barbe and W. P. McCormick,** Asymptotic expansions for infinite weighted convolutions of heavy tail distributions and applications, 2009

921 **Thomas Lehmkuhl,** Compactification of the Drinfeld modular surfaces, 2009

920 **Georgia Benkart, Thomas Gregory, and Alexander Premet,** The recognition theorem for graded Lie algebras in prime characteristic, 2009

919 **Roelof W. Bruggeman and Roberto J. Miatello,** Sum formula for SL_2 over a totally real number field, 2009

TITLES IN THIS SERIES

918 **Jonathan Brundan and Alexander Kleshchev,** Representations of shifted Yangians and finite W-algebras, 2008

917 **Salah-Eldin A. Mohammed, Tusheng Zhang, and Huaizhong Zhao,** The stable manifold theorem for semilinear stochastic evolution equations and stochastic partial differential equations, 2008

916 **Yoshikata Kida,** The mapping class group from the viewpoint of measure equivalence theory, 2008

915 **Sergiu Aizicovici, Nikolaos S. Papageorgiou, and Vasile Staicu,** Degree theory for operators of monotone type and nonlinear elliptic equations with inequality constraints, 2008

914 **E. Shargorodsky and J. F. Toland,** Bernoulli free-boundary problems, 2008

913 **Ethan Akin, Joseph Auslander, and Eli Glasner,** The topological dynamics of Ellis actions, 2008

912 **Igor Chueshov and Irena Lasiecka,** Long-time behavior of second order evolution equations with nonlinear damping, 2008

911 **John Locker,** Eigenvalues and completeness for regular and simply irregular two-point differential operators, 2008

910 **Joel Friedman,** A proof of Alon's second eigenvalue conjecture and related problems, 2008

909 **Cameron McA. Gordon and Ying-Qing Wu,** Toroidal Dehn fillings on hyperbolic 3-manifolds, 2008

908 **J.-L. Waldspurger,** L'endoscopie tordue n'est pas si tordue, 2008

907 **Yuanhua Wang and Fei Xu,** Spinor genera in characteristic 2, 2008

906 **Raphaël S. Ponge,** Heisenberg calculus and spectral theory of hypoelliptic operators on Heisenberg manifolds, 2008

905 **Dominic Verity,** Complicial sets characterising the simplicial nerves of strict ω-categories, 2008

904 **William M. Goldman and Eugene Z. Xia,** Rank one Higgs bundles and representations of fundamental groups of Riemann surfaces, 2008

903 **Gail Letzter,** Invariant differential operators for quantum symmetric spaces, 2008

902 **Bertrand Toën and Gabriele Vezzosi,** Homotopical algebraic geometry II: Geometric stacks and applications, 2008

901 **Ron Donagi and Tony Pantev (with an appendix by Dmitry Arinkin),** Torus fibrations, gerbes, and duality, 2008

900 **Wolfgang Bertram,** Differential geometry, Lie groups and symmetric spaces over general base fields and rings, 2008

899 **Piotr Hajłasz, Tadeusz Iwaniec, Jan Malý, and Jani Onninen,** Weakly differentiable mappings between manifolds, 2008

898 **John Rognes,** Galois extensions of structured ring spectra/Stably dualizable groups, 2008

897 **Michael I. Ganzburg,** Limit theorems of polynomial approximation with exponential weights, 2008

896 **Michael Kapovich, Bernhard Leeb, and John J. Millson,** The generalized triangle inequalities in symmetric spaces and buildings with applications to algebra, 2008

895 **Steffen Roch,** Finite sections of band-dominated operators, 2008

894 **Martin Dindoš,** Hardy spaces and potential theory on C^1 domains in Riemannian manifolds, 2008

For a complete list of titles in this series, visit the AMS Bookstore at **www.ams.org/bookstore/**.